Side Channel Attacks

Side Channel Attacks

Special Issue Editor

Seokhie Hong

MDPI • Basel • Beijing • Wuhan • Barcelona • Belgrade

MDPI

Special Issue Editor
Seokhie Hong
Korea University
Korea

Editorial Office
MDPI
St. Alban-Anlage 66
4052 Basel, Switzerland

This is a reprint of articles from the Special Issue published online in the open access journal *Applied Sciences* (ISSN 2076-3417) from 2018 to 2019 (available at: https://www.mdpi.com/journal/applsci/special_issues/Side_Channel_Attacks)

For citation purposes, cite each article independently as indicated on the article page online and as indicated below:

LastName, A.A.; LastName, B.B.; LastName, C.C. Article Title. *Journal Name* **Year**, *Article Number, Page Range*.

ISBN 978-3-03921-000-8 (Pbk)
ISBN 978-3-03921-001-5 (PDF)

Contents

About the Special Issue Editor

Seokhie Hong received his MA and PhD in mathematics from Korea University in 1997 and 2001, respectively. He worked for SECURITY Technologies Inc. from 2000 to 2004. From 2004 to 2005, he worked as a postdoctoral researcher in COSIC at K.U. Leuven, Belgium. Since 2005, he has been with Korea University, where he is now working in the School of Cyber Security. His specialty is information security, and his research interests include the design and analysis of symmetric-key cryptosystems, public-key cryptosystems, and side-channel attacks.

.

Preface to "Side Channel Attacks"

Cryptosystems are widely used in a growing number of embedded applications, such as smart cards, smartphones, Internet of Things (IoT) devices, and so on. Although these cryptosystems have been proven to be safe using mathematical tools, they could be susceptible to physical attacks that exploit additional sources of information, including timing information, power consumption, electromagnetic emissions (EM), sound, and so on. First introduced by Kocher, these types of attacks are referred to as side-channel attacks (SCAs). These attacks pose a very serious threat to embedded systems with cryptographic algorithms. For the past few years, there has been a great deal of effort put towards finding various SCAs and developing secure countermeasures This Special Issue provides an opportunity for researchers in the area of SCAs to highlight the most recent exciting technologies. The research papers published in this Special Issue represent recent progress in the field. The 13 papers in this Special Issue can be classified into the following four research themes: power analysis attacks and countermeasures, cache-based timing attacks, system-level countermeasures and their weaknesses, and recent technologies in the field of side-channel attacks. This Special Issue contains reaseach on various power analysis attacks and countermeasures on well-known cryptographic algorithms, such as elliptic curve cryptosystems (ECCs), block cipher SEED, and post-quantum cryptographies (PQCs). A new side-channel leakage of SEED in the financial IC cards of the Republic of Korea was detected, and new vulnerabilities were found using a single power consumption trace obtained in the elliptic curve scalar multiplication algorithm. Recently, PQCs, which refer to cryptographic algorithms executed on a classical computer that is expected to be secure against adversaries with quantum computers, have been actively researched. This Special Issue contains two papers about power analysis attacks on PQCs: the well-known NTRU algorithm and a cumulative distribution table (CDT) sampler used in lattice-based PQCs. This Special Issue also contains two research papers with regard to cache-based timing attacks utilizing the timing difference between cache hits and cache misses. One paper proposes a new constant-time method for the RSA modular exponentiation, which is resistant against fine-grained cache attacks. The second paper on cache-based timing attacks shows a non-access attack, a novel approach to exploit information gained from cache misses. Two research papers in this Special Issue introduce a new system-level countermeasure and a new vulnerability of the existing physically unclonable function (PUF). One paper deals with the re-keying scheme—the system-level countermeasure against SCAs—which makes attackers unable to collect enough power consumption traces for their analyses. The authors of this paper define a new security model and propose two provably secure re-keying schemes. The second paper on this topic shows that the PUF key can be derived from a chaotic circuit. This Special Issue also describes recent technologies in the field of SCAs, such as the machine-learning-based side-channel evaluation technique , the Merkle tree-based online data authentication technique with leakage resilience, the ID-based side-channel authentication technique, and a technique to distinguish ad-related network behavior. In summary, this Special Issue compiles excellent research, covering a wide area of SCAs.

Seokhie Hong
Special Issue Editor

applied
sciences

MDPI

Editorial

Special Issue on "Side Channel Attacks"

Seokhie Hong [1,2]

1 School of Cyber Security, Korea University, Seoul 02841, Korea; shhong@korea.ac.kr
2 Center for Information Security Technologies (CIST), Institute of Cyber Security and Privacy (ICSP), Korea University, Seoul 02841, Korea

Received: 25 April 2019; Accepted: 29 April 2019; Published: 8 May 2019

Cryptosystems are widely used in a growing number of embedded applications, such as smart cards, smart phones, Internet of Things (IoT) devices, and so on. Although these cryptosystems have been proven to be safe using mathematical tools, they are potentially susceptible to physical attacks which exploit additional sources of information, including timing information, power consumption, electromagnetic emissions (EM), and sound, amongst others. Introduced by Kocher, these types of attacks are referred to as side-channel attacks (SCAs) [1,2]. These attacks pose a very serious threat to embedded systems with cryptographic algorithms. There has been a great deal of effort put into finding various SCAs and developing secure counter-measures, recently [3–16].

This special issue has been organized to provide a possibility for researchers in the area of SCAs to highlight the most recent and exciting technologies. The research papers selected for this special issue represent recent progress in the field, including power analysis attacks [17–19], cache-based timing attacks [20–41], system-level counter-measures [42–48], and so on [49–60]. The thirteen papers in this special issue can be classified into the following four research themes:

Power analysis attacks and counter-measures: This special issue contains various power analysis attacks and counter-measures on well-known crypto algorithms: Elliptic curve cryptosystems (ECCs), the block cipher SEED, and the post-quantum cryptographies (PQCs). A new side channel leakage of the SEED in financial IC cards in the Republic of Korea was detected in [61]; and new vulnerabilities, using a single power consumption trace obtained in the elliptic curve scalar multiplication algorithm, were established in [62,63]. Recently, PQCs, cryptographic algorithms executed on a classical computer which are expected to be secure against adversaries with quantum computers, have been actively studied. This special issue contains two papers about power analysis attacks on PQCs: The well-known NTRU algorithm, and a cumulative distribution table (CDT) sampler used in the lattice-based PQCs [64,65].

Cache-based timing attacks: This special issue contains two research papers with regard to cache-based timing attacks, utilizing the timing difference between cache hits and cache misses. One paper proposes a new constant-time method for RSA modular exponentiation, which is resistant against fine-grained cache attacks [66]. The other one shows a non-access attack, a novel approach for exploiting the information gained from cache misses [67].

System-level counter-measures and their weaknesses: Two research papers in this special issue introduce a new system-level counter-measure and a new vulnerability of the existing physically un-clonable function (PUF), respectively. One paper deals with the re-keying scheme, a system-level counter-measure against SCAs, which makes attackers unable to collect enough power consumption traces for their analyses [68]. The authors of this paper define a new security model and propose two provably secure re-keying schemes. The other paper shows that a PUF key can be derived from a chaotic circuit [69].

Recent technologies in the field of Side Channel Attacks: This special issue also contains papers documenting recent technologies in the field of SCAs: A machine-learning based side-channel evaluation technique [70], a Merkle tree-based on-line data authentication technique with leakage resilience [71], an ID-based side-channel authentication technique [72], and a technique to distinguish ad-related network behavior [73].

In summary, this special issue contains many excellent studies, covering a wide range of SCA-related topics. This collection of 13 papers is highly recommended, and is believed to benefit readers in various aspects.

Funding: This work was supported by the Institute for Information and Communications Technology Promotion (IITP) grant, funded by the Korean government (MSIT) (No. 2017-0-00520, Development of SCR-Friendly Symmetric Key Cryptosystem and Its Application Modes).

Conflicts of Interest: The author declares no conflict of interest.

References

1. Kocher, P.C. Timing attacks on implementations of Diffie-Hellman, RSA, DSS, and other systems. In Proceedings of the Annual International Cryptology Conference (CRYPTO), Santa Barbara, CA, USA, 18–22 August 1996; pp. 104–113.
2. Kocher, P.; Jaffe, J.; Jun, B. Differential power analysis. In Proceedings of the Annual International Cryptology Conference (CRYPTO), Santa Barbara, CA, USA, 15–19 August 1999; pp. 388–397.
3. Gandolfi, K.; Mourtel, C.; Olivier, F. Electromagnetic analysis: Concrete results. In Proceedings of the International Workshop on Cryptographic Hardware and Embedded Systems, Paris, France, 14–16 May 2001; Springer: Berlin/Heidelberg, Germany, 2001; pp. 251–261.
4. Brier, E.; Clavier, C.; Olivier, F. Correlation power analysis with a leakage model. In Proceedings of the International Workshop on Cryptographic Hardware and Embedded Systems, Cambridge, MA, USA, 11–13 August 2004; Springer: Berlin/Heidelberg, Germany, 2004; pp. 16–29.
5. Gierlichs, B.; Batina, L.; Tuyls, P.; Preneel, B. Mutual information analysis. In Proceedings of the International Workshop on Cryptographic Hardware and Embedded Systems, Washington, DC, USA, 10–13 August 2008; Springer: Berlin/Heidelberg, Germany, 2008; pp. 426–442.
6. Chari, S.; Rao, J.R.; Rohatgi, P. Template attacks. In Proceedings of the International Workshop on Cryptographic Hardware and Embedded Systems, Redwood Shores, CA, USA, 13–15 August 2002; Springer: Berlin/Heidelberg, Germany, 2002; pp. 13–28.
7. Schindler, W.; Lemke, K.; Paar, C. A stochastic model for differential side channel cryptanalysis. In Proceedings of the International Workshop on Cryptographic Hardware and Embedded Systems, Edinburgh, UK, 29 August–1 September 2005; Springer: Berlin/Heidelberg, Germany, 2005; pp. 30–46.
8. Mangard, S.; Oswald, E.; Popp, T. *Power Analysis Attacks: Revealing the Secrets of Smart Cards*; Springer Science & Business Media: Berlin/Heidelberg, Germany, 2008; Volume 31.
9. Prouff, E.; Rivain, M.; Bevan, R. Statistical analysis of second order differential power analysis. *IEEE Trans. Comput.* **2009**, *58*, 799–811. [CrossRef]
10. Kim, H.S.; Hong, S. New type of collision attack on first-order masked AESs. *ETRI J.* **2016**, *38*, 387–396. [CrossRef]
11. Coron, J.S.; Goubin, L. On boolean and arithmetic masking against differential power analysis. In Proceedings of the International Workshop on Cryptographic Hardware and Embedded Systems, Worcester, MA, USA, 17–18 August 2000; Springer: Berlin/Heidelberg, Germany, 2000; pp. 231–237.
12. Goubin, L. A sound method for switching between boolean and arithmetic masking. In Proceedings of the International Workshop on Cryptographic Hardware and Embedded Systems, Paris, France, 14–16 May 2001; Springer: Berlin/Heidelberg, Germany, 2001; pp. 3–15.
13. Coron, J.S.; Tchulkine, A. A new algorithm for switching from arithmetic to boolean masking. In Proceedings of the International Workshop on Cryptographic Hardware and Embedded Systems, Cologne, Germany, 8–10 September 2003; Springer: Berlin/Heidelberg, Germany, 2003; pp. 89–97.
14. Tunstall, M.; Whitnall, C.; Oswald, E. Masking tables-an underestimated security risk. In Proceedings of the International Workshop on Fast Software Encryption, Washington, DC, USA, 11–13 March 2013; Springer: Berlin/Heidelberg, Germany, 2003; pp. 425–444.
15. Balasch, J.; Faust, S.; Gierlichs, B. Inner product masking revisited. In Proceedings of the Annual International Conference on the Theory and Applications of Cryptographic Techniques, Sofia, Bulgaria, 26–30 April 2015; pp. 486–510.

16. Bettale, L.; Coron, J.S.; Zeitoun, R. Improved high-order conversion from boolean to arithmetic masking. *IACR Trans. Cryptogr. Hardw. Embed. Syst.* **2018**, *2018*, 22–45.

17. Espitau, T.; Fouque, P.A.; Gérard, B.; Tibouchi, M. Side-Channel Attacks on BLISS Lattice-Based Signatures: Exploiting Branch Tracing Against strongSwan and Electromagnetic Emanations in Microcontrollers. In Proceedings of the 2017 ACM SIGSAC Conference on Computer and Communications Security, Dallas, TX, USA, 30 October–3 November 2017; ACM: New York, NY, USA, 2017; pp. 1857–1874.

18. Park, A.; Shim, K.A.; Koo, N.; Han, D.G. Side-Channel Attacks on Post-Quantum Signature Schemes based on Multivariate Quadratic Equations. *IACR Trans. Cryptogr. Hardw. Embed. Syst.* **2018**, *2018*, 500–523.

19. Saarinen, M.J.O. Arithmetic Coding and Blinding Countermeasures for Lattice Signatures. 2016. Available online: https://eprint.iacr.org/2016/276 (accessed on 7 May 2019).

20. Yarom, Y.; Falkner, K. FLUSH+RELOAD: A High Resolution, Low Noise, L3 Cache Side-Channel Attack. In Proceedings of the 23rd USENIX Security Symposium (USENIX Security 14), San Diego, CA, USA, 20–22 August 2014; pp. 719–732.

21. Liu, F.; Yarom, Y.; Ge, Q.; Heiser, G.; Lee, R.B. Last-Level Cache Side-Channel Attacks are Practical. In Proceedings of the 2015 IEEE Symposium on Security and Privacy, San Jose, CA, USA, 18–20 May 2015; pp. 605–622.

22. Lipp, M.; Schwarz, M.; Gruss, D.; Prescher, T.; Haas, W.; Fogh, A.; Horn, J.; Mangard, S.; Kocher, P.; Genkin, D.; et al. Meltdown: Reading kernel memory from user space. In Proceedings of the 27th USENIX Security Symposium (USENIX Security 18), Baltimore, MD, USA, 15–17 August 2018; pp. 973–990.

23. Kocher, P.; Genkin, D.; Gruss, D.; Haas, W.; Hamburg, M.; Lipp, M.; Mangard, S.; Prescher, T.; Schwarz, M.; Yarom, Y. Spectre attacks: Exploiting speculative execution. *arXiv* **2018**, arXiv:1801.01203.

24. Irazoqui, G.; Eisenbarth, T.; Sunar, B. Cross Processor Cache Attacks. In Proceedings of the 11th ACM on Asia Conference on Computer and Communications Security, Xi'an, China, 30 May–3 June 2016; pp. 353–364.

25. Xu, Y.; Cui, W.; Peinado, M. Controlled-channel attacks: Deterministic side channels for untrusted operating systems. In Proceedings of the 2015 IEEE Symposium on Security and Privacy, San Jose, CA, USA, 18–20 May 2015; pp. 640–656.

26. Zhang, Y.; Juels, A.; Reiter, M.K.; Ristenpart, T. Cross-tenant side-channel attacks in PaaS clouds. In Proceedings of the 2014 ACM SIGSAC Conference on Computer and Communications Security, Scottsdale, AZ, USA, 3–7 November 2014; pp. 990–1003.

27. Gruss, D.; Maurice, C.; Wagner, K.; Mangard, S. Flush+Flush: A fast and stealthy cache attack. In Proceedings of the International Conference on Detection of Intrusions and Malware, and Vulnerability Assessment, San Sebastián, Spain, 7–8 July 2016; Volume 9721, pp. 279–299.

28. Gruss, D.; Spreitzer, R.; Mangard, S. Cache template attacks: Automating attacks on inclusive last-level caches. In Proceedings of the 24th USENIX Security Symposium (USENIX Security 15), Washington, DC, USA, 12–14 August 2015; pp. 897–912.

29. Yarom, Y.; Genkin, D.; Heninger, N. CacheBleed: A timing attack on OpenSSL constant-time RSA. *J. Cryptogr. Eng.* **2017**, *7*, 99–112. [CrossRef]

30. Doychev, G.; Köpf, B.; Mauborgne, L.; Reineke, J. Cacheaudit: A tool for the static analysis of cache side channels. *ACM Trans. Inf. Syst. Secur.* **2015**, *18*, 4:1–4:32. [CrossRef]

31. Yarom, Y.; Benger, N. Recovering OpenSSL ECDSA Nonces Using the FLUSH+RELOAD Cache Side-channel Attack. *IACR Cryptol. ePrint Archi.* **2014**, *2014*, 140.

32. Lipp, M.; Gruss, D.; Spreitzer, R.; Maurice, C.; Mangard, S. ARMageddon: Cache attacks on mobile devices. In Proceedings of the 25th USENIX Security Symposium (USENIX Security 16), Austin, TX, USA, 10–12 August 2016; pp. 549–564.

33. Aldaya, A.C.; García, C.P.; Tapia, L.M.A.; Brumley, B.B. Cache-Timing Attacks on RSA Key Generation. *IACR Cryptol. ePrint Arch.* **2018**, *2018*, 367.

34. Deng, S.; Xiong, W.; Szefer, J. Analysis of Secure Caches and Timing-Based Side-Channel Attacks. *IACR Cryptol. ePrint Arch.* **2019**, *2019*, 167.

35. Irazoqui, G.; Guo, X. Cache Side Channel Attack: Exploitability and Countermeasures. *Black Hat Asia* **2017**, *2017*, 3.

36. Zhou, Z.; Reiter, M.K.; Zhang, Y. A software approach to defeating side channels in last-level caches. In Proceedings of the 2016 ACM SIGSAC Conference on Computer and Communications Security, Vienna, Austria, 24–28 October 2016; pp. 871–882.

37. Zhang, Y. Cache Side Channels: State of the Art and Research Opportunities. In Proceedings of the 2017 ACM SIGSAC Conference on Computer and Communications Security, Dallas, TX, USA, 30 October–3 November 2017; pp. 2617–2619.

38. Gruss, D.; Lettner, J.; Schuster, F.; Ohrimenko, O.; Haller, I.; Costa, M. Strong and efficient cache side-channel protection using hardware transactional memory. In Proceedings of the 26th USENIX Security Symposium (USENIX Security 17), Vancouver, BC, Canada, 16–18 August 2017; pp. 217–233.

39. Wang, S.; Wang, P.; Liu, X.; Zhang, D.; Wu, D. CacheD: Identifying cache-based timing channels in production software. In Proceedings of the 26th USENIX Security Symposium (USENIX Security 17), Vancouver, BC, Canada, 16–18 August 2017; pp. 235–252.

40. Dong, X.; Shen, Z.; Criswell, J.; Cox, A.L.; Dwarkadas, S. Shielding Software From Privileged Side-Channel Attacks. In Proceedings of the 27th USENIX Security Symposium (USENIX Security 18), Baltimore, MD, USA, 15–17 August 2018; pp. 1441–1458.

41. Gras, B.; Razavi, K.; Bos, H.; Giuffrida, C. Translation Leak-aside Buffer: Defeating Cache Side-channel Protections with TLB Attacks. In Proceedings of the 27th USENIX Security Symposium (USENIX Security 18), Baltimore, MD, USA, 15–17 August 2018; pp. 955–972.

42. Medwed, M.; Standaert, F.X.; Großschädl, J.; Regazzoni, F. Fresh Re-Keying: Security against Side-Channel and Fault Attacks for Low-Cost Devices. In *Progress in Cryptology—AFRICACRYPT 2010, Proceedings of the Third International Conference on Cryptology in Africa, Stellenbosch, South Africa, 3–6 May 2010*; Bernstein, D., Lange, T., Eds.; Springer: Berlin/Heidelberg, Germany, 2010; Volume 6055, pp. 279–296.

43. Medwed, M.; Petit, C.; Regazzoni, F.; Renauld, M.; Standaert, F.X. Fresh Re-keying II: Securing Multiple Parties Against Side-channel and Fault Attacks. In Proceedings of the 10th IFIP WG 8.8/11.2 International Conference on Smart Card Research and Advanced Applications, Stellenbosch, South Africa, 3–6 May 2011; Springer: Berlin/Heidelberg, Germany, 2011; pp. 115–132.

44. Rührmair, U.; Sölter, J.; Sehnke, F. On the Foundations of Physical Unclonable Functions. *IACR Cryptol. ePrint Arch.* **2009**, *2009*, 277.

45. Merli, D.; Schuster, D.; Stumpf, F.; Sigl, G. Side-channel analysis of PUFs and fuzzy extractors. In Proceedings of the International Conference on Trust and Trustworthy Computing, Pittsburgh, PA, USA, 22–24 June 2011; Volume 6740, pp. 33–47.

46. Tuyls, P.; Škorić, B.; Stallinga, S.; Akkermans, A.H.; Ophey, W. Information-theoretic security analysis of physical uncloneable functions. In Proceedings of the International Conference on Financial Cryptography and Data Security, Roseau, MN, USA, 28 February–3 March 2005; Volume 3570, pp. 141–155.

47. Rührmair, U.; Sehnke, F.; Sölter, J.; Dror, G.; Devadas, S.; Schmidhuber, J. Modeling attacks on physical unclonable functions. In Proceedings of the 17th ACM Conference on Computer and Communications Security, Chicago, IL, USA, 4–8 October 2010; pp. 237–249.

48. Škorić, B.; Tuyls, P.; Ophey, W. Robust key extraction from physical uncloneable functions. In Proceedings of the International Conference on Applied Cryptography and Network Security, New York, NY, USA, 7–10 June 2005; Volume 3531, pp. 407–422.

49. Lerman, L.; Bontempi, G.; Markowitch, O. Side channel attack: An approach based on machine learning. In Proceedings of the International Workshop on Constructive Side-Channel Analysis and Secure Design (COSADE), Darmstadt, Germany, 14 February 2011; pp. 29–41.

50. Hospodar, G.; Gierlichs, B.; De Mulder, E.; Verbauwhede, I.; Vewalle, J. Machine learning in side-channel analysis: A first study. *J. Cryptogr. Eng.* **2011**, *1*, 293–302. [CrossRef]

51. Bartkewitz, T.; Lemke-Rust, K. Efficient template attacks based on probabilistic multi-class support vector machines. In Proceedings of the International Conference on Smart Card Research and Advanced Applications (CARDIS), Graz, Austria, 28–30 November 2012; pp. 263–276.

52. Heuser, A.; Zohner, M. Intelligent machine homicide. In Proceedings of the International Workshop on Constructive Side-Channel Analysis and Secure Design (COSADE), Darmstadt, Germany, 3–4 May 2012; pp. 249–264.

53. Heyszl, J.; Ibing, A.; Mangard, S.; De Santis, F.; Sigl, G. Clustering Algorithms for Non-profiled Single-Execution Attacs on Exponentiations. In Proceedings of the International Conference on Smart Card Research and Advanced Applications (CARDIS), Paris, France, 5–7 November 2013; pp. 79–93.

54. Lerman, L.; Bontempi, G.; Markowitch, O. A machine learning approach against a masked AES. In Proceedings of the International Conference on Smart Card Research and Advanced Applications (CARDIS), Paris, France, 5–7 November 2013; pp. 61–75.

55. Specht, R.; Heyszl, J.; Kleinsteuber, M.; Sigl, G. Improving non-profiled attacks on exponentiations based on clustering and extracting leakage from multi-channel high-resolution EM measurements. In Proceedings of the International Workshop on Constructive Side-Channel Analysis and Secure Design (COSADE), Berlin, Germany, 13–14 April 2015; pp. 3–19.

56. Whitnall, C.; Oswald, E. Profiling DPA: Efficacy and efficiency trade-offs. In Proceedings of the International Workshop on Cryptographic Hardware and Embedded Systems (CHES), Santa Barbara, CA, USA, 20–23 August 2013; pp. 37–54.

57. Maghrebi, H.; Portigliatti, T.; Prouff, E. Breaking cryptographic implementations using deep learning techniques. In Proceedings of the International Conference on Security, Privacy, and Applied Cryptography Engineering (SPACE), Hyderabad, India, 14–18 December 2016; pp. 3–26.

58. Cagli, E.; Dumas, C.; Prouff, E. Convolutional neural networks with data augmentation against jitter-based countermeasures. In Proceedings of the International Workshop on Cryptographic Hardware and Embedded Systems (CHES), Taipei, Taiwan, 25–18 September 2017; pp. 45–68.

59. Picek, S.; Samiotis, I.P.; Kim, J.; Heuser, A.; Bhasin, S.; Legay, A. On the performance of convolutional neural networks for side-channel analysis. In Proceedings of the International Conference on Security, Privacy, and Applied Cryptography Engineering (SPACE), Goa, India, 13–17 December 2018; pp. 157–176.

60. Carbone, M.; Conin, V.; Cornélie, M.A.; Dassance, F.; Dufresne, G.; Dumas, C.; Prouff, E.; Venelli, A. Deep Learning to Evaluate Secure RSA Implementations. *IACR Trans. Cryptogr. Hardw. Embed. Syst.* **2019**, *2*, 132–161.

61. Won, Y.S.; Lee, J.; Han, D.G. Side Channel Leakages Against Financial IC Card of the Republic of Korea. *Appl. Sci.* **2018**, *8*, 2258. [CrossRef]

62. Sim, B.Y.; Kang, J.; Han, D.G. Key Bit-Dependent Side-Channel Attacks on Protected Binary Scalar Multiplication. *Appl. Sci.* **2018**, *8*, 2168. [CrossRef]

63. Cho, S.M.; Jin, S.; Kim, H. Side-Channel Vulnerabilities of Unified Point Addition on Binary Huff Curve and Its Countermeasure. *Appl. Sci.* **2018**, *8*, 2002. [CrossRef]

64. Kim, S.; Hong, S. Single Trace Analysis on Constant Time CDT Sampler and Its Countermeasure. *Appl. Sci.* **2018**, *8*, 1809. [CrossRef]

65. An, S.; Kim, S.; Jin, S.; Kim, H.; Kim, H. Single Trace Side Channel Analysis on NTRU Implementation. *Appl. Sci.* **2018**, *8*, 2014. [CrossRef]

66. Shin, Y. Fast and Secure Implementation of Modular Exponentiation for Mitigating Fine-Grained Cache Attacks. *Appl. Sci.* **2018**, *8*, 1304. [CrossRef]

67. Briongos, S.; Malagón, P.; de Goyeneche, J.M.; Moya, J.M. Cache Misses and the Recovery of the Full AES 256 Key. *Appl. Sci.* **2019**, *9*, 944. [CrossRef]

68. Komano, Y.; Hirose, S. Re-Keying Scheme Revisited: Security Model and Instantiations. *Appl. Sci.* **2019**, *9*, 1002. [CrossRef]

69. Gołofit, K.; Wieczorek, P.Z. Chaos-Based Physical Unclonable Functions. *Appl. Sci.* **2019**, *9*, 991. [CrossRef]

70. Mukhtar, N.; Mehrabi, M.A.; Kong, Y.; Anjum, A. Machine-Learning-Based Side-Channel Evaluation of Elliptic-Curve Cryptographic FPGA Processor. *Appl. Sci.* **2019**, *9*, 64. [CrossRef]

71. Koo, D.; Shin, Y.; Yun, J.; Hur, J. Improving Security and Reliability in Merkle Tree-Based Online Data Authentication with Leakage Resilience. *Appl. Sci.* **2018**, *8*, 2532. [CrossRef]

72. Li, Y.; Kasuya, M.; Sakiyama, K. Comprehensive Evaluation on an ID-Based Side-Channel Authentication with FPGA-Based AES. *Appl. Sci.* **2018**, *8*, 1898. [CrossRef]

73. Su, M.Y.; Wei, H.S.; Chen, X.Y.; Lin, P.W.; Qiu, D.Y. Using Ad-Related Network Behavior to Distinguish Ad Libraries. *Appl. Sci.* **2018**, *8*, 1852. [CrossRef]

applied
sciences

MDPI

Article

Single Trace Analysis on Constant Time CDT Sampler and Its Countermeasure

Suhri Kim and Seokhie Hong *

Center for Information Security Technologies (CIST), Korea University, Seoul 02841, Korea; suhrikim@gmail.com
* Correspondence: shhong@korea.ac.kr

Received: 17 September 2018; Accepted: 29 September 2018; Published: 3 October 2018

Abstract: The Gaussian sampler is an integral part in lattice-based cryptography as it has a direct connection to security and efficiency. Although it is theoretically secure to use the Gaussian sampler, the security of its implementation is an open issue. Therefore, researchers have started to investigate the security of the Gaussian sampler against side-channel attacks. Since the performance of the Gaussian sampler directly affects the performance of the overall cryptosystem, countermeasures considering only timing attacks are applied in the literature. In this paper, we propose the first single trace power analysis attack on a constant-time cumulative distribution table (CDT) sampler used in lattice-based cryptosystems. From our analysis, we were able to recover every sampled value in the key generation stage, so that the secret key is recovered by the Gaussian elimination. By applying our attack to the candidates submitted to the National Institute of Standards and Technology (NIST), we were able to recover over 99% of the secret keys. Additionally, we propose a countermeasure based on a look-up table. To validate the efficiency of our countermeasure, we implemented it in Lizard and measure its performance. We demonstrated that the proposed countermeasure does not degrade the performance.

Keywords: post-quantum cryptography; lattice-based cryptography; Gaussian sampling; CDT sampling; side-channel attack; single trace analysis

---

1. Introduction

The security of currently used public-key cryptosystems is based on the hardness of mathematical problems such as integer factorization or a discrete logarithm problem over a finite field. For example, RSA, proposed in 1977, is based on the hardness of factoring large integers. Elliptic curve cryptography (ECC), proposed in 1985, is based on the hardness of solving discrete logarithm problem on an elliptic curve (ECDLP). Since the currently best-known algorithms for solving integer factorization and ECDLP take sub-exponential time and exponential time, respectively, RSA and ECC are believed to be secure when using large enough parameters. However, due to the seminal work of Peter Shor in 1994, these problems can be solved in polynomial time when a quantum computer is built running the Shor algorithm [1]. For this reason, quantum-resistant cryptographic schemes have become an active area of research due to the increase in concerns over the security of current public-key cryptosystems.

Post-quantum cryptography (PQC) refers to cryptographic algorithms executed on a classical computer that is expected to be secure against adversaries with quantum computers. PQC is considered to be secure against quantum computers since there is no known quantum algorithm that can solve security base problems more efficiently than classical algorithms. Five categories are mainly considered in post-quantum cryptography: multivariate-based cryptography, code-based cryptography, lattice-based cryptography, hash-based digital signature, and isogeny-based cryptography. Multivariate-based cryptography is based on the hardness of solving a system of multivariate equations, while code-based cryptography is based on the hardness of decoding general linear code. Lattice-based cryptography is based on the hardness of solving lattice problems and hash-based digital signature is based on the security of cryptographic hash functions. Lastly, isogeny-based cryptography is based on the hardness of finding isogeny between two given elliptic curves. Each cryptosystem in its category has both pros and cons, and no cryptosystem is known to be better than the others. After the National Institute of Standards and Technology (NIST) announced a standardization project for PQC in 2016, its first submission ended in November 2017. Among the 69 submissions, 26 were lattice-based cryptosystems that ranked the highest. Lattice-based cryptography was first proposed by Ajtai and Dwork in 1997 [2]. Later on, due to the groundbreaking work of Regev in 2005 and 2010, lattice-based cryptography became one of the most promising candidates in PQC because of its efficiency [3,4]. By using the learning with errors (LWE) problem over an ideal lattice as a security base, the key size and ciphertext size have decreased significantly compared with Ajtai and Dwork's proposal. Therefore, most cryptographic schemes in lattice-based cryptography nowadays are based on the LWE problem.

Intuitively, the LWE problem aims to find a solution s given a sequence of "approximate" random linear equations on s. That is, given a set of linear equations on s—e.g., $As = B$, where A and B are known, we insert an error e to make the problem significantly more difficult. In this regard, the attacker now needs to solve $As + e = B$, where e is unknown to him. Components of this error vectors are selected from a Gaussian distribution. If an attacker knows all the values of error vectors, the system can be easily broken. In this regard, the closer the sampler is to the actual Gaussian distribution, the more secure the cryptosystem is. At the same time, since the Gaussian sampler is one of the main modules in lattice-based cryptography, it must be efficient to avoid performance degradation. Therefore, previous works have been conducted on efficiency and accuracy (i.e., closeness to the ideal Gaussian distribution) trade-off for Gaussian sampling algorithms [5–8]. Recently, security against side-channel attacks on a Gaussian sampler has been recognized as an important problem.

Just as side-channel analysis has been thoroughly performed on classical cryptosystems such as RSA or ECC, we also need to examine the case in lattice-based cryptography. A side-channel attack proposed by Kocher et al. in 1996 is an attack based on exploiting information gained from the physical implementation of a cryptographic system [9]. In order to substitute RSA or ECC with post-quantum cryptography, resistance against side-channel analysis is required. Hence, researchers are examining possible side-channel attacks on lattice-based cryptography. Particularly, side-channel attacks on Gaussian sampling have been proposed [7,10,11]. Since the Gaussian sampler uses different random values each time of execution, side-channel attacks that exploit multiple iterations of an algorithm such as differential power analysis (DPA) cannot be performed. Moreover, since efficiency is as important as security when implementing the Gaussian sampler, countermeasures that significantly degrade the performance cannot be used. In this regard, current implementation on Gaussian samplers mainly considers timing attacks so that constant-time algorithms on Gaussian sampling have been proposed [7,12–14]. However, single-trace attacks targeting the Gaussian sampler have not been analyzed so far.

The goal of this work is to examine the vulnerability of the current implementation of CDT sampler and to provide its countermeasure. Due to its efficiency, the CDT sampler is one of the most widely used algorithms. As stated above, a constant-time CDT sampler is implemented to withstand timing side-channel attacks. The following list details the main contributions of this work.

- We performed the first single trace analysis (STA) on a constant-time CDT sampler. We theoretically and experimentally analyzed that existing a constant-time CDT algorithm is vulnerable to STA. Our attack is simple as it does not have any restriction on the attack environment. We exploit the gap of Hamming weight between the positive and negative value to recover the sampled value through a single power consumption trace. When the errors are known, the attacker can reveal the secret key by solving $As = B - e$. Therefore, every lattice-based cryptosystem including key exchange is vulnerable to STA. Details of our attack are presented in Section 3.

- We proposed an efficient countermeasure based on look-up table schemes. Especially for the Gaussian sampling algorithm used in lattice-based cryptography, the countermeasures must be as efficient as possible since every entry of an error matrix used in the cryptosystem requires executions of the sampler individually. In this regard, at least 2000 samplings are required for single executions of an algorithm proposed nowadays. Our countermeasure is efficient since it is faster than the constant-time CDT sampler, with the use of ROM. However, the additional memory usage is minimal—only about 256 to 2048 bytes, depending on the choice of parameter. To validate the efficiency of our countermeasure, we implemented it in FrodoKEM and Lizard and compared its speed.
- We present an implementation result to validate the efficiency of our proposed countermeasure. We implemented our countermeasure in Lizard, one of the submitted candidates in NIST standardization projects [15]. From our result, we demonstrated that our countermeasure does not degrade the performance of the constant-time CDT sampler. The result of our implementation is given in Section 4.

This paper is organized as follows: Section 2 recalls the definition of a lattice and describes lattice-based cryptography and a CDT sampler. In Section 3, our attack on constant-time CDT sampler is described. We propose our countermeasure in Section 4 and conclude in Section 5.

2. Preliminaries

In this section, we briefly introduce lattice and lattice-based cryptography. Next, we describe the CDT sampling method. Although there are several sampling methods in the literature, we focused on the CDT sampler since it is most widely used due to its efficiency. In the last subsection, we describe timing a side-channel attack performed in the CDT sampler and its countermeasure.

2.1. Lattice-Based Cryptography

Intuitively, a lattice can be defined as a discrete subgroup of $\mathbb{R}^n$ or a set of integer combinations of basis. Below is the formal definition of a lattice.

Definition 1. *Let $B = \{b_1, ..., b_d\} \subset \mathbb{Z}^n$ be a set of d independent vectors. Then, the set*

$$L(\{b_1, ..., b_d\}) = \left\{ \sum_{i=1}^{d} x_i \cdot b_i \middle| x_i \in \mathbb{Z} \right\} \tag{1}$$

is a lattice and B is called a basis of L.

Given two bases B and B', B and B' generate the same lattice if and only if there exists a unimodular matrix U such that $B = UB'$. It is known that there are infinitely many bases for a given lattice L, and some bases are considered as "good" and many are considered as "bad". Using a "good" basis, one can solve lattice problems more efficiently. Below are definitions of some of the computational problems on lattices.

Definition 2. *(Shortest Vector Problem, SVP) Given an arbitrary basis B of some lattice $L = L(B)$, find a shortest nonzero lattice vector.*

Definition 3. *(Closest Vector Problem, CVP) Given an arbitrary basis B of some lattice $L = L(B)$, and a target vector $t \in \mathbb{R}^n$, find some $v \in L$ such that $||t - v|| \leq \gamma \cdot dist(t, L)$, where $dist(L, t) = min_{x \in L}||x - t||$ and $\gamma \geq 1$ is an approximation factor.*

The above problems are known to be NP-hard, and several cryptosystems based on these problems have been proposed. Since there is no quantum algorithm to solve these problems efficiently, such lattice-based cryptosystems are secure even against quantum computers. The first public-key cryptosystem based on lattice was proposed by Ajtai and Dwork in 1997 [2]. Although their system was theoretically secure, it was inefficient due to large public-key, private key, and ciphertext sizes. Meanwhile, NTRU was proposed in 1996 by Hoffstein, Pipher, and Silverman [16]. Their system was based on the closest vector problem and has remained secure until now. NTRU is famous for faster operation compared with RSA at equivalent cryptographic strength.

In 2005, Regev proposed that the LWE problem is as hard to solve as several worst-case lattice problems [3]. He showed a quantum reduction from worst-case lattice problems such as SVP to LWE. This LWE problem can be considered as an extension of the "learning parity with noise" problem to higher moduli. Regev further proposed a public-key cryptosystem based on this problem, which is significantly more efficient than the previous lattice-based cryptosystems [3]. The public-key is of size $\tilde{O}(n^2)$ and encrypting a message increases its size by a factor of $\tilde{O}(n)$, where previous cryptosystems were $\tilde{O}(n^4)$ and $\tilde{O}(n^2)$, respectively. Before describing the LWE problem, we first define the LWE distribution.

Definition 4. *(LWE distribution) For a vector $s \in \mathbb{Z}_q^n$, called the secret, the LWE distribution $A_{s,\chi}$ over $\mathbb{Z}_q^n \times \mathbb{Z}_q$ is sampled by choosing $a \in \mathbb{Z}_q^n$ uniformly at random, choosing $e \leftarrow \chi$ and generating $(a, b = <s, a> + e \mod q)$.*

In the above definition, χ is an error distribution over $\mathbb{Z}$, usually taken to be a discrete Gaussian distribution. There are two versions of the LWE problem: the search-LWE problem that aims to find the secret vector given LWE samples and the decision-LWE problem that aims to distinguish between LWE samples and uniformly random samples. Below are definitions of search-LWE and decision-LWE problems [17]. In 2010, Lyubashevsky, Peikert, and Regev proposed a ring-base analog of LWE problems, known as the ring LWE problem (RLWE) [4]. Due to its structure and the usage of number theoretic transform (NTT) as an underlying computation, RLWE further improved lattice-based cryptography.

Definition 5. *(Search-LWE problem) Given m independent samples $(a_i, b_i) \in \mathbb{Z}_q^n \times \mathbb{Z}_q$ drawn from $A_{s,\chi}$ for a uniformly random $s \in \mathbb{Z}_q^n$ (fixed for all samples), find s.*

Definition 6. *(Decision-LWE problem) Given m independent samples $(a_i, b_i) \in \mathbb{Z}_q^n \times \mathbb{Z}_q$ where every sample is distributed according to either (1) $A_{s,\chi}$ for a uniformly random $s \in \mathbb{Z}_q^n$ (fixed for all samples), or (2) the uniform distribution, which distinguish which is the case (with non-negligible advantage).*

Note that, without errors, both problems are easy to solve using Gaussian elimination. Errors smudge the lattice points and somewhat erase the structure of the lattice, making problems challenging to solve. After Regev's work, most of the lattice-based cryptosystems are based on LWE. Below is the encryption algorithm proposed by Regev.

- **Key Generation:** Sample $A \in \mathbb{Z}_q^{m \times n}$ and $s \in \mathbb{Z}_q^n$, chosen uniformly at random. Sample $e \in \chi^m$ from Gaussian distribution. Compute $b = As + e$. The public-key is $(A, b) \in \mathbb{Z}_q^{(n+1) \times m}$ and the secret key is $s \in \mathbb{Z}_q^n$.
- **Encryption:** For given message M, let $t \leftarrow \{0,1\}^m$. Compute $C_1 = t^T A$ and $C_2 = t^T b + \lfloor \frac{q}{2} \rceil \cdot M$. The ciphertexts are (C_1, C_2).
- **Decryption:** Compute $d = C_2 - C_1 \cdot s$. If the result is close to 0, output 0, else output 1.

As error vectors play an important role in the security of the LWE-based cryptosystem, the importance of implementing the Gaussian sampler has been raised.

2.2. Discrete Gaussian Sampling

Before describing the Gaussian sampling algorithms, the discrete Gaussian distribution on a lattice is defined below.

Definition 7. *(Discrete Gaussian Distribution) Let $L \subset \mathbb{Z}^n, c \in \mathbb{R}^n, \sigma \in \mathbb{R}^+$. Define:*

$$\rho_{\sigma,c}(x) = exp\left(-\pi \frac{\|x-c\|^2}{\sigma^2} \right) \text{ and } \rho_{\sigma,c}(L) = \sum_{x \in L} \rho_{\sigma,c}(x).$$

The discrete Gaussian distribution over lattice L with center c and parameter σ is defined as

$$\forall x \in L, D_{L,\sigma,c}(x) = \frac{\rho_{\sigma,c}(x)}{\rho_{\sigma,c}(L)}.$$

For $c = 0$, denote $\rho_{\sigma,0}$ as ρ_σ and $D_{L,\sigma,0}$ as $D_{L,\sigma}$.

In lattice-based cryptography, one has to sample a vector from a discrete Gaussian distribution on a lattice $L \subset \mathbb{Z}^n$. This reduces to sampling n number of integers from a discrete Gaussian distribution on $\mathbb{Z}$. There are several methods for sampling numbers from Gaussian distributions. Gaussian sampling algorithms aim to produce a distribution of random numbers that are statistically close to the ideal Gaussian distribution within a certain bound, depending on the security parameter. The distance between the two distributions can be measured by statistical distance. The statistical distance between the distribution generated by Gaussian sampler X and ideal Gaussian distribution Y is defined as follows:

$$\Delta(X, Y) = \frac{1}{2} \sum_{x \in L} |Pr(X = x) - Pr(Y = x)|.$$

Given the security parameter λ, we aim to produce random numbers from X with certain precision so that its statistical distance between ideal Gaussian distribution Y is $\Delta(X, Y) < 2^\lambda$.

Recently, Bai et al. proposed that Rényi divergence can be used as an alternative to the statistical distance in security proofs for lattice-based cryptography [8]. The definition of Rényi divergence is as follows.

Definition 8. *(Rényi divergence of order a) For any two discrete probability distribution P and Q, where $Supp(P) \subseteq Supp(Q)$, and $a \in (0, \infty)$, Rényi divergence of order a is defined as*

$$R_a(P\|Q) = \left(\sum_{x \in Supp(P)} \frac{P(x)^a}{Q(x)^{a-1}} \right)^{\frac{1}{a-1}}.$$

Bai et al. showed that using Rényi divergence in place of statistical distance can result in less precision in the implementation of the security analysis for lattice-based cryptography [8]. In some cases, using Rényi divergence leads to security proofs allowing for taking smaller parameters in the cryptographic schemes. For more information, please refer to [8]. The work of Bai et al. is evidence that Gaussian sampling is an integral part of lattice-based cryptography that affects both security and efficiency. In [13], Micciancio and Walter also stated that the discrete Gaussian sampling can be the main hurdle in implementation, and a bottleneck to achieve good performance. For example, one parameter set of FrodoKEM-640 requires $640 \times 8 = 5,120$ entries in the error matrix, which is equal to the number of samplings needed [18]. Therefore, one has to consider the characteristics of an algorithm and a minimum number of bits required to satisfy the desired security level for efficient implementation. Before presenting the Gaussian sampling methods used in this field, the parameters needed in Gaussian sampling are as follows [19].

- **The standard deviation:** The standard deviation σ determines the shape of the Gaussian distribution. The larger the standard deviation, the more difficult the implementation. That is, the required bits and table sizes increase when the standard deviation is large.
- **The precision parameter:** The precision parameter λ determines how close the Gaussian sampler is to the ideal Gaussian distribution.
- **The tailcut parameter:** Given a target security of λ bits, the target distance from the ideal Gaussian distribution should not be no less than 2^λ. Hence, there is no need to sample from $x \in (-\infty, \infty)$. Instead, x is selected from $x \in (-\tau\sigma, \tau\sigma)$ for some tailcut parameter τ, depending on λ.

Cumulative Distribution Table Sampling

There are many sampling algorithms such as rejection sampling and Knuth–Yao sampling [20,21]. However, since the Gaussian sampler plays a crucial role in the performance of LWE-based cryptosystems, CDT sampling is one of the widely chosen sampling algorithms due to its efficiency. CDT sampling is an instantiation of inversion sampling that requires a precomputed cumulative density function table.

The cumulative distribution function (CDF) of a random variable X is defined as follows.

Definition 9. *(Cumulative Distribution Function (CDF)) The CDF F_X of a random variable X is the function given by*

$$F_X = P(X \leq x).$$

The right-hand side of the equation represents the probability of the random variable X takes on a value less than or equal to x. The CDT sampler first constructs the CDT table evaluated at some positive integer points including zero. Given a random number x, sampled from uniform distribution, the idea of the CDT sampler is to return the smallest index i of the precomputed CDT table Ψ such that $\Psi[i] < x \leq \Psi[i+1]$. The length of the table depends on the Gaussian parameters, tailcut, and precision. An example of sampling by CDT is as follows:

Suppose we want to sample from some distribution f, by sampling eight bits uniformly. The first seven bits correspond to a uniform random integer in $[0, 128)$, and the last eighth bit is used to determine the sign of the sampled value. Suppose that, by sampling $0, \pm1, \pm2$, and ±3, we can obtain the desired statistical distance between f and our CDT sampler. Then, we first obtain the probability of $0, 1, 2$ and 3: $P[i] = \{0.4, 0.3, 0.2, 0.1\}$ according to f. Then, we map this probability to integers between $[0, 128)$: $\boldsymbol{P}[i] = \{51, 38, 26, 13\}$. From this probability density table, we obtain: $\Psi[i] = \{26, 64, 90, 103\}$. Note that we halved the probability of 0 since $-0 = +0$ so that 0 is sampled twice. By sampling $x \in [0, 128)$, CDT sampler returns the smallest $i \in [0, 3]$ such that $x \leq \Psi[i]$. The previous implementation used a binary search to find i efficiently.

Note that the greater the number of bits used to sample, the more accurate the sampler is to the Gaussian distribution. To conclude the section, CDT sampling is a popular choice for implementing lattice-based cryptographic schemes due to their efficiency. Rejection sampling requires high-precision floating-point arithmetic [6]. Knuth–Yao sampling is fast, but it requires large precomputed tables when the standard deviation is large. Therefore, the CDT sampler is widely chosen when implementing lattice-based cryptography.

2.3. Timing Attacks on CDT Sampler

As stated repetitively, the Gaussian sampler is the core element in implementing lattice-based cryptography. If an attacker knows every sampled value from the Gaussian sampler, then the secret key is revealed by simple linear algebra. Hence, resistance against a potential side-channel attack is necessary. In this section, we mainly focus on timing attack performed in the CDT sampler and its countermeasures.

The timing attack proposed by Kocher et al. is a side-channel attack performed by measuring the time of operation of a cryptographic device [22]. The attack exploits the data-dependent characteristic of an implementation that leads to the revelation of the secret key. For example, if an implementation terminates or executes an algorithm depending on the input data, this may be vulnerable to the timing attack. A timing attack on a CDT sampler is performed as follows. Recall that steps 4 to 6 in Algorithm 1 are repeated until it finds k such that $\Psi[k]$ is smaller than the selected random S. Let n be the length of a CDT table. By measuring its time, an attacker can classify the timing of an algorithm to n different groups, where the shortest sampling time will most likely have output zero. This means that naive implementation allows an attacker to determine the sampled value which can lead to the secret key. Hence, Poppelmann et al. proposed a constant-time CDT sampler which guarantees that no information is leaked through timing analysis [7]. Below is the pseudocode of the constant-time CDT sampler implemented in cryptosystems such as FrodoKEM [18].

Algorithm 1 CDT sampling

Require: CDT table Ψ, σ, τ
Ensure: Sampled value S
 1: $S \leftarrow [0, \tau\sigma)$
 2: $sign \leftarrow [0, 1] \cap \mathbb{Z}$ uniformly at random
 3: $k \leftarrow 0$
 4: while $(S > \Psi[k])$
 5:    $k{+}{+}$
 6: end while
 7: $S \leftarrow ((-sign) \wedge k) + sign$
 8: return S

Note that, instead of terminating the algorithm when the index is found, Algorithm 2 searches through the whole table. Algorithm 2 adds 0 to a variable S when the selected random value is less than the table value and adds 1 when the selected random value is greater than the table value. Therefore, the returned value can be equal to the index i such that $rnd < \Psi[i+1]$ even though the whole table has been searched. The timing attack can be eliminated since the execution time of the algorithm is independent from the selected random value.

Algorithm 2 Constant-time CDT sampling

Require: CDF table Ψ of length ℓ, σ, τ
Ensure: Sampled value S
 1: $S \leftarrow 0$
 2: $rnd \leftarrow [0, \tau\sigma) \cap \mathbb{Z}$ uniformly at random
 3: $sign \leftarrow [0, 1] \cap \mathbb{Z}$ uniformly at random
 4: For $i = 0$ up to $\ell - 1$ do
 5:    $S \mathrel{+}= (\Psi[i] - rnd) \gg 15$
 6: $S \leftarrow ((-sign) \wedge S) + sign$
 7: return S

---

3. Single Trace Analysis on a CDT Sampler

STA is a statistical power analysis attack using a single power consumption trace. The STA performs by statistically analyzing the distribution of an operation related to a secret key. For example, Amiel et al. recovered the secret exponent in RSA by exploiting the average Hamming weight difference between squaring and multiplication operations [23]. Since STA uses a single power consumption trace, blinded exponent or randomization is meaningless. More examples of these kinds of power analysis attacks include Horizontal Correlation Analysis (HCA) [24], Horizontal Collision Correlation Analysis (HCCA) [25], Recovery of Secret Exponent by Triangular Trace Analysis (ROSETTA) [26], and Hanley et al.'s attack [27]. In this section, we propose that the constant-time CDT sampler is not secure against STA. We theoretically and experimentally analyzed that an attacker can reveal every value of an error by exploiting a single power consumption trace. This indicates that lattice-based cryptosystems that use a constant-time CDT sampler are vulnerable since an attacker can reveal the secret key by using the recovered error vectors from STA.

This section is organized as follows: first, we justify our attack by examining raw traces of a CDT sampler. Then, we show that our attack works on the simulated CDT sampler. Finally, we show how to recover the secret key given the sampled values. We demonstrated that any other lattice-based cryptosystems that use the constant-time CDT sampler as in Algorithm 2 is vulnerable to STA.

3.1. Examining Raw Traces

The targets of attack are steps 4 to 6 in Algorithm 2. Due to the application of timing attack countermeasures, step 5 in Algorithm 2 is repeated *len* times regardless of the selected random value. However, note that 1 is added to S when the selected random is larger than the table value. When the selected random is larger than the table value, the result of a subtraction is negative so that the Hamming weight of the value increases. More specifically, note that the significant bits are 1 s when the negative numbers are expressed in two's complements. Therefore, the Hamming weight of the subtraction result is at least 16 bits minus the number of bits used for sampling. On the other hand, when the result of a subtraction is positive, then 0 is added to the value S. Consequently, its Hamming weight is at most the number of bits used for sampling. For a numerical example, suppose the values are expressed in 16-bit integers, and nine bits are used to sample as in FrodoKEM [18]. When the result of a subtraction is negative, the most significant seven bits are 1, while half of the remaining bits are 1 on average. Therefore, negative values will have Hamming weight of 11.5 on average. In a similar manner, when the result of a subtraction is positive, the most significant seven bits are zero, while half of the remaining bits are 1 on average. Hence, positive numbers will have Hamming weight of 4.5 on average.

Due to the gap of Hamming weight between positive and negative number, the sign of the subtracted value (i.e., $\Psi[i] - rnd$) is visible in the power consumption trace. Furthermore, this sign is related to adding 1 or 0 to the S. As a result, instantaneous power consumption increases when 1 is added to the value compared to when 0 is added. We claim that adding 1 and 0 in step 5 can be detected through STA, which allows an attacker to find out which number was sampled by counting this information.

In order to support our claim, we sampled using a CDT sampler that does not cast the subtracted value to an unsigned type. We implemented the CDT sampler in ATmega128 (Atmel, San Jose, CA, USA) and the measurement devices used to obtain the power consumption traces involved a LeCroy HDO6104A oscilloscope (Testforce, Pickering, ON, Canada) sampling at 1 GS/s. The red line in Figure 1a illustrates the power consumption trace when the subtraction result is negative, and the red line in Figure 1b illustrates the power consumption trace when the subtraction result is positive. As depicted in Figure 1, the result was expected as described above. Due to the large Hamming weight of the negative value, an attacker can deduce whether the subtracted result is positive or negative.

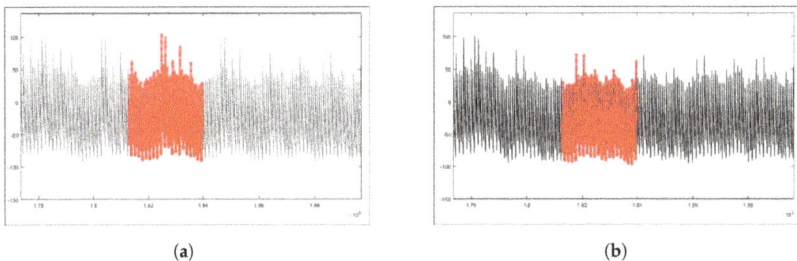

| (a) | (b) |

Figure 1. (**a**) power consumption trace when subtraction result is negative; (**b**) power consumption trace when subtraction result is positive.

3.2. Attack on the CDT Sampler

This section describes how to recover the sampled values through a single power consumption trace by using the result from the previous section. Although Algorithm 2 casts the value to an unsigned integer, the very moment when the difference between the sampled value and the table value is calculated can be captured in power consumption trace. In other words, high power consumption indicates that the negative value is calculated and hence 1 is added to the variable S. On the contrary, low power consumption indicates that the positive value is calculated so that 0 is added to the variable. For the experiment, a CDT sampler in Algorithm 2 with the table of length 9 was implemented in ATmega128. We sampled 100 times by selecting 100 different random value. In order to clearly capture when the CDT is processed, Algorithm 2 was repeated 10 times with the selected random number as input. The power consumption traces were collected using a LeCroy HDO6104A oscilloscope at a sampling rate of 250 M/s. We used a low pass filter on the collected power consumption traces to reduce noise. Figure 2 illustrates the entire power consumption trace of our implementation, which is then zoomed into the region of interest.

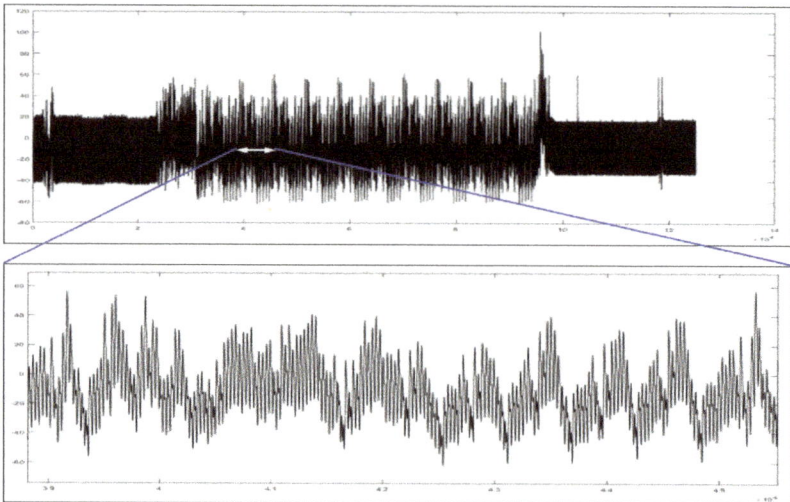

Figure 2. Part of a single execution of the algorithm.

As depicted in Figure 2, we can see 10 identical iterations where its unit is marked with a white arrow line. To better represent the power consumption trace of one execution of the CDT sampler, we cut the power consumption trace into each execution of an algorithm. The bottom of Figure 2 illustrates the power consumption trace when a single value is sampled.

If the power consumption of sampling is identical to every random value, that is, if the algorithm resists against STA, then the power consumption trace of sampling with certain random values will be identical to the power consumption trace of sampling with any other random values. However, this is not the case for the currently used constant-time CDT sampler. Due to the gap of the Hamming weight of the negative and positive values, the sign of the subtracted result is reflected in the power consumption trace. Figure 3a,b illustrates when the subtracted result is negative and positive, respectively. The power consumption trace when 1 or 0 was added is marked as a black line over gray-colored trace as shown in the figure below. As illustrated in Figure 3, the sign of the subtracted result can be detected through power consumption trace.

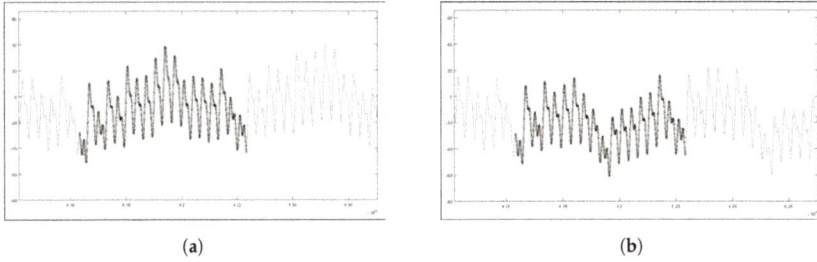

(a)   (b)

Figure 3. (**a**) power consumption trace when the subtracted result is negative; (**b**) power consumption trace when the subtracted result is positive.

The sign of the sampled bit can also be revealed through STA. Step 6 in Algorithm 2 depends on the sign bit selected in step 3. If 0 is selected, then the result of the sampled value is positive, and negative otherwise. Step 6 in Algorithm 2 flips the variable S when the sign bit is 1 and remains the same when the sign bit is 0. Therefore, this process is visible through power consumption trace. Figure 4 illustrates the power consumption trace when the sign bit is 1 and 0.

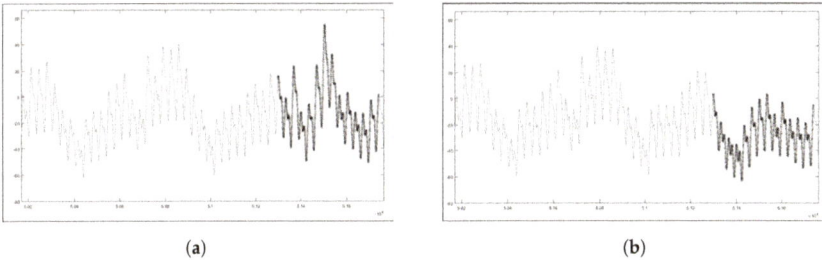

(a)   (b)

Figure 4. (**a**) power consumption trace when the sign bit is 1; (**b**) power consumption trace when the sign bit is 0.

For a clear comparison, Figure 5 is the overlapped power consumption trace of Figure 4a,b. The black power consumption trace illustrates when the sign bit is 1 and the red power consumption trace illustrates when the sign bit is 0.

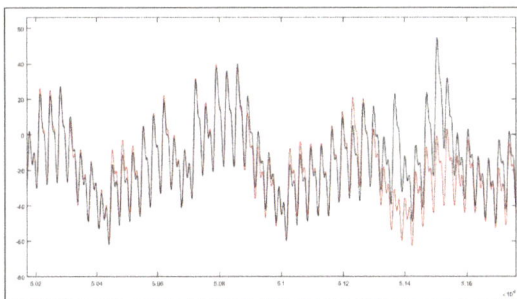

Figure 5. Overlapped power consumption trace of Figure 4a,b. The black line illustrates when the sign bit is 1 and the red line illustrates when the sign bit is 0.

Combining the information in Figures 3 and 4, an attacker can determine the sampled values exactly through a single power consumption trace. To justify our observation, we constructed a tool to

15

extract the sampled value automatically given a power consumption trace. First, we divided the power consumption trace into one execution of the algorithm. In our experiment, each execution consists of 6160 points. For each partitioned trace, eight iterations were visible, with each iteration consisting of 577 points on average. For example, step 4 starts approximately at 1,154, 1731, 2307, 2883, 3460, 4036, 4615, and 5192.

Figure 6 illustrates the part of the single execution of the algorithm, where the red lines indicate when step 4 starts. In order to visualize the difference of the graph when the subtraction result is positive or negative, power consumption traces when 6 and 2 are sampled through Algorithm 2 are overlapped in Figure 6. The blue power consumption trace illustrates when 6 is sampled and black power consumption trace illustrates when 2 is sampled. In our implementation, the sampled value is equal to the number of the subtractions that result in a negative value out of eight subtractions. Therefore, compared to the case when two are sampled, four more of the subtracted result is negative when six is sampled and is visualized as shown in the blue shaded rectangle as in Figure 6. Particularly, the difference between the power trace when the subtracted result is positive and negative was most noticeable after 157 points for each iteration on average. An example of such difference is illustrated in the blue shaded rectangle in Figure 6, which consists of 284 points on average. This means that the difference between the average power consumption when 0 or 1 is distinct so that an attacker can set a threshold to divide the gap. From our experiment, when the average is greater than –1, we assumed that the subtracted result was negative so that 1 was added to the variable S.

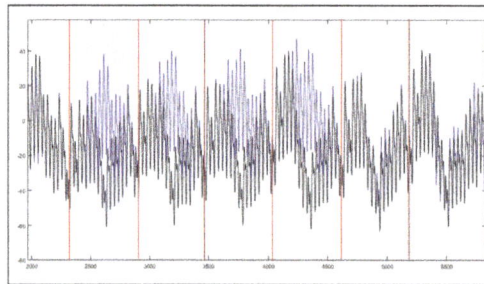

Figure 6. Part of a single execution of the algorithm. The black line is the power consumption trace when 6 is sampled, and red line is the power consumption trace when 2 is sampled. The blue area indicates the difference between the power consumption traces.

Let $x(i)$ be the point where the i-th loop starts. By calculating the average between $x(i) + 157$ and $x(i) + 441$ and comparing it with –1, we were able to deduce whether 1 or 0 was added to the variable. In this way, we were able to recover 1000 sampled values out of 1000 number of power consumption traces in Figure 1 and 100 sampled values out of 100 number of power consumption traces in Figure 2. Since the difference between the average power consumption of negative and positive is apparent, we believe that full recovery of the sampled value is possible even with a large number of samples.

To summarize the attack, STA on a CDT sampler can be performed as follows: suppose that an attacker obtains a power consumption trace that samples 2000 values. The attacker first divides the power consumption trace into each execution of the algorithm, which results in 2000 number of sub-traces. For each sub-trace, an attacker identifies loops of an algorithm. The number of loops must equal the number of table length. By overlapping the sub-traces, an attacker approximates the location where step 5 of the Algorithm 2 is processed. Next, by averaging the power consumption where one or zero adds, an attacker sets a threshold t. For each loop of the sub-trace, an attacker averages the power consumption trace and compares it with the threshold. If the average is greater than t, then the subtracted result is negative so that one is added to S. If the average is smaller than t, then the subtracted result is positive so that zero is added. An attacker can recover the sampled value since the summation of the number of 1s added is equal to the value sampled. The sign of the sampled value can be revealed with the similar process. Note that the CDT sampler is more vulnerable to such an attack if casting to an unsigned type is not used in step 5 of Algorithm 2. If the right shift of the difference is not unsigned, then −1 is added to the variable S making the gap wider. This results in more visibility in the power consumption graph.

3.3. Remarks on the Proposed Attack

As denoted in Section 2, for the public-key (A, b), a standard LWE-based cryptosystem takes s as the secret key where $b = As + e$ for an error matrix e sampled from the discrete Gaussian distribution. If e is revealed, then $As = b - e$ so that s can be solved in polynomial time using the Gaussian elimination. As we demonstrated above, any lattice-based cryptosystem that uses the constant time CDT sampler is vulnerable to our attack. This includes the FrodoKEM proposed by Bos et al. and Lizard proposed by Cheon et al. [15,18]. The secret key of FrodoKEM is $(s \| pk, S)$, where pk is the public-key, s is a random seed chosen uniformly at random, and S is a sample through the Gaussian sampler [18]. Although we can directly obtain S through our STA, the secret key cannot fully be recovered since there is no information relating to s. However, our attack reduces the substantial amount of entropy of the secret key. For a numerical example, consider FrodoKEM-640, which aims at the brute-force security of AES-128. The size of S is equal to $640 \times 8 \times 2 = 10240$ bytes, while the size of s equals to 16 bytes. Considering the public-key, we can recover 99% of the secret key through our attack. On the other hand, the secret key of Lizard is S, where S satisfies the equation $B = -AS + E$ for the public-key (A, B) and E sampled from the Gaussian distribution [15]. Therefore, by performing our STA to recover E, we were able to obtain S through Gaussian elimination.

4. Proposed Countermeasure

Due to the gap of the Hamming weight of negative and positive value, constant-time CDT sampler is vulnerable to STA. In Algorithm 2, when the selected random value is subtracted from the table value, 1 or 0 is added to the variable S if the result was negative or positive, respectively. The negative values will have higher power consumption than the positive values since negative values have larger Hamming weight. Therefore, an attacker can deduce whether 1 or 0 was added by analyzing the power consumption trace. By counting how many 1s were added to the S and recognizing the additional peak after the loop to determine the sign, an attacker can determine the exact sampled value. In this section, we propose an algorithm that resists against STA described in Section 3. Note that we also need to consider the efficiency of the countermeasure, given the fact that numerous samplings are conducted for one execution of the cryptosystem. Our countermeasure is efficient since it is faster than the unprotected algorithm with the use of ROM.

4.1. Table-Based CDT Sampler

Note that the output of the CDT sampler is determined from the moment the random value is selected from the uniform distribution. Let $\Psi[i] = \{26, 64, 90, 103\}$ be the corresponding CDT as in Section 2. The CDT sampler outputs 0 when the random value selected is between $[0, 26)$ and outputs 1 when the random value selected is between $[26, 64)$ and so on. Thus, we can make the look-up table as follows:

From the above table, we can construct the CDF table as $\Phi[i] = \{0, \ldots, 0, 1, \ldots 1, 2, \ldots, 4\}$. A similar method was implemented in [28], where the look-up table was constructed from the start and end addresses of the linear feedback shift register. Sampling values from this reconstructed CDF table are as denoted in Algorithm 3.

Unlike the constant-time CDT sampler described in Algorithm 2, the execution of Algorithm 3 does not depend on the sampled random value. In Algorithm 2, whether the selected random value is greater than the CDT table value was visible in the power consumption trace due to the difference in Hamming weight of positive and negative values. This vulnerability is eliminated in our countermeasure by precalculating what output of the sampler will be given the random input value. As soon as a random value is selected (step 2 of Algorithm 3) Algorithm 3 uses the random value as the index of the table Φ and outputs the corresponding table value. Since there is no Hamming weight difference or the difference in the execution of an algorithm depending on the selected random value, an attacker cannot extract information from a single power consumption trace.

Algorithm 3 STA-resistant CDT sampling

Require: CDF look-up table Φ, σ, τ
Ensure: Sampled value S
 1: $sign \leftarrow [0,1] \cap \mathbb{Z}$ uniformly at random
 2: $rnd \leftarrow [0, \tau\sigma) \cap \mathbb{Z}$ uniformly at random
 3: return $S \leftarrow (-sign) \wedge \Phi[rnd] + sign$

However, Algorithm 3 exposes the sign bit through STA in the same manner as in Algorithm 2. In order to prevent the exposure of the signed bit, one can additionally sample one more bit from the uniform distribution and double the length of the table instead of sampling the sign bit separately. For example, suppose that the original algorithm selects eight bits uniformly at random, where the first seven bits correspond to a uniform random integer in $[0, 128)$ and the last bit is used to determine the sign of the sampled value. To use the look-up table for sampling, first, sample 8 bits, and map to integer $[0, 128)$ when the most significant bit is 0 and to $(-128, 0]$ when the least significant bit is 1. Then, Table 1 changes to Table 2 with the length doubled.

Table 1. Sampled values given random input value.

x	0	1	...	25	26	...	63	64	...	127
Output	0	0	...	0	1	...	1	2	...	4

Table 2. Sampled values given random input value including the sign.

x	0	1	...	25	26	...	127	128	...	154	...	255
Output	0	0	...	0	1	...	4	0	...	−1	...	−4

One disadvantage of this algorithm is that it needs to store a look-up table. However, since the output of the Gaussian sampler does not exceed one byte, for λ-bit of precision, we need a table size of 2^{λ} bytes for Table 1 and $2^{\lambda+1}$ bytes when using Table 2. When applying the look-up table method to lattice-based cryptosystem, table size is approximately 128–1024 bytes for Table 1 and 256–2048 bytes for Table 2. Specifically, for FrodoKEM-640, since the CDT sampler uses a 7-bit random value, the size of Table 1 is 128 bytes, and the size of Table 2 is 256 bytes. Considering the Knuth–Yao table (1080 bytes for LOTUS-128 that aims the brute-force security of AES-128 [29]) used in the lattice-based cryptosystem, we believe that this storage size is acceptable.

4.2. Implementation

To validate the efficiency of our proposed countermeasure, we implemented our table-based CDT sampler on Lizard. The selected parameters for Lizard we used were CCA_CATEGORY1_N536, CCA_CATEGORY3_N816, and CCA_CATEGORY3_N952, which aim at the brute-force security of AES-128, AES-192, and AES-192, respectively. For specific parameters, please refer to [15]. The test was performed on one core of an Intel Core i7-6700 (Skylake) at 3.40 GHz, running Ubuntu 16.04 LTS. The key generation for each parameter was repeated 10^4 times, and its average is given in Table 3.

The CDT table length of CCA_CATEGORY1_N536, CCA_CATEGORY3_N816, and CCA_CATEGORY3_N952 are 9, 5, and 6, respectively. As shown in Table 3, our countermeasure becomes more efficient when the table length of the constant-time CDT sampler is long. In addition, as in Table 3, our countermeasure does not degrade the performance of the cryptosystem since applying our countermeasure does not make any difference in speed compared to the original implementation. Overall, the proposed countermeasure protects the STA without efficiency loss.

Table 3. Comparison of the running time of key generation stage of the original cryptosystem verses countermeasure-implemented cryptosystem represented in milliseconds.

	Lizard		
	CCA_CATEGORY1_N536	CCA_CATEGORY3_N816	CCA_CATEGORY3_N952
Original	213.687	335.144	361.328
Countermeasure (Table 1)	207.979	333.235	356.383
Countermeasure (Table 2)	206.739	330.509	353.166

5. Conclusions

In this paper, we showed that the existing constant-time CDT sampling method for Gaussian sampler is vulnerable against STA and proposed its countermeasure. Since the knowledge of the sampled value can reveal the secret key by using the Gaussian elimination, security regarding the implementation has been recognized as an important problem. However, as efficiency is as important as security when implementing the Gaussian sampler, only constant-time algorithms for Gaussian samplers have been proposed in order to prevent the timing side-channel attacks. However, we proposed that STA was possible for the constant-time CDT sampling method. The constant-time CDT sampling is vulnerable to STA due to the gap of Hamming weight of negative and positive values. This value can be related to adding one and zero to some variable, and, consequently, the sampled value can be revealed. We also recovered the sign bit of the sampled value in a similar way. Therefore, we were able to recover the exact sampled values each time the algorithm was executed. In order to resist STA, we proposed an alternative algorithm for the CDT-based Gaussian sampler. The idea is to use a look-up table that precomputes the sampled value from a given input. Unlike calculating the differences in the constant-time CDT sampler, the proposed countermeasure uses the random value as the index of the precomputed table and outputs the corresponding table value. Since there is no Hamming weight difference or the difference in the execution of an algorithm depending on the selected random value, an attacker cannot extract information from a single power consumption trace. In addition, we implemented our countermeasure in Lizard and check its performance. The proposed countermeasure does not degrade the performance since it works by directly referencing table values.

Author Contributions: S.K. performed the experiments, analyzed the data, and wrote the paper. S.H. analyzed the data and verified the validity of results.

Funding: This research was supported by the MSIT (Ministry of Science and ICT), Korea, under the ITRC (Information Technology Research Center) support program (IITP-2018-2015-0-00385) supervised by the IITP (Institute for Information & communications Technology Promotion).

Conflicts of Interest: The authors declare no conflict of interest.

References

1. Shor, P.W. Polynomial-time algorithms for prime factorization and discrete logarithms on a quantum computer. *SIAM Rev.* **1999**, *41*, 303–332. [CrossRef]
2. Ajtai, M.; Dwork, C. A public-key cryptosystem with worst-case/average-case equivalence. In Proceedings of the Twenty-Ninth Annual ACM Symposium on Theory of Computing, El Paso, TX, USA, 4–6 May 1997; pp. 284–293.
3. Regev, O. On lattices, learning with errors, random linear codes, and cryptography. *J. ACM* **2009**, *56*, 34. [CrossRef]

4. Lyubashevsky, V.; Peikert, C.; Regev, O. On ideal lattices and learning with errors over rings. In Proceedings of the Annual International Conference on the Theory and Applications of Cryptographic Techniques, French Riviera, France, 30 May–3 June 2010; Springer: Berlin/Heidelberg, Germany, 2010; pp. 1–23.

5. Peikert, C. An efficient and parallel Gaussian sampler for lattices. In Proceedings of the Annual Cryptology Conference, Santa Barbara, CA, USA, 15–19 August 2010; Springer: Berlin/Heidelberg, Germany, 2010; pp. 80–97.

6. Galbraith, S.D.; Dwarakanath, N.C. Efficient sampling from discrete Gaussians for lattice-based cryptography on a constrained device. *Preprint* **2012**.

7. Pöppelmann, T.; Güneysu, T. Towards practical lattice-based public-key encryption on reconfigurable hardware. In Proceedings of the International Conference on Selected Areas in Cryptography, Burnaby, BC, Canada, 14–16 August 2013; Springer: Berlin/Heidelberg, Germany, 2013; pp. 68–85.

8. Bai, S.; Lepoint, T.; Roux-Langlois, A.; Sakzad, A.; Stehlé, D.; Steinfeld, R. Improved security proofs in lattice-based cryptography: Using the Rényi divergence rather than the statistical distance. *J. Cryptol.* **2018**, *31*, 610–640. [CrossRef]

9. Kocher, P.; Jaffe, J.; Jun, B. Differential power analysis. In Proceedings of the Annual International Cryptology Conference, Santa Barbara, CA, USA, 15–19 August 1999; Springer: Berlin/Heidelberg, Germany, 1999; pp. 388–397.

10. Bruinderink, L.G.; Hülsing, A.; Lange, T.; Yarom, Y. Flush, Gauss, and Reload—A cache attack on the BLISS lattice-based signature scheme. In Proceedings of the International Conference on Cryptographic Hardware and Embedded Systems, Santa Barbara, CA, USA, 17–19 August 2016; Springer: Berlin/Heidelberg, Germany, 2016; pp. 323–345.

11. Espitau, T.; Fouque, P.A.; Gérard, B.; Tibouchi, M. Side-channel attacks on BLISS lattice-based signatures: Exploiting branch tracing against strongswan and electromagnetic emanations in microcontrollers. In Proceedings of the 2017 ACM SIGSAC Conference on Computer and Communications Security, Dallas, TX, USA, 30 October–3 November 2017; pp. 1857–1874.

12. Karmakar, A.; Roy, S.S.; Reparaz, O.; Vercauteren, F.; Verbauwhede, I. Constant-time Discrete Gaussian Sampling. *IEEE Trans. Comput.* **2018**. [CrossRef]

13. Micciancio, D.; Walter, M. Gaussian sampling over the integers: Efficient, generic, constant-time. In Proceedings of the Annual International Cryptology Conference, Santa Barbara, CA, USA, 20–24 August 2017; Springer: Cham, Switzerland, 2017; pp. 455–485.

14. Roy, S.S.; Reparaz, O.; Vercauteren, F.; Verbauwhede, I. Compact and Side Channel Secure Discrete Gaussian Sampling. *IACR Cryptol. ePrint Arch.* **2014**, *2014*, 591.

15. Cheon, J.H.; Kim, D.; Lee, J.; Song, Y. Lizard: Cut Off the Tail! A Practical Post-quantum Public-Key Encryption from LWE and LWR. In Proceedings of the International Conference on Security and Cryptography for Networks, Amalfi, Italy, 5–7 September 2018; Springer: Cham, Switzerland, 2018; pp. 160–177.

16. Hoffstein, J.; Pipher, J.; Silverman, J.H. NTRU: A ring-based public key cryptosystem. In Proceedings of the International Algorithmic Number Theory Symposium, Portland, OR, USA, 21–25 June 1998; Springer: Berlin/Heidelberg, Germany, 1998; pp. 267–288.

17. Peikert, C. A decade of lattice cryptography. *Found. Trends® Theor. Comput. Sci.* **2016**, *10*, 283–424. [CrossRef]

18. Bos, J.; Costello, C.; Ducas, L.; Mironov, I.; Naehrig, M.; Nikolaenko, V.; Raghunathan, A.; Stebila, D. Frodo: Take off the ring! practical, quantum-secure key exchange from LWE. In Proceedings of the 2016 ACM SIGSAC Conference on Computer and Communications Security, Vienna, Austria, 24–28 October 2016; pp. 1006–1018.

19. Du, C.; Bai, G. Towards efficient discrete Gaussian sampling for lattice-based cryptography. In Proceedings of the 2015 25th International Conference on Field Programmable Logic and Applications (FPL), London, UK, 2–4 September 2015; pp. 1–6.

20. Gentry, C.; Peikert, C.; Vaikuntanathan, V. Trapdoors for hard lattices and new cryptographic constructions. In Proceedings of the Fortieth Annual ACM Symposium on Theory of Computing, Victoria, BC, Canada, 17–20 May 2008; pp. 197–206.

21. Knuth, D.; Yao, A. The complexity of nonuniform random number generation. In *Algorithms and Complexity: New Directions and Recent Results*; Academic Press: Cambridge, MA, USA, 1976.

22. Kocher, P.C. Timing attacks on implementations of Diffie-Hellman, RSA, DSS, and other systems. In Proceedings of the Annual International Cryptology Conference, Santa Barbara, CA, USA, 18–22 August 1996; Springer: Berlin/Heidelberg, Germany, 1996; pp. 104–113.

23. Amiel, F.; Feix, B.; Tunstall, M.; Whelan, C.; Marnane, W.P. Distinguishing multiplications from squaring operations. In Proceedings of the International Workshop on Selected Areas in Cryptography, Sackville, NB, Canada, 14–15 August 2008; Springer: Berlin/Heidelberg, Germany, 2008; pp. 346–360.

24. Clavier, C.; Feix, B.; Gagnerot, G.; Roussellet, M.; Verneuil, V. Horizontal correlation analysis on exponentiation. In Proceedings of the International Conference on Information and Communications Security, Barcelona, Spain, 15–17 December 2010; Springer: Berlin/Heidelberg, Germany, 2010; pp. 46–61.

25. Bauer, A.; Jaulmes, E.; Prouff, E.; Reinhard, J.R.; Wild, J. Horizontal collision correlation attack on elliptic curves. *Cryptogr. Commun.* **2015**, *7*, 91–119. [CrossRef]

26. Clavier, C.; Feix, B.; Gagnerot, G.; Giraud, C.; Roussellet, M.; Verneuil, V. ROSETTA for single trace analysis. In Proceedings of the International Conference on Cryptology in India, Kolkata, India, 9–12 December 2012; Springer: Berlin/Heidelberg, Germany, 2012; pp. 140–155.

27. Hanley, N.; Kim, H.; Tunstall, M. Exploiting collisions in addition chain-based exponentiation algorithms using a single trace. In Proceedings of the Cryptographers' Track at the RSA Conference, San Francisco, CA, USA, 20–24 April 2015; Springer: Cham, Switzerland, 2015; pp. 431–448.

28. Göttert, N.; Feller, T.; Schneider, M.; Buchmann, J.; Huss, S. On the design of hardware building blocks for modern lattice-based encryption schemes. In Proceedings of the International Workshop on Cryptographic Hardware and Embedded Systems, Leuven, Belgium, 9–12 September 2012; Springer: Berlin/Heidelberg, Germany, 2012; pp. 512–529.

29. Le Trieu Phong, T.H.; Aono, Y.; Moriai, S. LOTUS: Algorithm Specifications and Supporting Documentation. Available online: https://csrc.nist.gov/Projects/Post-Quantum-Cryptography/Round-1-Submissions (accessed on 17 September 2018)

applied
sciences

MDPI

Article

Single Trace Side Channel Analysis on NTRU Implementation

Soojung An [1], Suhri Kim [1], Sunghyun Jin [1], HanBit Kim [1] and HeeSeok Kim [2,*]

[1] Graduate School of Information Security and Institute of Cyber Security & Privacy (ICSP), Korea University, Seoul 02841, Korea; soojung02@korea.ac.kr (S.A.); suhrikim@gmail.com (S.K.); sunghyunjin@korea.ac.kr (S.J.); luz_damoon@naver.com (H.K.)

[2] Department of Cyber Security, College of Science and Technology, Korea University, 2511 Sejong-Ro, Sejong 30019, Korea

* Correspondence: 80khs@korea.ac.kr; Tel.: +82-044-860-1383

Received: 17 September 2018; Accepted: 17 October 2018; Published: 23 October 2018

Abstract: As researches on the quantum computer have progressed immensely, interests in post-quantum cryptography have greatly increased. NTRU is one of the well-known algorithms due to its practical key sizes and fast performance along with the resistance against the quantum adversary. Although NTRU has withstood various algebraic attacks, its side-channel resistance must also be considered for secure implementation. In this paper, we proposed the first single trace attack on NTRU. Previous side-channel attacks on NTRU used numerous power traces, which increase the attack complexity and limit the target algorithm. There are two versions of NTRU implementation published in succession. We demonstrated our attack on both implementations using a single power consumption trace obtained in the decryption phase. Furthermore, we propose a countermeasure to prevent the proposed attacks. Our countermeasure does not degrade in terms of performance.

Keywords: side channel analysis; single trace analysis; post quantum cryptography; NTRU

1. Introduction

The currently used public key cryptography (PKC) such as RSA and Elliptic Curve Cryptography (ECC) are no longer secure if the quantum computer is developed running the Shor algorithm [1–3]. Due to the recent advances in quantum computing, post-quantum cryptography (PQC) is an active area of research. Moreover, the national institute of standards and technology (NIST) announced a project to define new standards for the PKC [4]. There are five categories studied for PQC: lattice-based cryptography, multivariate-based cryptography, hash-based cryptography, code-based cryptography, and isogeny-based cryptography. Among those categories, lattice-based cryptography is one of the prominent candidates due to the fast performance with a practical key size. In the same context, the largest number of candidates submitted to NIST project belong to the lattice-based cryptography. One of the well known lattice-based cryptography is NTRU, which is an abbreviation of N-th degree truncated polynomial ring.

NTRU proposed in 1996 by Hoffstein et al. [5] is an encryption algorithm based on the shortest vector problem. NTRU has attracted much attention to the researchers due to the faster speed than classical cryptosystems by more than two orders of magnitude on the same security level. Regarding the implementation, only the encryption code was open to the public until 2017. After the patent release in 2017, all of the source code was available. Currently, two kinds of implementation are proposed in the literature, one released on GitHub in 2017 and the other submitted on NIST standardization project. We distinguish the prior one as NTRU Open Source and the latter as NTRUEncrypt, and also the algorithm itself as NTRU. Although NTRU has withstood various mathematical attacks, the security of its implementation is an open question.

After the proposal of the side channel analysis (SCA) by Kocher et al. [6] in 1996, most of the cryptosystems consider a SCA as a de facto standard in nowadays. Additionally, since the resistance of a SCA is included as a requirement in FIPS140-2 (Federal Information Processing Standard publication 140-2), NTRU should consider the resistance against SCA to substitute the RSA and ECC. Moreover, the NIST standardization project suggests a resistance to SCA for the submitted candidates. The SCA is an attack using additional information such as time, sound, and power consumption during the operation of a cryptographic device. Among these methods, power analysis attack such as the differential power analysis (DPA) and simple power analysis (SPA) is known to be the most practical method. The SPA performs by analyzing a single power consumption trace of the device. The DPA is a statistical approach by exploiting a number of power traces related to secret data. Even though the cryptographic algorithm is theoretically safe, the private key or secret message can be exposed by the side-channel leakage when executing the algorithm. In this regard, there are previous studies on SCA on NTRU by Lee et al. [7] and Zheng et al. [8]. They performed DPA on NTRU and revealed the secret key. However, whether other types of power analysis can be performed has not been analyzed so far.

1.1. Our Contribution

In this paper, we propose the first single trace side channel analysis (STA) against on both NTRU Open Source and NTRUEncrypt with experimental results, and propose a countermeasure. Previous SCA on NTRU [7–9] targeted the polynomial multiplication between the cipher-text and the secret key. However, since NTRU was patented until 2017, existing SCAs on NTRU are based on the assumption that publicized polynomial multiplications are used in the decryption process. In this paper, we performed STA on the decryption algorithm used in [10] and on the version submitted in NIST standardization project [11].

Based on the proposed analysis, every NTRU based cryptosystem can be vulnerable to our attack. Moreover, as the PKC is mostly used to exchange the session key between two parties, there might be the case where power consumption trace can only be obtained by one execution of the algorithm. Since we recover with a single power consumption trace, our attack is indeed a threat to these implementations whereas existing DPA cannot be applied in this circumstances. We implement the algorithm on the ATmega128 8-bit processor of the KLA-SCARF AVR [12] and applied the proposed attack. The details of our attack are presented in Section 3.

We propose two versions of countermeasure against the proposed analysis. Although the previous DPA target the different implementation, their method can still be applied on NTRU Open Source and NTRUEncrypt. In this paper, we propose a countermeasure that prevents not only our proposed attack but also the DPA. The proposed countermeasure on NTRU Open Source does not increase the computational cost. Furthermore, the proposed countermeasure on NTRUEncrypt reduces the computational cost approximately by half. The description of our countermeasure is presented in Section 4.

1.2. Organization

This paper is organized as follows. In Section 2, we describe NTRU and its implementation. Also, previous SCAs on NTRU are described. In Section 3, we propose our single trace attack and show experimental results. Next, proposed countermeasures and computation comparisons are in Section 4. We make our conclusion in Section 5.

2. Background

2.1. Algorithm of NTRU

The NTRU is a PKC based on the shortest vector problem whose computational complexity is exponential even in the presence of a quantum computer. The encryption and decryption scheme use polynomial rings in $\mathcal{R} = \mathbb{Z}[X]/(x^N - 1)$, which consist of all polynomials with degree less than

N and coefficients in $\mathbb{Z}$. Thus, an element $f \in \mathcal{R}$ can be written as $f = \sum_{i=0}^{N-1} f_i x^i$. The polynomial multiplication in $\mathcal{R}$ is denoted as $\cdot$, and is performed as in Equation (1).

$$h = f \cdot g, \; f, g, h \in \mathcal{R}$$

$$h_k = \sum_{i=0}^{k} f_i g_{k-i} + \sum_{i=k+1}^{N-1} f_i g_{N+k-i} = \sum_{i+j \equiv k \pmod{N}} f_i g_j \tag{1}$$

Let $\mathcal{L}_f$ be a set of $f \in \mathcal{R}$ with $d_f + 1$ coefficients equal to 1 and d_f coefficients to -1 and let $\mathcal{B}_g$ be a set of $g \in \mathcal{R}$ with d_g coefficients equal to 1 and -1, where d_f and d_g are fixed parameter. We express the polynomials in $\mathcal{L}_f$ and $\mathcal{B}_g$ as a trinary polynomial because they consist of only three number of coefficient. The modulus values of integers p and q are used and they satisfy the conditions $gcd(p,q) = 1$ and $p \ll q$. We define f_p as a polynomial in $\mathbb{Z}_p[X]/(x^N - 1)$ obtained by reducing the coefficients of $f \in \mathcal{R}$ modulo p. The inverse of f_p in $\mathbb{Z}_p[X]/(x^N - 1)$ is denoted as f_p^{-1}. The f_q and f_q^{-1} are defined in the same manner.

Key Generation

The private key f is a trinary polynomial selected from $\mathcal{L}_f$ and the public key h satisfies $h = pf_q^{-1} \cdot g \pmod{q}$, $g \in \mathcal{B}_g$. The public key is used in the data encryption and private key is used in the data decryption.

Encryption

The purpose of encryption is to transport the data by converting a message using the public key of the receiver. Then only an owner of proper private key can decrypt the message. To encrypt a plain-text $m \in \mathcal{B}_m$, we first choose a random polynomial r in $\mathcal{B}_r$ and compute the cipher-text e as Equation (2).

$$e = r \cdot h + m \pmod{q} \tag{2}$$

The modulus q in the above equation means that each coefficient in a polynomial is reduced modulo q.

Decryption

Decryption is used to recover the message from sender. The received data is usually called as cipher-text. The cipher-text e is decrypted by computing the following equations.

$$a = f \cdot e \pmod{q} \tag{3}$$
$$m = a \cdot f_p^{-1} \pmod{p} \tag{4}$$

The correctness of the decryption is confirmed by the Equations (5) and (6).

$$
\begin{aligned}
a &= f \cdot e \pmod{q} \\
&= r \cdot h \cdot f + m \cdot f \\
&= pr \cdot g + m \cdot f \pmod{q}
\end{aligned}
\tag{5}
$$

$$
\begin{aligned}
a \cdot f_p^{-1} &= (pr \cdot g + m \cdot f) \cdot f_p^{-1} \pmod{p} \\
&= m \cdot f \cdot f_p^{-1} \pmod{p} \\
&= m
\end{aligned}
\tag{6}
$$

Note that by choosing the private key f as $pF + 1$ where $F \in \mathcal{L}_f$, then f_p^{-1} is equal to 1 so that the Equation (4) can be omitted [13]. Both target of this paper (NTRUEncrypt, NTRU Open Source) use this optimization.

2.2. Side Channel Analysis and Related Work

Although an algorithm is mathematically secure, naive implementation can make cryptosystem vulnerable to various attacks. The most important considerations in implementation are random number generator and leakage of the secret information. In most cryptosystems, the quality of the random numbers used directly determines the security of the system. Therefore, a predictable random value (i.e., low entropy source) may weaken the system. The studies for the randomness have done in the respect of entropy [14–19]. However, the analysis herein discusses the implementation in the side of SCA.

The SCA is first introduced by Kocher et al. in 1996 [6]. Subsequently, power analysis attacks such as the simple power analysis (SPA), differential power analysis (DPA), and correlation power analysis (CPA) [20,21] have been proposed. Nowadays, any attack that exploits information gained from the implementation is considered as SCA. This includes cache attack, EM analysis, and attacks that exploit hardware vulnerabilities [22–25]. However, we mainly focus on the power analysis attack. The power analysis attacks rely on the dependency between the power consumption of the device and the operated data during the execution of an algorithm. The SCA is an actual threat since it can recover the private key of the cryptosystem in practical time. To prevent these type of attack, masking and hiding are studied [26]. Masking refers to a method of computing secret information with random values, so that the actual value is unused during the encryption and decryption. Hiding removes the relationship between power consumption and the data. Hiding is one of the hardware level countermeasure which is focused on the security during the operation.

Additionally, the Internet of Things (IoT) devices are advanced nowadays, the security against low-power design is essential [27]. However, the conventional PKC is difficult to implement on the resource-constrained environment. Therefore, there is a research on the physical unclonable function (PUF) as a light-weight authentication security primitive. For example, side-channel resistant PUF was intensively studied in [28].

2.3. Previous Side Channel Analysis on NTRU

The first studied SCA on the NTRU was timing attacks in 2007 [29]. In 2010, Lee et al. introduced a SPA and CPA on NTRU and proposed a countermeasure against the attack [7]. The idea behind the proposed SPA in [7] is that there exists a difference in the power consumption when adding non-zero with zero values and non-zero with non-zero values. They also performed CPA on the multiplication using 1000 traces. Also, a second order CPA is proposed in [7] using 10,000 traces. For the description of the attack, please refer to [7]. To prevent the SPA, they proposed to initialize the temporary buffer with a non-zero value and to randomize the order of computation and data. They also provided countermeasures against CPA such as masking and shuffling.

In 2013, a first-order collision attack was proposed in [8] with the purpose of incapacitating the countermeasure proposed in [7]. Their attack against the first-order countermeasure is an improvement in [8] since the attack is performed with 5,000 traces. The target of the attack was when the same registers are loaded during the multiplication. Overall, the decryption code used for the analysis in [7] and [8] was not an official implementation. Although the proposed attacks can be applied to official implementation, the attack environment is restricted to the case where multiple executions of NTRU with the same key is possible.

3. Proposed Single Trace Side Channel Analysis on NTRU Implementation

In this section, we propose our STA on the two cases of NTRU implementation. For each case, we first describe the implementation and then suggest our STA. Lastly, we present the experimental results on our attack. The purpose of our attack is to recover the private key. Therefore, only the implementation of decryption is introduced in this paper.

3.1. NTRU Open Source

The integral parts of NTRU implementation are the way to store polynomials and a polynomial multiplication.

Representing Polynomials

To store a polynomial f of the private key, NTRU Open Source stores the degree of indeterminant x whose coefficient is -1 or 1. Because the addition is computed according to the degree of -1 and 1, it is possible to operate without the degree of 0. Thus, the private key array first stores all the degree whose coefficient is 1 and then it stores all the degree where its coefficient is -1 in an array. For example, if $f = x^3 - x + 1$, then the array of f would be $\{0, 3, 1\}$. The polynomial in general, is stored such that the coefficient of the xth degree is the xth element in an array. For example, the polynomial $e = 3x^4 - x^2 + 9x - 5$ represent as $\{-5, 9, -1, 0, 3\}$.

Polynomial Multiplication

For efficiency, the private key is set as $f = pF + 1$ and F is divided into three trinary polynomials $F = F_1 \cdot F_2 + F_3$, F_1, F_2, and $F_3 \in \mathcal{L}_F$. The advantage of splitting F, is that it lowers the hamming weight of polynomials so that the multiplication could be speed up [13,30]. Consequently, the decryption of NTRU Open Source performs as in Equation (7) considering the order of multiplication.

$$a = f \cdot e = (1 + pF) \cdot e = (1 + p(F_1 \cdot F_2 + F_3)) \cdot e = e + p(((e \cdot F_1) \cdot F_2) + (e \cdot F_3)) \tag{7}$$

Computation of Equation (7) is represented in Algorithm 1 and algorithm for polynomial multiplication is in Algorithm 2.

Algorithm 1 Decryption in NTRU Open Source

Require: The trinary polynomials F_1, F_2, F_3 and $e \in R$ with degree N ▷ F_1, F_2, F_3 is a private key
     polynomial satisfied $f = 1 + p(F_1 \cdot F_2 + F_3)$
Ensure: message $m = f \cdot e \pmod q = (1 + (F_1 \cdot F_2 + F_3)) \cdot e \pmod q$
  1:   $t \leftarrow$ Algorithm 2(F_1, e) ▷ Algorithm 2 is polynomial multiplication
  2:   $t \leftarrow$ Algorithm 2(F_2, t)
  3:   $u \leftarrow$ Algorithm 2(F_3, e)
  4: **for** $0 \leq i < N$ **do**
  5:     $v_i \leftarrow (t_i + u_i) \pmod q$ ▷ add t and u
  6: **end for**
  7: **for** $0 \leq i < N$ **do**
  8:     $m_i \leftarrow (e_i + p * v_i) \pmod q$ ▷ $*$ is a word multiplication
  9: **end for**
10: **return** m

The input b of Algorithm 2 is formed in a way such that the degree having coefficient 1 is stored in ascending order and then degree having -1 is stored. The polynomial multiplication starts with the smallest degree where its coefficient equals to -1 and add cipher-text to the initialized array. Since the result must be reduced modulo $(x^N - 1)$, this implementation performs the addition from the beginning to $(N - 1)$ and restarts for the 0th element in an array when the degree exceeds N. After the modular operation, the sign is reversed and the same steps are repeated on for the degree having coefficient 1. Lastly, the $\pmod q$ operation is performed by AND$(\wedge)$ $(q - 1)$ since the q is set as power of 2.

Algorithm 2 Polynomial Multiplication during NTRU Open Source Decryption

---

Require: Polynomial $e \in R$ with degree N and Private key array b ▷ let b be a information of private
    key F
Ensure: $H = F \cdot e \pmod{q}$
  1: **for** $i = 0; i < N; i{++}$ **do**
  2:    $t_i \leftarrow 0$
  3: **end for**
  4: **for** $j = d_F + 1; j < 2d_F + 1; j{++}$ **do**   ▷ private key has d_F coefficients equal -1
  5:    $k \leftarrow b_j$
  6:    **for** $i = 0; k < N; i{++}, k{++}$ **do**
  7:        $t_k \leftarrow t_k + e_i$
  8:    **end for**
  9:    **for** $k = 0; i < N; i{++}, k{++}$ **do**
10:        $t_k = t_k + e_i$
11:    **end for**
12: **end for**
13: **for** $i = 0; i < N; i{++}$ **do**   ▷ This step is because the above process is for -1
14:    $t_i \leftarrow -t_i$
15: **end for**
16: **for** $j = 0; j < d_F + 1; j{++}$ **do**   ▷ private key has $d_F + 1$ coefficients equal 1
17:    $k \leftarrow b_j$
18:    **for** $i = 0; k < N; i{++}, k{++}$ **do**
19:        $t_k \leftarrow t_k + e_i$
20:    **end for**
21:    **for** $k = 0; i < N; i{++}, k{++}$ **do**
22:        $t_k = t_k + e_i$
23:    **end for**
24: **end for**
25: **for** $i = 0; i < N; i{++}$ **do**
26:    $H_i \leftarrow t_i \pmod{q}$ ▷ in the case of q is powering of 2, $\wedge(q-1)$ works for mod q
27: **end for**
28: **return** H

3.1.1. Proposed Method

The idea behind the attack is that the correlation between power consumption traces obtained when performing the same operations is higher than the power consumption trace obtained when performing different operations. Let the power trace obtained during the addition operation be taken as a reference trace R. Let O be the subtraces of the power consumption trace in Algorithm 2. When calculating the correlation between R and O, the correlation coefficient will be obtained when computing Algorithm 2. When plotting the gained coefficients values, then a graph appear like Figure 1. There are peaks, called as high peak herein, which signify the affinity between R and O. Then, we recover the private key polynomial by calculating the distance between the high peaks.

As in Algorithm 2, the additions in steps 4 to 12 and steps 16 to 24 depend on the private value. For example, suppose $N = 11$ and let 5 be the smallest degree when its coefficient equals to -1. Then the steps 6 to 8 are repeated 6 times and steps 9 to 11 are repeated 5 times. Note that, there is a moment when the loop passes to the next loop, then the distance between high peaks is different at that moment. Thus, if the real value is x, so that the interval between $(N - x)$th and $(N - x + 1)$th high peak is different from the others. Therefore, we can recover the whole value by applying the same steps for the coefficients -1 and 1.

3.1.2. Experiment

Figure 2 is a full trace of the NTRU Open Source porting on an KLA-SCARF AVR, captured in Lecroy HDO6104A oscilloscope with 250 M sampling rate [12,31]. The parameters for the experiment are $N = 50$, $d_{F_1} = 8$, $d_{F_2} = 8$, $d_{F_3} = 6$ and the private key is as follows.

$$b = \{0x03, 0x01, 0x1e, 0x11, 0x05, 0x06, 0x1a, 0x0e, 0x13, 0x01, 0x28, 0x23, 0x10, 0x29, 0x22, 0x0c,$$
$$0x07, 0x08, 0x0b, 0x15, 0x1b, 0x25, 0x2e, 0x2c, 0x18, 0x21, 0x17, 0x2f, 0x19, 0x04, 0x30, 0x00,$$
$$0x02, 0x0f, 0x27, 0x2d, 0x12, 0x2a, 0x2b, 0x14, 0x1c, 0x1f, 0x26, 0x20\}$$

We choose these values as we considered them to be suitable in the experimental environment. The first 16 entries of b represent F_1, the next 16 values represent F_2 and the rest of the values represent F_3.

The first step for analysis is discovering a reference trace R by SPA (Figure 3). The length of R is calculated by dividing the full trace length by the total number of operations. After that, the correlation coefficient can be calculated from the trace using the reference. Figure 1 is a part of the result containing the high peaks and the following intervals. There are two indices tagged on each peak, one represents an order of the high peak and the other is a distance between the previous high peak. The 31th peak has different distance than others, so the first degree where coefficient is -1 is $50 - 31 = 19 = 0x13$. With this process, we can recover F_1, F_2, F_3, and the private key.

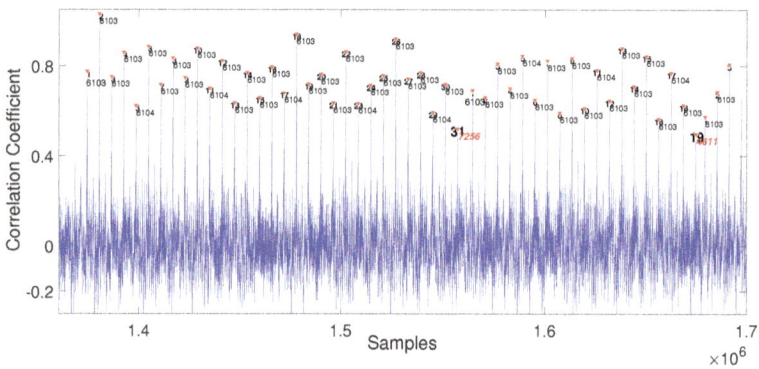

Figure 1. Result: High peaks of the addition and a distance value between the peaks.

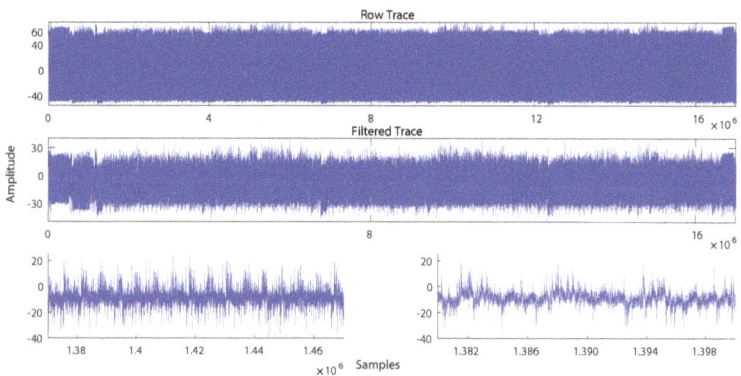

Figure 2. Full trace of NTRU Open Source: The top figure is a raw trace, the middle figure is a filtered trace, and the bottom two are enlarged figure of the filtered trace.

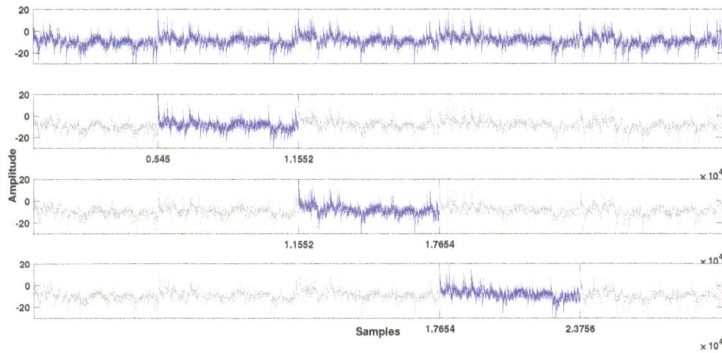

Figure 3. Figuring the reference trace: The three set of an addition operation.

3.2. NTRUEncrypt

Representing Polynomials

In the NTRUEncrypt, the polynomial is represented as the coefficients in order. For example, $F(x) = x^3 + x - 1$ stored as $F = \{-1, 1, 0, 1\}$. Before the polynomial multiplication of cipher-text and private key, there are steps to compute $f = pF + 1$.

Polynomial Multiplication

The the Equation (3) operates using the grade school multiplication. Unlike NTRU Open Source, the polynomial multiplication operates separately. These steps are described in Algorithm 3.

Algorithm 3 Decryption in NTRUEncrypt

Require: Trinary polynomial $F \in \mathcal{L}_f$, cipher-text $e \in \mathcal{R}$
Ensure: message $m = f \cdot e \pmod{q}$
1: **for** $0 \leq i < N$ **do**
2:     $f_i \leftarrow F_i \times p$
3: **end for**
4: $f_0 \leftarrow f_0 + 1$
5: **for** $0 \leq j < N$ **do**
6:     $t_j \leftarrow e_0 \times f_j$
7: **end for**
8: **for** $1 \leq i < N$ **do**
9:     $t_{i+N-1} \leftarrow 0$
10:     **for** $0 \leq j < N$ **do**
11:         $t_{i+j} \leftarrow t_{i+j} + e_i \times f_j$
12:     **end for**
13: **end for**
14: $t_{2N-1} \leftarrow 0$
15: **for** $0 \leq i < N$ **do**
16:     $m_i \leftarrow (t_i + t_{i+N}) \pmod{q}$
17: **end for**
18: **return** m

3.2.1. Proposed Method

The proposed method exploits the power consumption of steps 1 to 3 and steps 5 to 13 in Algorithm 3 to recover the trinary polynomial F. When F get recovered, the private key polynomial f is computed by $f = pF + 1$, where p is a public value. The relative order of coefficients -1 is discovered by analyzing the steps 1 to 3 operation. Because F is a trinary polynomial, a constant value

p is multiplied by three values $-1, 0,$ and 1. Since most of the processor apply 2's complement method to express negative value, a hamming weight of -1 is bigger than others. Thus we can observe the high peaks in the power consumption trace when the -1 is operated. Note that, the proposed analysis depends on the operation of the processor. Thus, if the processor uses another method to represent negative value, the proposed analysis should consider such circumstances.

The next step, the relative orders of the coefficient 0 are known from 5 to 13 steps which are the polynomial multiplication of cipher-text e and private key f. The power consumption when calculating the coefficient of the cipher-text and 0 will be lower than other calculation processes. This portion where the power consumption is low is referred to as low peak. Therefore, after finding the relative position from 0, 1 to -1, and combining this result with the information of 0s then F is completed recovered. Finally, we can get f by computing $pF + 1$.

3.2.2. Experiment

Figure 4a is a full trace of the NTRUEncrypt porting on the KLA-SCARF AVR and is captured with a Lecroy HDO6104A oscilloscope at a 250 M sampling rate [12,31]. The parameters for the experiment are $N = 49$, $p = 3$, $q = 2048$, and a private key is as follows.

$$f = \{1, 1, 1, 1, 1, 0, 1, 0, -1, -1, 1, 0, 0, 1, 1, -1, -1, 1, 0, 0, 0, -1, 0, -1, -1, 1, -1, 1, 0, 0, 0, -1, 0, 0, -1,$$
$$0, 1, 0, 1, -1, 0, 0, -1, -1, 1, 1\}$$

The p and q follow the proposed parameter but N is smaller than the standard because of the experimental environment.

Figure 4c depicts the power consumption of steps 1 to 3 in Algorithm 3. As mentioned above, the high peaks represent the moment when p is multiplied by -1. Also, in the Figure 4c, there are the low peaks related to the coefficient 0 and 1. Thus the relative orders of -1 and others can be recovered by analyzing Figure 4c.

The following process is to recover the coefficients 0. For each coefficient of the cipher-text, there are N multiplications with the private key. During the N operations, the operation of the private key 0 appears in the same order, so the low peaks appear regularly on the whole power trace (Figure 4a). To recover the degree, we should classify a set of multiplications by SPA among the trace. The multiplication between cipher-text and private key occurs after computing pF, and the total recovered number of multiplications is N^2. To reduce the noise, one can average multiple power consumption trace. Figure 5 illustrates the average of 10 traces. Figure 4b is an enlarged plot of four low peaks to deduce that peaks are identified. Lastly, with the three coefficients recovered from the analysis, the private key f is obtained.

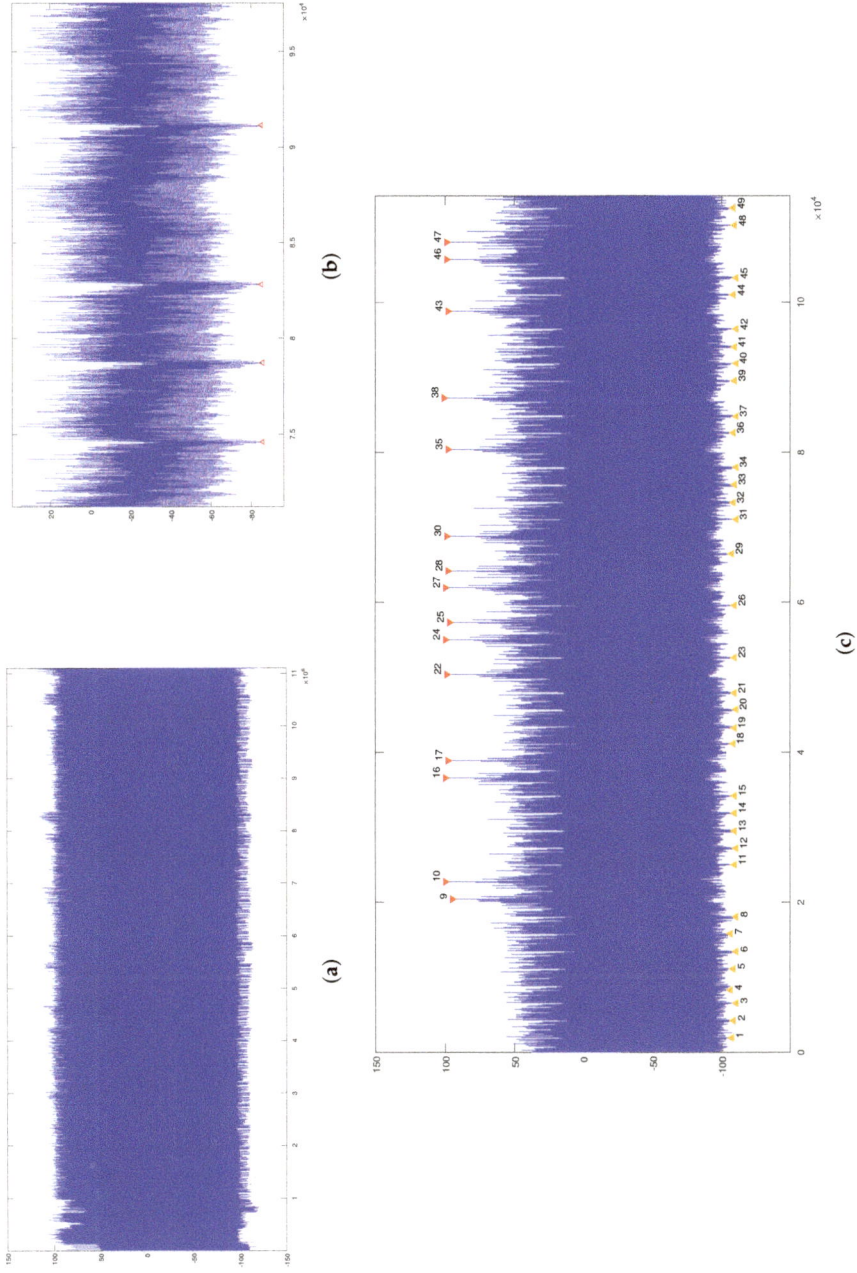

Figure 4. (a) Full trace of NTRUEncrypt; (b) Enlarged Low Peaks; (c) Result of SPA against $f = pF + 1$.

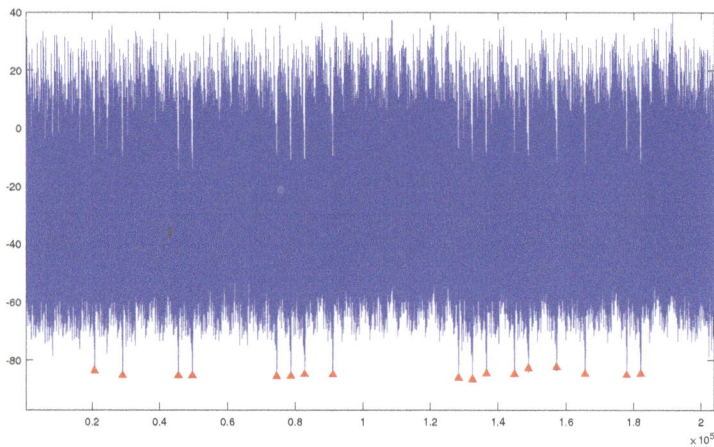

Figure 5. The Average of 10 Power Consumption Traces of Polynomial Multiplication using Grace School Multiplication Method.

4. Countermeasure

In this section, we propose a countermeasure for each of the two implementations. The proposed analysis on NTRU Open Source and NTRUEncrypt does not depend on the data information. Since we used a single power consumption trace, countermeasures to prevent DPA such as adding dummy operation and shuffling cannot prevent our attack.

4.1. Countermeasure against NTRU Open Source Implementation

Since the advantage of the original implementation is that computes both modular reduction $(x^N - 1)$ and polynomial multiplication, simultaneously, the countermeasure we propose also process both of the operation at the same time. Furthermore, the modified implementation has the same number of polynomial coefficient addition as the original implementation. The Algorithm 4 is a countermeasure for the polynomial multiplications described in Algorithm 2.

The Algorithm 4 is a method that precomputes the index i, where the cipher-text polynomial e_i will be added. For example, let 9 be the degree of the polynomial and let 2 be a coefficient of degree 0, then the original cipher-text polynomial coefficient addition performs as in Figure 6a. During the original iteration, the additions when $i = 0$ to 6 operate with different loop when $i = 7$ to 8. However, the addition in the proposed method operates in the same loop so there is no leakage for side channel analysis. The proposed method first finds the index of cipher-text which is added to the middle index of the result array, then the addition operates simultaneously as in Figure 6b.

Algorithm 4 Countermeasure of NTRU Open Source Proeject

Require: cipher-text polynomial $e \in R$ and coefficient location indices of private key b

Ensure: $H = F \cdot e \pmod{q}$

1: **for** $i = 0; i < N; i$++ **do**
2: $\quad t_i \leftarrow r$ $\qquad\qquad\qquad\qquad\qquad\qquad\qquad\qquad\qquad$ $\triangleright r$ is a random value
3: **end for**
4: **for** $j = d_f + 1; j < 2d_f + 1; j$++ **do**
5: $\quad k \leftarrow b_j$
6: $\quad x \leftarrow \frac{N-1}{2} - k, y \leftarrow N - k$
7: $\quad t_{\frac{N-1}{2}} \leftarrow t_{\frac{N-1}{2}} + e_x$
8: $\quad x \leftarrow x + 1$
9: $\quad$ **for** $i = 0; i < \frac{N-1}{2}; i$++, x++,y++ **do**
10: $\quad\quad t_{\frac{N}{2}+i+1} \leftarrow t_{\frac{N}{2}+i+1} + e_x$
11: $\quad\quad t_i \leftarrow t_i + e_y$
12: $\quad$ **end for**
13: **end for**
14: **for** $i = 0; i < N; i$++ **do**
15: $\quad t_i \leftarrow -t_i$
16: **end for**
17: **for** $j = 0; j < d_F + 1; j$++ **do**
18: $\quad k \leftarrow b_j$
19: $\quad x \leftarrow \frac{N-1}{2} - k, y \leftarrow N - k$
20: $\quad t_{\frac{N-1}{2}} \leftarrow t_{\frac{N-1}{2}} + e_x$
21: $\quad x \leftarrow x + 1$
22: $\quad$ **for** $i = 0; i < \frac{N-1}{2}; i$++, x++,y++ **do**
23: $\quad\quad t_{\frac{N}{2}+i+1} \leftarrow t_{\frac{N}{2}+i+1} + e_x$
24: $\quad\quad t_i \leftarrow t_i + e_y$
25: $\quad$ **end for**
26: **end for**
27: **for** $i = 0; i < N; i$++ **do**
28: $\quad H_i \leftarrow t_i - r \pmod{q}$
29: **end for**
30: **return** H

---

t_0	t_1	t_2	t_3	t_4	t_5	t_6	t_7	t_8
+	+	+	+	+	+	+	+	+
e_7	e_8	e_0	e_1	e_2	e_3	e_4	e_5	e_6
$i=7$	$i=8$	$i=0$	$i=1$	$i=2$	$i=3$	$i=4$	$i=5$	$i=6$

(a)Original Implementation Iteration

t_0	t_1	t_2	t_3	t_4	t_5	t_6	t_7	t_8
+	+	+	+	+	+	+	+	+
e_7	e_8	e_0	e_1	e_2	e_3	e_4	e_5	e_6
$i=0$	$i=1$	$i=2$	$i=3$	step 7	$i=0$	$i=1$	$i=2$	$i=3$

(b)Proposed Implementation Iteration

Figure 6. Iteration of Addition.

33

4.2. Countermeasure against NTRUEncrypt Implementation

The proposed countermeasure in this paper uses three tables initialized with a random number. The countermeasure prevents not only proposed attack in this paper but also the other previous attacks with a decreased number of computations compared to the submitted implementation. The NTRUEncrypt first calculates the private key $f(= pF + 1)$ then multiplies $f \cdot e$. However, the proposed countermeasure uses a trinary polynomial F, and computes $p \times F \cdot e$ and adds e to decrypt the cipher-text. This is expressed in Equation (8).

$$m = f \cdot e = (pF + 1) \cdot e = pe \cdot F + e \tag{8}$$

To compute $f \times e$, the proposed countermeasure first computes $p \times e$ and temporarily save the value. After that, it updates the output of computation in the three tables according to $-1, 0, 1$ from the trinary polynomial F. Then the table of the coefficient -1 is subtracted from the table of the coefficient 1. During the accumulation, a start index is sought by $(i + j) \pmod{N}$ to process $\pmod{x^N - 1}$ at the same time for increased efficiency.

For the side channel resistance, by encoding the trinary polynomial of the private key $F' = enc(F)$ at the storage step, it relieves the difference in power consumption coming from loading -1 and $0, 1$. The encoding function enc is chosen by considering the physical property. The proposed countermeasure uses an encoding function as $enc(-1) = 1, enc(0) = 2$, and $enc(1) = 4$, to have the same hamming weight. Then the trinary polynomial would be represented with $1, 2$, and 4 at the proposed algorithm. Algorithm 5 describes the above procedure.

Algorithm 5 Countermeasure Applied Decryption of NTRUEncrypt

Require: Trinary polynomial F' an encoding of $F \in \mathcal{L}_f$, cipher-text $e \in \mathcal{R}$
Ensure: message $m = f \cdot e \pmod{q}$
1: **for** $0 \le i < N$ **do**
2:      $PE_i \leftarrow p \times e_i$
3:      $T_i[1] \leftarrow r$ ▷ r is a random
4:      $T_i[2] \leftarrow r$
5:      $T_i[4] \leftarrow r$
6: **end for**
7: **for** $0 \le i < N$ **do**
8:      **for** $0 \le j < N$ **do**
9:          $T_{i+j \ (mod \ N)}[F'_i] \leftarrow T_{i+j \ (mod \ N)}[F'_i] + PE_j$
10:      **end for**
11: **end for**
12: **for** $0 \le i < N$ **do**
13:      $m_i \leftarrow (T_i[4] - T_i[1] + e_i) \pmod{q}$
14: **end for**
15: **return** m

At the final step 13, a subtraction of 1 and -1, an addition of e and $\pmod{q}$ could be processed at once. Since q is a power of 2 in the proposed parameters, $\pmod{q}$ can be operated as AND ($\wedge$). In Algorithm 5, different tables are accessed according to the coefficient of F. However, since the same operation is performed regardless of the coefficients, it can be considered that there is no difference in the power consumption which depends on $-1, 0$, and 1. Furthermore, as the three tables are initialized with the same random number (steps 3 to 5), the algorithm also prevents SPA and CPA proposed in [7]. SPA can be prevented by initializing the array to hold the result with non-zero values and removing the operations that add zero and non-zero values. Also, choosing the non-zero value for initial as random, the algorithm is protected from CPA because the intermediate value cannot be guessed. When three tables initialized with the same random non-zero value, the random values can be removed without additional computation, as in step 13.

4.3. Implementation of Countermeasure

4.3.1. Comparison of the NTRUEncrypt

The computational cost of the proposed countermeasure on NTRU Open Source is similar to the unprotected version, only with additional precomputations. Moreover, it is hard to compare the NTRU Open Source and NTRUEncrypt because of the different process of the private key multiplication. Thus, we only compare the computational cost and memory of the protected and unprotected version of NTRUEncrypt. Table 1 includes the number of initialization, addition, and multiplication costs when the degree is N. Since the computational cost of subtraction is similar to that of the addition, we include both the numbers of subtractions and the numbers of additions. The computation to find the start degree $(i + j) \pmod N$ is not included.

Table 1. Comparison of Operation between Unprotected and Protected NTRUEncrypt.

	Unprotected	Protected
Initial	N	$3N$
Add/Sub	N^2	$N^2 + 2N$
Mul	$N^2 + N$	N

As shown in Table 1, the total number of computational steps necessary to calculate $f \cdot e$ is reduced when applying our countermeasure. Moreover, the number of multiplications is reduced to square root of the original. Also, the comparison of the memory usage is presented in Table 2. Since NTRUEncrypt use 16 bit as a word size, Table 2 refers to the multiplication of a word size and the number of used arrays.

Table 2. Comparison of the Memory Usage between Unprotected and Protected NTRUEncrypt.

	Unprotected	Protected
RAM (bytes)	$6N$	$12N$

Although the number of used arrays is doubled compared to that used in the unprotected NTRUEncrypt, the total computational cost(number of computational steps) is reduced from $2N^2 + 2N$ to $N^2 + 6N$, where N is at least 443. Consequently, considering the computational costs and the memory size, Algorithm 5 is more efficient compared to Algorithm 3 and has side channel resistance.

4.3.2. Result of the Countermeasure Implementation

Figure 7a is the full trace of the Algorithm 5 implemented in the same environment as the analyzed traces. Since the countermeasure uses a table for all coefficients in the private key, the power consumption difference depending on the private key is not exposed and this is observed in Figure 7b.

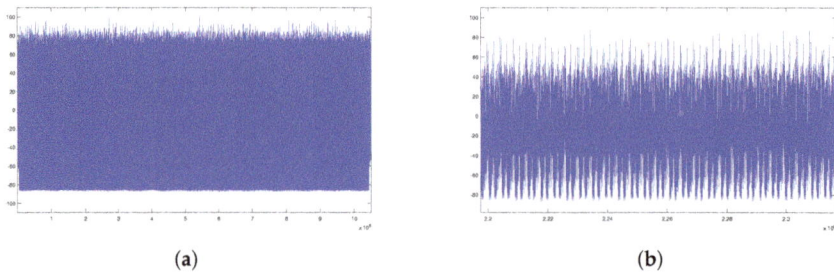

(a)   (b)

Figure 7. Iteration of Addition. (a) Full Trace of Countermeasure; (b) Enlarged Trace of Countermeasure.

5. Conclusions

Although the cryptosystem is proven to be secure, the security of its implementation must be considered as the devices may expose side channel leakages. Since PQC cryptosystems are implemented in classical computers, side channel must be considered. In this paper, we analyzed the two versions of NTRU implementation – NTRUEncrypt and NTRU Open Source. By using a single power consumption obtained in the decryption, we were able to recover the private key on both implementations. Our attack is practical and powerful since it can be applied without constraints of the environment. We also proposed countermeasures for our attack. Our countermeasures not only prevent our proposed attack but also prevents the previous attack. Moreover, our countermeasures do not degrade its performance. In addition, as the NIST standardization project is still in process, every algorithm including NTRU may provide an updated optimized implementation. The proposed analysis is based on the implementation up to now but more optimized version might appear in the future.

Acknowledgments: This research was supported by Basic Science Research Program through the National Research Foundation of Korea (NRF) funded by the Ministry of Science and ICT (NRF-2017R1C1B2004583).

Author Contributions: All authors have contributed to this work. Soojung An and Suhri Kim analyzed the algorithm and drafted and revised the manuscript. Soojung An and Sunghyun Jin performed the experiment and analyzed the result. Soojung An, Suhri Kim, Sunghyun Jin, and HanBit Kim devised the countermeasure and HeeSeok Kim verified the analytical methods and supervised this work.

Conflicts of Interest: The authors declare no conflict of interest.

References

1.	Shor, P.W. Polynomial-time algorithms for prime factorization and discrete logarithms on a quantum computer. *SIAM Rev.* **1999**, *41*, 303–332. [CrossRef]
2.	Rivest, R.L.; Shamir, A.; Adleman, L. A method for obtaining digital signatures and public-key cryptosystems. *Commun. ACM* **1978**, *21*, 120–126. [CrossRef]
3.	Koblitz, N. Elliptic curve cryptosystems. *Math. Comput.* **1987**, *48*, 203–209. [CrossRef]
4.	Chen, L.; Chen, L.; Jordan, S.; Liu, Y.K.; Moody, D.; Peralta, R.; Perlner, R.; Smith-Tone, D. *Report on Post-Quantum Cryptography*; US Department of Commerce, National Institute of Standards and Technology: Gaithersburg, MD, USA, 2016.
5.	Hoffstein, J.; Pipher, J.; Silverman, J.H. NTRU: A ring-based public key cryptosystem. In *Algorithmic Number Theory, Proceedings of the Third International Symposiun, ANTS-III, Portland, OR, USA, 21–25 June 1998*; Springer: Berlin, Germany, 1998; Volume 1423, pp. 267–288.
6.	Kocher, P.C. Timing attacks on implementations of Diffie-Hellman, RSA, DSS, and other systems. In *Advances in Cryptology—CRYPTO '96, Proceedings of the 16th Annual International Cryptology Conference, Santa Barbara, CA, USA, 18–22 August 1996*; Springer: Berlin, Germany, 1996; Volume 1109, pp. 104–113.
7.	Lee, M.K.; Song, J.E.; Choi, D.; Han, D.G. Countermeasures against power analysis attacks for the NTRU public key cryptosystem. *IEICE Trans. Fundam. Electron. Commun. Comput. Sci.* **2010**, *93*, 153–163. [CrossRef]

8. Zheng, X.; Wang, A.; Wei, W. First-order collision attack on protected NTRU cryptosystem. *Microprocess. Microsyst.* **2013**, *37*, 601–609. [CrossRef]
9. Song, J.E.; Han, D.G.; Lee, M.K.; Choi, D.H. Power analysis attacks against NTRU and their countermeasures. *J. Korea Inst. Inf. Secur. Cryptol.* **2009**, *19*, 11–21.
10. Whyte, W.; Etzel, M. NTRU Open Source. 2017. Available online: https://github.com/NTRUOpenSourceProject/ntru-crypto (accessed on 19 October 2018).
11. Zhang, Z.; Chen, C.; Hoffstein, J.; Whyte, W. NTRUEncrypt. 2017. Available online: https://csrc.nist.gov/Projects/Post-Quantum-Cryptography/Round-1-Submissions (accessed on 19 October 2018).
12. Choi, Y.; Choi, D.; Ryou, J. Implementing Side Channel Analysis Evaluation Boards of KLA-SCARF system. *J. Korea Insti. Inf. Secur. Cryptol.* **2014**, *24*, 229–240. [CrossRef]
13. Hoffstein, J.; Silverman, J. Optimizations for NTRU. In *Public-Key Cryptography and Computational Number Theory*; Walter de Gruyter: Warsaw, Poland, 2001; pp. 77–88.
14. Guariglia, E. Entropy and fractal antennas. *Entropy* **2016**, *18*, 84. [CrossRef]
15. Zmeskal, O.; Dzik, P.; Vesely, M. Entropy of fractal systems. *Comput. Math. Appl.* **2013**, *66*, 135–146. [CrossRef]
16. Zanette, D.H. Generalized Kolmogorov entropy in the dynamics of multifractal generation. *Phys. Stat. Mech. Its Appl.* **1996**, *223*, 87–98. [CrossRef]
17. Guariglia, E. Harmonic Sierpinski Gasket and Applications. *Entropy* **2018**, *20*, 714. [CrossRef]
18. Berry, M.V.; Lewis, Z.; Nye, J.F. On the Weierstrass-Mandelbrot fractal function. *Proc. R. Soc. Lond. A* **1980**, *370*, 459–484. [CrossRef]
19. Guariglia, E. Spectral analysis of the Weierstrass-Mandelbrot function. In Proceedings of the 2017 2nd International Multidisciplinary Conference on Computer and Energy Science (SpliTech), Split, Croatia, 12–14 July 2017; pp. 1–6.
20. Brier, E.; Clavier, C.; Olivier, F. Correlation power analysis with a leakage model. In *Cryptographic Hardware and Embedded Systems—CHES 2004, Proceedingds of the 6th International Workshop, Cambridge, MA, USA, 11–13 August 2004*; Springer: Berlin, Germany, 2004; Volume 3156, pp. 16–29.
21. Kocher, P.; Jaffe, J.; Jun, B. Differential power analysis. In *Advances in Cryptology—CRYPTO' 99, Proceedingds of the 19th Annual International Cryptology Conference, Santa Barbara, CA, USA, 15–19 August 1999*; Springer: Berlin, Germany, 1999; Volume 1666, pp. 388–397.
22. Yarom, Y.; Falkner, K. FLUSH+ RELOAD: A High Resolution, Low Noise, L3 Cache Side-Channel Attack. In Proceedings of the 23rd USENIX Security Symposium, San Diego, CA, USA, 20–22 August 2014; Volume 1, pp. 22–25.
23. Inci, M.S.; Gulmezoglu, B.; Irazoqui, G.; Eisenbarth, T.; Sunar, B. Cache attacks enable bulk key recovery on the cloud. In *Cryptographic Hardware and Embedded Systems—CHES 2016, Proceedings of the 18th International Conference, Santa Barbara, CA, USA, 17–19 August 2016*; Springer: Berlin, Germany, 2016; Volume 9813, pp. 368–388.
24. Martinasek, Z.; Zeman, V.; Trasy, K. Simple electromagnetic analysis in cryptography. *Int. J. Adv. Telecommun. Electrotech. Signals Syst.* **2012**, *1*, 13–19. [CrossRef]
25. Rooney, C.; Seeam, A.; Bellekens, X. Creation and Detection of Hardware Trojans Using Non-Invasive Off-The-Shelf Technologies. *Electronics* **2018**, *7*, 124. [CrossRef]
26. Coron, J.S.; Goubin, L. On boolean and arithmetic masking against differential power analysis. In *Cryptographic Hardware and Embedded Systems—CHES 2000, Proceedings of the Second International Workshop, Worcester, MA, USA, 17–18 August 2000*; Springer: Berlin, Germany, 2000; Volume 1965, pp. 231–237.
27. Yuan, J.S.; Lin, J.; Alasad, Q.; Taheri, S. Ultra-Low-Power Design and Hardware Security Using Emerging Technologies for Internet of Things. *Electronics* **2017**, *6*, 67. [CrossRef]
28. Cao, Y.; Zhao, X.; Ye, W.; Han, Q.; Pan, X. A Compact and Low Power RO PUF with High Resilience to the EM Side-Channel Attack and the SVM Modelling Attack of Wireless Sensor Networks. *Sensors* **2018**, *18*, 322. [CrossRef] [PubMed]
29. Silverman, J.H.; Whyte, W. Timing attacks on NTRUEncrypt via variation in the number of hash calls. In *Topics in Cryptology—CT-RSA 2007, Proceedings of the Cryptographers' Track at the RSA Conference 2007, San Francisco, CA, USA, 5–9 February 2007*; Springer: Berlin, Germany, 2007; Volume 4377, pp. 208–224.

30. Howgrave-Graham, N.; Silverman, J.H.; Singer, A.; Whyte, W. NAEP: Provable Security in the Presence of Decryption Failures. *IACR Cryptol. ePrint Arch.* **2003**, *2003*, 172.

31. LeCroy, T. HDO6000A High Definition. Available online: http://cdn.teledynelecroy.com/files/pdf/hdo6000a-oscilloscopes-datasheet.pdf (accessed on 11 October 2018).

*applied
sciences*

MDPI

Article

Side-Channel Vulnerabilities of Unified Point Addition on Binary Huff Curve and Its Countermeasure

Sung Min Cho [1], Sunghyun Jin [1] and HeeSeok Kim [2,*]

[1] Center for Information Security Technologies (CIST), Korea University, Seoul 02841, Korea;
muji0828@korea.ac.kr (S.M.C.); sunghyunjin@korea.ac.kr (S.J.)

[2] Department of Cyber Security, College of Science and Technology, Korea University, Sejong 30019, Korea

* Correspondence: 80khs@korea.ac.kr

Received: 18 September 2018; Accepted: 19 October 2018; Published: 22 October 2018

check for
updates

Abstract: Unified point addition for computing elliptic curve point addition and doubling is considered to be resistant to simple power analysis. Recently, new side-channel attacks, such as recovery of secret exponent by triangular trace analysis and horizontal collision correlation analysis, have been successfully applied to elliptic curve methods to investigate their resistance to side-channel attacks. These attacks turn out to be very powerful since they only require leakage of a single power consumption trace. In this paper, using these side-channel attack analyses, we introduce two vulnerabilities of unified point addition on the binary Huff curve. Also, we propose a new unified point addition method for the binary Huff curve. Furthermore, to secure against these vulnerabilities, we apply an equivalence class to the side-channel atomic algorithm using the proposed unified point addition method.

Keywords: unified point addition; binary Huff curve; recovery of secret exponent by triangular trace analysis; horizontal collision correlation analysis

1. Introduction

Side-channel attacks (SCAs) are major threats to the security of cryptographic embedded devices. Power analysis, the most actively researched SCA technique, can be used to find secret information by using the power consumption data extracted during the cryptographic operations of embedded devices. Power analysis attacks on elliptic curve cryptosystems (ECCs) are classified into two types: simple power analysis (SPA) and differential power analysis (DPA) [1]. SPA exposes secret information by observing the power consumption of a single execution of a cryptographic algorithm. For example, a secret key can be easily extracted from the binary scalar multiplication algorithm by differentiating the point addition signal from the point doubling signal. On the other hand, DPA reveals secret information by statistically analyzing many executions of the same algorithm with different inputs without the physical decapsulation of the target device, even if it is impossible to apply SPA. DPA utilizes a correlation between power consumption and specific key-dependent bits that appear at the cryptographic computations. Among the representative countermeasures against DPA are randomization techniques, e.g., scalar/message blinding methods and randomized projective coordinates, which make it impossible to guess the specified values [2]. The countermeasures against SPA can be divided into two main categories. The first strategy is to perform point addition and point doubling, regardless of the secret bit value, such as the double-and-add-always method and Montgomery ladder algorithm [2,3]. The second approach is to make basic operations indistinguishable, such as side-channel atomicity and unified point addition [4,5].

Recently, two new SCAs using only one power consumption trace—recovery of secret exponent by triangular trace analysis (ROSETTA) and horizontal collision correlation analysis (HCCA)—have been proposed to analyze various countermeasures against DPA and SPA [6,7]. While ROSETTA can find secret information by distinguishing whether the operands of a field multiplication are the same or different, HCCA can find it by distinguishing whether the two field multiplications have at least one operand in common. These two attacks do not require any prior knowledge of the input operands of the field multiplications.

Unified point addition is useful for resisting ECCs to SPA. This technique, by which point addition and point doubling use the same sequence of field operations, was first introduced by Brier and Joye in affine and projective coordinates [5]. After that, various unified point addition formulae were proposed for their application to many kinds of elliptic curves, such as Edwards curves, binary Huff curves, and so on. Recently, unified point addition for the binary Huff curve was proposed by Debigne and Joye at the CT-RSA 2011 conference [8]. However, at the CHES 2013 conference, S. Ghosh et al. showed that unified point addition was insecure against SPA. They further proposed a modified unified point addition formula for the binary Huff curve which would provide resistance to SPA [9].

In this paper, we demonstrate two vulnerabilities of unified point addition on the binary Huff curve using ROSETTA and HCCA. Unified point addition operates with an identical sequence of field operations, regardless of the input points. However, some field multiplications of the unified point addition computation can be affected by investigating whether the two input points are equal or not. If two input points of the unified point addition operation are equal, field multiplications are computed with the same operands (i.e., squaring). Also, there are some field multiplication pairs with common operands. Hence, unified point addition can be exposed to the risk of these vulnerabilities using ROSETTA and HCCA. In order to show that unified point addition actually has these weaknesses, we implemented unified point addition on a binary Huff curve on an ARM cortex-m4 processor that performs field multiplications depending on the secret bit value, repeatedly. Then, we analyzed a power consumption trace collected from the implementation by using our attack methods. As a result of the actual experiments, we were able to find secret bit values more than 94% of the time, which proves that this unified point addition operation is indeed vulnerable to our attacks, and the single trace attack is a practical threat.

To provide security against our attack methods, we propose a new countermeasure using an equivalence class for unified point addition. By using the equivalence class, even though two input points of the unified point addition operation are in the same class, the two points can be different projective coordinate values. In addition, to provide perfect security against our attack methods, we reconfigured the operations of the unified point addition formula. The proposed unified point addition method for the binary Huff curve using the equivalence class is just about 2~4.4% slower than the existing unified point addition method from [8,9]. In addition, the proposed method is about 8.5~17.5% faster than an existing countermeasure that provides same security, i.e., unified point addition using blinding operands of a field multiplication [10]. We applied the aforementioned attacks to the unified point addition formulae of other elliptic curves and confirmed that most unified point addition formulae have these vulnerabilities.

This paper is organized as follows. Section 2 introduces basic knowledge of binary Huff curves and a description of ROSETTA and HCCA. In Sections 3 and 4, we explain the vulnerabilities of the unified point addition formulae and describe the experimental results of applying these methods. Section 5 proposes our method to make unified point addition secure against our attacks. In Section 6, we compare the proposed method with previous methods. Finally, Section 7 addresses our conclusions. In addition, we explain the vulnerabilities of several unified addition formulae and their countermeasures in the Appendix A.

2. Preliminaries

2.1. Binary Huff Curve and Unified Point Addition

At CT-RSA in 2011, a Huff curve for the binary field was proposed by Devigne and Joye. Instead of providing general point addition, this construction provides a unified point addition operation to resist side-channel attacks. However, at CHES in 2013, Ghosh et al. demonstrated that the unified point addition method from CT-RSA 2011 was insecure against SPA. Even though both point addition and point doubling are computed with the same formula and executed by the same sequence of finite field operations, they demand different amounts of power consumption. Specifically, point doubling with unified point addition produces a zero value in some intermediate operations. However, point addition does not. Such zero values in point doubling are used in some field multiplications in unified point addition. Apparently, the outputs are also zero. The power consumption of these multiplications with zero and nonzero inputs are significantly different. Therefore, it is possible to distinguish between point doubling and point addition. Hence, they proposed a new unified point addition formula which is secure against SPA. Here, we provide a brief description.

Definition 1 ([11]). *A generalized binary Huff curve is the set of projective points* $(X : Y : Z) \in \mathbb{P}^2(\mathbb{F}_{2^m})$ *satisfying the equation*

$$E/\mathbb{F}_{2^m} : aX(Y^2 + fYZ + Z^2) = bY(X^2 + fXZ + Z^2), \tag{1}$$

where $a, b, f \in (F)^*_{2^m}$ *and* $a \neq b$.

There are three points at infinity that satisfy the curve equation, namely, $(a : b : 0)$, $(1 : 0 : 0)$, and $(0 : 1 : 0)$. Let $P_1 = (X_1 : Y_1 : Z_1)$ and $P_2 = (X_2 : Y_2 : Z_2)$; then, we get $P_1 + P_2 = (X_3 : Y_3 : Z_3)$ with unified point addition [8]:

$$
\begin{cases}
X_3 = (Z_1Z_2 + Y_1Y_2)((X_1Z_2 + X_2Z_1)(Z_1^2Z_2^2 + X_1X_2Y_1Y_2) + \\
\quad \alpha X_1X_2Z_1Z_2(Z_1Z_2 + Y_1Y_2)) \\
Y_3 = (Z_1Z_2 + X_1X_2)((Y_1Z_2 + Y_2Z_1)(Z_1^2Z_2^2 + X_1X_2Y_1Y_2) + \\
\quad \beta Y_1Y_2Z_1Z_2(Z_1Z_2 + X_1X_2)) \\
Z_3 = (Z_1Z_2 + X_1X_2)(Z_1Z_2 + Y_1Y_2)(Z_1^2Z_2^2 + X_1X_2Y_1Y_2),
\end{cases}
\tag{2}
$$

where $\alpha = (a + b)/b$ and $\beta = (a + b)/a$. The unified point addition formula in Equation (2) can be evaluated as described in [9]:

$m_1 = X_1X_2, \quad m_2 = Y_1Y_2, \quad m_3 = Z_1Z_2,$
$m_4 = (X_1 + Z_1)(X_2 + Z_2), \quad m_5 = (Y_1 + Z_1)(Y_2 + Z_2),$
$m_6 = m_1m_3, \quad m_7 = m_2m_3, \quad m_8 = m_1m_2 + m_3^2,$
$m_9 = m_6(m_2 + m_3)^2, m_{10} = m_7(m_1 + m_3)^2, \quad m_{11} = m_8(m_2 + m_3),$
$Z_3 = m_{11}(m_1 + m_3), \quad X_3 = m_4m_{11} + \alpha m_9 + Z_3,$
$Y_3 = m_5m_8(m_1 + m_3) + \beta m_{10} + Z_3.$

The above operation needs 17 field multiplications, which is exactly the same as in the original one. Since point doubling does not have a zero value in any intermediate operation, it is secure against SPA. Recently, however, SCAs such as SPA using only one power consumption trace have been proposed [6,7]. Therefore, security analysis of the unified point addition formula should be considered not only for SPA but also for other analyses. Using these analyses, we present the vulnerabilities of the unified point addition method from [9] and report our experimental results in Sections 3 and 4, respectively.

2.2. ROSETTA and HCCA

Recovery of secret exponent by triangular trace analysis (ROSETTA) [7] and horizontal collision correlation attack (HCCA) [6] are based on the observations of the power consumption of the cryptosystems during the executions of field multiplications. They are powerful attacks on elliptic curve cryptosystems since they use only one power consumption trace for SPA. ROSETTA and HCCA can be used to reveal secret information by analyzing the correlation between the secret bit value and the power consumption of field multiplications without any prior knowledge of the inputs. Details of the analyses are as follows.

ROSETTA. Clavier's attack needs a single power consumption trace to recover secret information. For each field multiplication, ROSETTA detects whether the operation is $x \cdot x$ (squaring) or $x \cdot y$ (multiplication). Let $x = (x_{m-1}, x_{m-2}, ..., x_0)_{2^w}$ and $y = (y_{m-1}, y_{m-2}, ..., y_0)_{2^w}$. A w-bit multiplication $x_i \cdot y_j$ can be identified from the specific pattern in side-channel power consumption. ROSETTA considers the observation O_1 and O_2 extracted from the multiplication $x_i \cdot y_j$ for all $i \neq j$:

$(O_1) : x \cdot y$ s.t $x = y \Rightarrow \text{Prob}(x_i \cdot y_j = x_j \cdot y_i) = 1$ for all $i \neq j$

$(O_2) : x \cdot y$ s.t $x \neq y \Rightarrow \text{Prob}(x_i \cdot y_j = x_j \cdot y_i) \approx 0$ for all $i \neq j$

From the observations O_1 and O_2, collisions between $x_i \cdot y_j$ and $x_j \cdot y_i$ for all $i \neq j$ can be used to identify squarings from multiplications. To identify these collisions of field multiplication trace, ROSETTA exploits a triangle trace analysis which uses a Euclidean distance distinguisher relying on a collision correlation technique.

HCCA. Bauer et al. introduced this method to extract keys using the collision of field multiplications in a single power consumption trace. The core idea of this attack is that collision occurs during two field multiplication computations when the same operands are used, which can be detected by HCCA. When performed in a horizontal setting, the observations O_1 and O_2 are extracted from the two field multiplications.

$(O_1) : x_1 \cdot y_1$ and $x_2 \cdot y_2$ s.t $x_1 = x_2$ and $y_1 = y_2 \Rightarrow \text{Prob}((x_1)_i \cdot (y_1)_j = (x_2)_i \cdot (y_2)_j) = 1$ for all i, j

$(O_2) : x_1 \cdot y_1$ and $x_2 \cdot y_2$ s.t $x_1 \neq x_2$ and $y_1 \neq y_2 \Rightarrow \text{Prob}((x_1)_i \cdot (y_1)_j \neq (x_2)_i \cdot (y_2)_j) \approx 0$ for all i, j

The correlation between the two observations is then estimated by Pearson's coefficient in order to determine whether the two operands of the field multiplications are the same or different.

The advantage of these analyses is that the inputs of field multiplication can remain unknown since the adversary does not need to compute intermediate values. Countermeasures against ROSETTA and HCCA include shuffling the operands and blinding the operands of a field multiplication [10]. For n-bit field multiplication, the blinding operand method requires $t^2 + 2t + 1$ w-bit multiplications, where $t = \lceil n/w \rceil$. Unified point addition using blinding operands requires a great additional computational cost. Therefore, for efficiency, we propose a suitable and efficient countermeasure for the unified point addition operation, and we compare and analyze the proposed method with the existing unified point addition method using blinding operands on the binary Huff curve.

3. Vulnerabilities of Unified Point Addition

Many methods have been proposed to prevent SPA, such as unified point addition and the Montgomery ladder algorithm. Since unified point addition can compute point addition and point doubling with the same formula, it is secure against SPA. In addition, it can be applied to various algorithms easily. In this section, we define two types of vulnerabilities of unified point addition and find vulnerabilities of unified point addition of the binary Huff curve in [9].

3.1. Vulnerabilities of Unified Point Addition

We describe the vulnerabilities of unified point addition considering ROSETTA and HCCA. Both are analyses using the correlation between the input data and operations. ROSETTA can determine whether the operands of a field multiplication are equal (squaring) or different (multiplication). HCCA can determine whether two field multiplications have the same or different operands. We defined the two types of vulnerabilities exposed by these analyses.

Type 1. (Vulnerability by ROSETTA): The unified point addition operation can compute the point doubling and point addition with the same formula. However, depending on the input points of unified point addition, field multiplications can be performed as squaring or multiplication. For example, let $P_1 = (X_1 : Y_1 : Z_1)$ and $P_2 = (X_2 : Y_2 : Z_2)$ be the two input points of the unified point addition formula. Note that there exists the operation $X_1 \cdot X_2$ in unified point addition. If $P_1 = P_2$, then this operation computes to $X_1 \cdot X_1$. If $P_1 \neq P_2$, then this operation computes to $X_1 \cdot X_2$. Then, this operation becomes a vulnerability that is exploitable by ROSETTA.

Type 2. (Vulnerability by HCCA): Considering two field multiplications, if they have at least one common operand, they can be distinguished by HCCA. In unified point addition, the two different multiplications can be identically computed according to the inputs. For example, the operations $X_1 \cdot Y_1$ and $X_2 \cdot Y_2$ exist in unified point addition. If $P_1 = P_2$, then $X_1 \cdot Y_1$ will be computed twice. If $P_1 \neq P_2$, then $X_1 \cdot Y_1$ and $X_2 \cdot Y_2$ will be computed. Then, these operations become a vulnerability that is exploitable by HCCA.

3.2. Vulnerabilities of Binary Huff Curve

In this section, we find Type 1 and Type 2 vulnerabilities of unified point addition on the binary Huff curve from [9] during the computations of $P_1 + P_2$ for $P_1 \neq P_2$ and $P_1 = P_2$. Let $P_1 = (X_1 : Y_1 : Z_1)$ and $P_2 = (X_2 : Y_2 : Z_2)$. In each case, the unified point addition formula can be evaluated as shown in Table 1.

Table 1. Unified point addition on binary Huff curve.

Out	$P_1 = P_2$	$P_1 \neq P_2$
m_1	$[X_1] \cdot [X_1]$	$[X_1] \cdot [X_2]$
m_2	$[Y_1] \cdot [Y_1]$	$[Y_1] \cdot [Y_2]$
m_3	$[Z_1] \cdot [Z_1]$	$[Z_1] \cdot [Z_2]$
m_4	$[(X_1 + Z_1)] \cdot [(X_1 + Z_1)]$	$[(X_1 + Z_1)] \cdot [(X_2 + Z_2)]$
m_5	$[(Y_1 + Z_1)] \cdot [(Y_1 + Z_1)]$	$[(Y_1 + Z_1)] \cdot [(Y_2 + Z_2)]$
$m_6 = m_1 \cdot m_3$	$[X_1^2] \cdot [Z_1^2]$	$[X_1 X_2] \cdot [Z_1 Z_2]$
$m_7 = m_2 \cdot m_3$	$[Y_1^2] \cdot [Z_1^2]$	$[Y_1 Y_2] \cdot [Z_1 Z_2]$
$m_8 = m_1 \cdot m_2 + m_3^2$	$[X_1^2] \cdot [Y_1^2] + (Z_1^2)^2$	$[X_1 X_2] \cdot [Y_1 Y_2] + (Z_1 Z_2)^2$
$m_9 = m_6 \cdot (m_2 + m_3)^2$	$[X_1^2 Z_1^2] \cdot [(Y_1^2 + Z_1^2)^2]$	$[X_1 X_2 Z_1 Z_2] \cdot [(Y_1 Y_2 + Z_1 Z_2)^2]$
$m_{10} = m_7 \cdot (m_1 + m_3)^2$	$[Y_1^2 Z_1^2] \cdot [(X_1^2 + Z_1^2)^2]$	$[Y_1 Y_2 Z_1 Z_2] \cdot [(X_1 X_2 + Z_1 Z_2)^2]$
$m_{11} = m_8 \cdot (m_2 + m_3)$	$[(X_1^2 Y_1^2 + Z_1^4)] \cdot [(Y_1^2 + Z_1^2)]$	$[(X_1 X_2 Y_1 Y_2 + (Z_1 Z_2)^2)] \cdot [(Y_1 Y_2 + Z_1 Z_2)]$
$Z_3 = m_{11} \cdot (m_1 + m_3)$	$[(X_1^2 Y_1^2 + Z_1^4)(Y_1^2 + Z_1^2)] \cdot [(X_1^2 + Z_1^2)]$	$[(X_1 X_2 Y_1 Y_2 + (Z_1 Z_2)^2)(Y_1 Y_2 + Z_1 Z_2)] \cdot [(X_1 X_2 + Z_1 Z_2)]$
$X_3 = m_4 \cdot m_{11}$	$[(X_1^2 + Z_1^2)] \cdot [(X_1^2 Y_1^2 + Z_1^4)(Y_1^2 + Z_1^2)]$	$[(X_1 + Z_1)(X_2 + Z_2)] \cdot [(X_1 X_2 Y_1 Y_2 + (Z_1 Z_2)^2)(Y_1 Y_2 + Z_1 Z_2)]$
$+\alpha \cdot m_9 + Z_3$	$+[\alpha] \cdot [X_1^2 Z_1^2 (Y_1^2 + Z_1^2)^2] + Z_3$	$+[\alpha] \cdot [X_1 X_2 Z_1 Z_2 (Y_1 Y_2 + Z_1 Z_2)^2] + Z_3$
$Y_3 =$	$[(Y_1^2 + Z_1^2)] \cdot [(X_1^2 Y_1^2 + Z_1^4)]$	$[(Y_1 + Z_1)(Y_2 + Z_2)]$
$m_5 \cdot m_8 \cdot (m_1 + m_3)$	$\cdot [(X_1^2 + Z_1^2)]$	$\cdot [(X_1 X_2 Y_1 Y_2 + (Z_1 Z_2)^2)] \cdot [(X_1 X_2 + Z_1 Z_2)]$
$+\beta \cdot m_{10} + Z_3$	$+[\beta] \cdot [Y_1^2 Z_1^2 (X_1^2 + Z_1^2)^2] + Z_3$	$+[\beta] \cdot [Y_1 Y_2 Z_1 Z_2 (X_1 X_2 + Z_1 Z_2)^2] + Z_3$

Type 1 vulnerability: Let us consider the computation of $m_1 = [X_1] \cdot [X_2]$. In this formula, it is computed as $[X_1] \cdot [X_1]$ for $P_1 = P_2$, whereas it is computed as $[X_1] \cdot [X_2]$ for $P_1 \neq P_2$. Similarly, for $P_1 = P_2$, for m_2, m_3, m_4, and m_5, these are computed as $[Y_1] \cdot [Y_1]$, $[Z_1] \cdot [Z_1]$, $[(X_1 + Z_1)] \cdot [(X_1 + Z_1)]$, and $[(Y_1 + Z_1)] \cdot [(Y_1 + Z_1)]$, respectively. Thus, an adversary can distinguish between $P_1 = P_2$ and $P_1 \neq P_2$.

Type 2 vulnerability: In Table 1, let us consider the computations of $[m_{11}] \cdot [(m_1 + m_3)]$, $[m_4] \cdot [m_{11}]$, and $[m_5 m_8] \cdot [(m_1 + m_3)]$ for Z_3, X_3, and Y_3, respectively. If $P_1 = P_2$, then $m_4 = (X_1 +$

$Z_1)(X_1 + Z_1) = X_1 X_1 + Z_1 Z_1 + X_1 Z_1 + X_1 Z_1$. Since the value of $X_1 Z_1 + X_1 Z_1$ is zero in $\mathbb{F}_{2^m}$, then $m_4 = X_1 X_1 + Z_1 Z_1 = m_1 + m_3$ for $P_1 = P_2$. Also, $m_5 m_8 = (Y_1^2 + Z_1^2)(X_1^2 Y_1^2 + Z_1^4) = m_{11}$ for $P_1 = P_2$. Thus, the operands of $[m_{11}] \cdot [(m_1 + m_3)]$, $[m_4] \cdot [m_{11}]$, and $[m_5 m_8] \cdot [(m_1 + m_3)]$ for Z_3, X_3, and Y_3 are the same for $P_1 = P_2$ but different for $P_1 \neq P_2$. Similarly, consider $[m_8] \cdot [(m_2 + m_3)]$ and $[m_5] \cdot [m_8]$ for m_{11} and Y_3. Since $m_5 = (Y_1 + Z_1)(Y_2 + Z_2) = Y_1 Y_1 + Z_1 Z_1 = m_2 + m_3$, $[m_8] \cdot [(m_2 + m_3)]$ and $[m_5] \cdot [m_8]$ have the same inputs for $P_1 = P_2$ but different inputs for $P_1 \neq P_2$. Therefore, they can be distinguished between $P_1 = P_2$ and $P_1 \neq P_2$.

In this section, we have defined the two types of vulnerabilities and highlighted them in unified point addition on the binary Huff curve. These vulnerabilities can also be found in unified point additions on other elliptic curves. We explain how to find these vulnerabilities of unified point addition on other elliptic curves in the Appendix A.

4. Experiments

In this section, we provide experimental results showing that unified point addition on the binary Huff curve is vulnerable to HCCA and ROSETTA. For this, we implemented a field multiplication for unified point addition on the binary Huff curve on an ARM cortex-m4 processor on the ChipWhisperer CW308 UFO evaluation board [12]. The scheme of the experimental setup used for measuring the power consumption is shown in Figure 1.

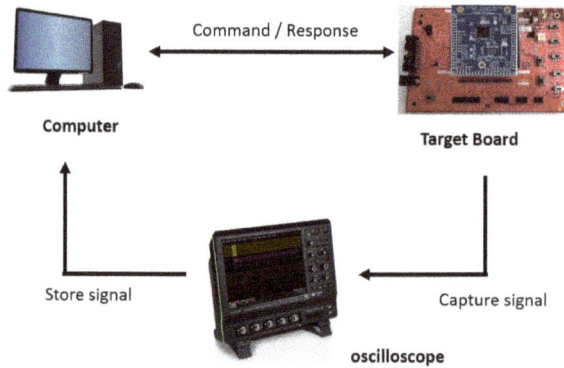

Figure 1. The scheme of the experimental setup used for measuring power consumption.

We collected a power consumption trace which is measured when 192 field multiplications are performed. We randomly selected whether the two operands of the two multiplications of each pair are identical or not for HCCA. Also, we randomly selected whether the operands of the multiplication are identical or not for ROSETTA. The power consumption trace was acquired using a Lecroy HDO oscilloscope with a sampling rate of 5 GS/s. We preprocessed the power consumption trace with a 168 MHz low-pass filter and 3-point maximum compression only for ROSETTA. Figure 2 shows a power consumption trace of field multiplications for unified point addition on the binary Huff curve. Using SPA and a cross-correlation technique, we identified each w-bit multiplication in a field multiplication and separated these into subtraces which correspond to each w-bit multiplication, as shown in Figure 3. For the experiment, we divided them into 96 pairs of subtraces of field multiplications for $(x_1) \cdot (y_1)$ and $(x_2) \cdot (y_2)$ for HCCA. Similarly, we separated a power consumption trace into subtraces of 192 field multiplications for $(x) \cdot (y)$ for ROSETTA. To perform HCCA and ROSETTA, each subtrace was classified into two groups appropriately according to each analysis method. To find a pairwise collision, we separated the subtraces into two groups based on the following fact. Since HCCA determines whether a collision occurs during two field multiplications or not, we divided the subtraces of the w-bit multiplications $(x_1)_i \cdot (y_1)_j$ and $(x_2)_i \cdot (y_2)_j$ for all i, j of the two multiplications $(x_1) \cdot (y_1)$ and $(x_2) \cdot (y_2)$ into each group. In the case of ROSETTA, similar to HCCA, we divided the subtraces of the w-bit multiplications $(x)_i \cdot (y)_j$ and $(x)_j \cdot (y)_i$ for all $i \neq j$ of a field multiplication $x \cdot y$ into each group.

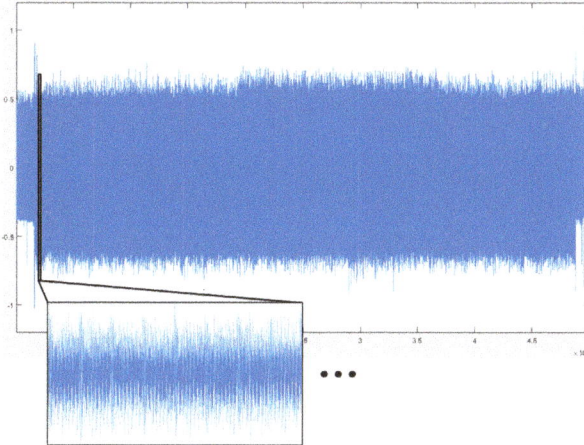

Figure 2. A single power consumption trace of field multiplications for binary Huff curve software implementation on an ARM cortex-m4 processor. The power consumption trace is composed of subtraces corresponding to field multiplications.

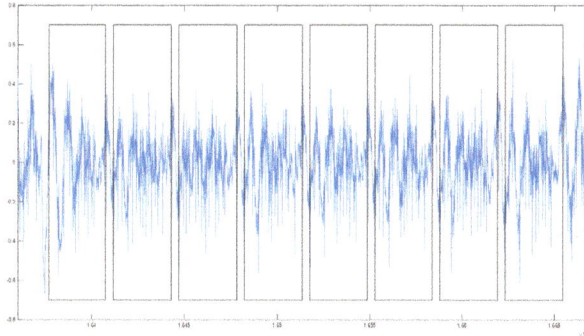

Figure 3. Beginning of a field multiplication power consumption trace. Each w-bit multiplication subtrace in a field multiplication can be identified using simple power analysis (SPA) and cross-correlation.

To find points of interest (POIs), i.e., those having the most collision-related leakage information, we calculated the sum of squared pairwise t-differences (SOST), which is Welch's t-test of two groups, using the following:

$$\left(\frac{m_1 - m_2}{\sqrt{\frac{\sigma_1^2}{n_1} + \frac{\sigma_2^2}{n_2}}} \right)^2 \tag{3}$$

where m_i is the mean trace of group i, and σ_i^2 is the variance trace of group i [13,14]. SOST is a tool mainly used to identify side-channel leakage and is discussed in the SCA literature [15–17]. Because SOST is computed depending on the group's statistics and each group is separated based on the operand of w-bit multiplication, points having high SOST indicate POIs. Since HCCA uses both the inputs and the output of w-bit multiplication, we selected points having a SOST value higher than some heuristic threshold. However, ROSETTA uses the output of w-bit multiplication, and we selected points having leakage of manipulating the output, considering the sequence of the multiplication. The SOST results and POIs for HCCA and ROSETTA are shown in Figure 4a,b, respectively.

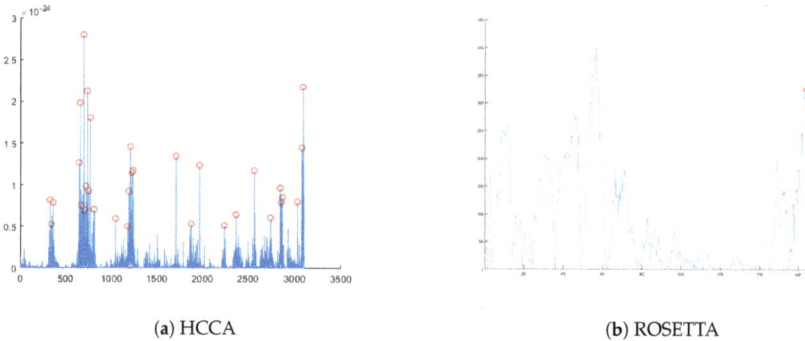

(a) HCCA   (b) ROSETTA

Figure 4. Squared pairwise t-differences (SOST; line) and points of interest (POIs; red circle). (a) Points having higher SOST values than the heuristic threshold are chosen for HCCA's POIs. (b) Unlike HCCA, ROSETTA's POIs, upon which the output value of w-bit multiplication is processed, are chosen heuristically.

We checked for a collision between subtraces corresponding to each group. The occurrence of a collision was determined by calculating Pearson's correlation coefficients. For this, we reconstructed all subtraces composed of values of POIs only. Then, Pearson's correlation coefficients were calculated between subtraces corresponding to each group over every point. Then, correlation coefficients corresponding to the same field multiplications and the same groups were averaged over the points. The values of the correlation coefficient sequences indicating a collision were averaged. As a result, this averaged value became a criterion for determining whether a collision occurs or not. We set the threshold by averaging all final values, which were the criteria for each collision check, and confirmed collisions by comparing the magnitude of each value and threshold. If a value was higher than the threshold, we guessed that collision occurs; otherwise, the collision was assumed not to occur. The analysis results of HCCA and ROSETTA are shown in Figure 5a,b, respectively. As a result, the success rates of HCCA and ROSETTA are 97.92% and 94.79%, respectively. These results prove that the aforementioned HCCA and ROSETTA vulnerabilities are real.

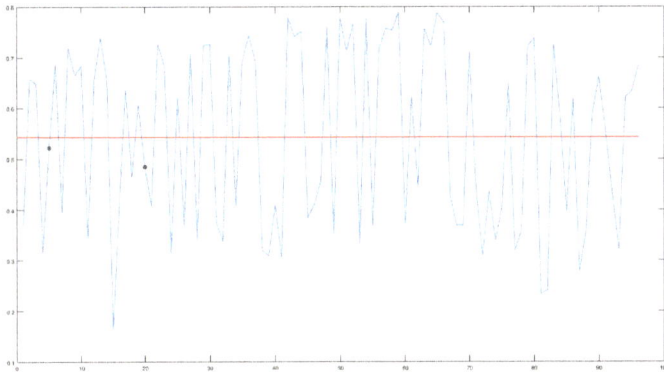

(a) HCCA

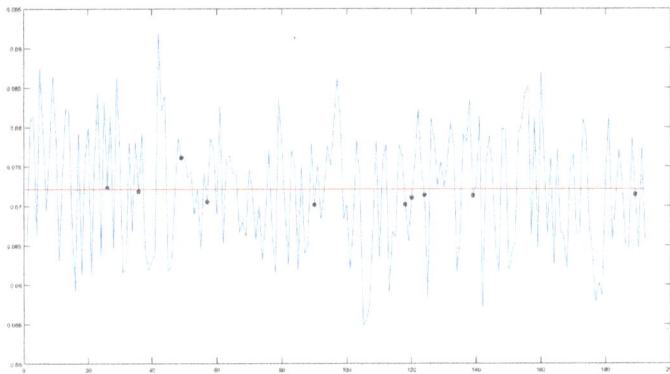

(b) ROSETTA

Figure 5. Results of the secret bit value guess by (a) HCCA and (b) ROSETTA. The blue line is the secret bit value guess and the horizontal red line is the threshold value for the secret bit value discrimination; points with a black circle indicate where the attack failed.

5. Countermeasures

5.1. Countermeasures

As for the two types of vulnerabilities considered in this paper, we introduce the following interesting properties: they make use of a single power consumption trace, yet they do not require knowledge of the inputs to the unified point addition formula for the binary Huff curve. Due to these properties, the application of classical blinding countermeasures (point blinding, scalar blinding, random projective coordinates) is not recommended. We propose new countermeasures against these vulnerabilities of unified point addition.

Type 1 and Type 2 vulnerabilities are due to two problems in unified point addition on the binary Huff curve. The first is that each coordinate of input points of the unified point addition operation has the same value. This problem can be solved by using the equivalence class of projective coordinates [18]. Let $\mathbb{F}$ be a finite field. In a binary Huff curve, the equivalence class containing (X, Y, Z) is

$$(X : Y : Z) = \{(rX, rY, rZ) : r \in \mathbb{F}\}. \tag{4}$$

Notice that if $(X', Y', Z') \in (X : Y : Z)$, then $(X' : Y' : Z') = (X : Y : Z)$. Let $P = (X : Y : Z)$ and $P' = (X' : Y' : Z')$ be the equivalence class, where $X' = rX, Y' = rY$, and $Z' = rZ, r \neq 1$. Then, $(X : Y : Z) = (X' : Y' : Z')$. When considering $P_3 = P + P'$ and $P_4 = P + P$, each coordinate of input points of P and P' has a different value, but $P_3 = P_4$. The equivalence class has been used in random projective coordinates (RPCs), which is a countermeasure of DPA [19]. However, RPCs are generally applied only to the input P of the elliptic curve scalar multiplication. Of course, RPCs can be applied to every execution or after each unified point addition. Unfortunately, in this case, the computational cost is disadvantageously increased for RPCs. Since we only need to convert P to a different coordinate of the same equivalence class, the bit size of r need not be the same as the bit size of the finite field. Therefore, for computational efficiency, we propose a w-bit random projective coordinate (wRPC) that limits the size of r to w bits. The proposed wRPC for the binary Huff curve is depicted in Algorithm 1.

Algorithm 1: A w-bit random projective coordinate for the binary Huff curve (wRPC)

Require: $P = (X : Y : Z)$
Ensure: $P' = (X' : Y' : Z')$
1: Generate a w-bit random number r with $r \neq 1$
2: $X' \leftarrow rX; Y' \leftarrow rY; Z' \leftarrow rZ$
3: return P'

In Algorithm 1, w is the bit size of a word multiplication for a field multiplication. In this work, we only considered the application of wRPC on a side-channel atomic algorithm using unified point addition [4]. The side-channel atomic algorithm using wRPC is described by Algorithm 2. We show the additional cost of Algorithm 2 in Section 5.

Algorithm 2: Side-channel atomic algorithm using wRPC

Require: $P = (X : Y : Z), k = (k_{n-1}...k_0)_2$
Ensure: kP
1: $R_0 \leftarrow O; R_1 \leftarrow P; R_2 \leftarrow O; i \leftarrow n - 1;$
2: $k' \leftarrow 0$
3: while $(i \geq 1)$ do
4:     $R_0 \leftarrow wRPC(R_2)$
5:     $R_1 \leftarrow wRPC(R_1)$
6:     $R_2 \leftarrow R_2 + R_{k'}$
7:     $k' \leftarrow k' \oplus k_i$
8:     $i \leftarrow i - \neg k$
9: end while
10: return R_0

Although Algorithm 2 using unified point addition is secure against Type 1 vulnerabilities, it is still insecure against Type 2. We show in the next subsection that it is not secure against Type 2 vulnerabilities. To be secure against Type 2 vulnerabilities, it is necessary to reconstruct the calculation process of unified point addition. For this reason, we propose a new unified point addition formula for the binary Huff curve as follows:

$$m_1 = X_1 X_2, \quad m_2 = Y_1 Y_2, \quad m_3 = Z_1 Z_2,$$
$$m_4 = (X_1 + Z_1)(X_2 + Z_2), \quad m_5 = (Y_1 + Z_1)(Y_2 + Z_2),$$
$$m_6 = m_1 m_3, \quad m_7 = m_2 m_3, \quad m_8 = m_1 m_2 + m_3^2,$$
$$m_9 = m_6(m_2 + m_3)^2, m_{10} = m_7(m_1 + m_3)^2, \quad m_{11} = m_8(m_2 + m_3),$$
$$m_{12} = m_8(m_1 + m_3),$$
$$Z_3 = m_{11}(m_1 + m_3),$$
$$X_3 = (m_4 + m_{11})m_{11} + m_{11}^2 + \alpha m_9 + Z_3,$$
$$Y_3 = (m_5 + m_{12})m_{12} + m_{12}^2 + \beta m_{10} + Z_3.$$

The proposed unified point addition operation is based on masking by m_4 and m_5. To use the advantage of almost no computational cost for squaring in a binary field, we configured the calculation of masking m_4 and m_5 by squaring. Thus, the proposed method needs 17 field multiplications, which is exactly the same as in [9]. Furthermore, we explain Type 1 and Type 2 vulnerabilities of several unified point addition formulae and propose countermeasures in the Appendix A.

5.2. Security Analysis of the Proposed Method

In this section, we analyze Type 1 and Type 2 vulnerabilities of Algorithm 2 using the proposed unified point addition method. Let the input $R_2 = (X_1 : Y_1 : Z_1)$ in step 4 and let the input $R_1 = (X_2 : Y_2 : Z_2)$ in step 5. Then, in step 6, the two inputs R_2 and $R_{k'}$ of the proposed unified point addition are $P_1 = R_2, P_2 = R_0$ if $k' = 0$ and $P_1 = R_2, P_2 = R_1$ if $k' = 1$. The two inputs are expressed as follows:

$$\begin{cases} P_1 = (X_1 : Y_1 : Z_1), P_2 = (r_1 X_1 : r_1 Y_1 : r_1 Z_1) & \text{If } k' = 0, \\ P_1 = (X_1 : Y_1 : Z_1), P_2 = (r_2 X_2 : r_2 Y_2 : r_2 Z_2) & \text{If } k' = 1. \end{cases} \tag{5}$$

where $r \neq 1$. The proposed unified point addition method can be evaluated as shown in Table 2.

Table 2. The proposed unified point addition method on the binary Huff curve in Algorithm 2.

Out	$P_1 = P_2(k' = 0)$	$P_1 \neq P_2(k' = 1)$
m_1	$[X_1] \cdot [r_1 X_1]$	$[X_1] \cdot [r_2 X_2]$
m_2	$[Y_1] \cdot [r_1 Y_1]$	$[Y_1] \cdot [r_2 Y_2]$
m_3	$[Z_1] \cdot [r_1 Z_1]$	$[Z_1] \cdot [r_2 Z_2]$
m_4	$[(X_1 + Z_1)] \cdot [(r_1 X_1 + r_1 Z_1)]$	$[(X_1 + Z_1)] \cdot [(r_2 X_2 + r_2 Z_2)]$
m_5	$[(Y_1 + Z_1)] \cdot [(r_1 Y_1 + r_1 Z_1)]$	$[(Y_1 + Z_1)] \cdot [(r_2 Y_2 + r_2 Z_2)]$
$m_6 = m_1 \cdot m_3$	$[r_1 X_1^2] \cdot [r_1 Z_1^2]$	$[r_2 X_1 X_2] \cdot [r_2 Z_1 Z_2]$
$m_7 = m_2 \cdot m_3$	$[r_1 Y_1^2] \cdot [r_1 Z_1^2]$	$[r_2 Y_1 Y_2] \cdot [r_2 Z_1 Z_2]$
$m_8 =$ $m_1 \cdot m_2 + m_3^2$	$[r_1 X_1^2] \cdot [r_1 Y_1^2] + (r_1 Z_1^2)^2$	$[r_2 X_1 X_2] \cdot [r_2 Y_1 Y_2] + (r_2 Z_1 Z_2)^2$
$m_9 =$ $m_6 \cdot (m_2 + m_3)^2$	$[r_1^2 X_1^2 Z_1^2] \cdot [(r_1 Y_1^2 + r_1 Z_1^2)^2]$	$[r_2^2 X_1 X_2 Z_1 Z_2] \cdot [(r_2 Y_1 Y_2 + r_2 Z_1 Z_2)^2]$
$m_{10} =$ $m_7 \cdot (m_1 + m_3)^2$	$[r_1^2 Y_1^2 Z_1^2] \cdot [(r_1 X_1^2 + r_1 Z_1^2)^2]$	$[r_2^2 Y_1 Y_2 Z_1 Z_2] \cdot [(r_2 X_1 X_2 + r_2 Z_1 Z_2)^2]$
$m_{11} =$ $m_8 \cdot (m_2 + m_3)$	$[r_1^2 (X_1^2 Y_1^2 + Z_1^4)] \cdot [r_1 (Y_1^2 + Z_1^2)]$	$[r_2^2 (X_1 X_2 Y_1 Y_2 + (Z_1 Z_2)^2)] \cdot [r_2 (Y_1 Y_2 + Z_1 Z_2)]$
$m_{12} =$ $m_8 \cdot (m_1 + m_3)$	$[r_1^2 (X_1^2 Y_1^2 + Z_1^4)] \cdot [r_1 (X_1^2 + Z_1^2)]$	$[r_2^2 (X_1 X_2 Y_1 Y_2 + (Z_1 Z_2)^2)] \cdot [r_2 (X_1 X_2 + Z_1 Z_2)]$
$Z_3' =$ $m_{11} \cdot (m_1 + m_3)$	$[r_1^3 (X_1^2 Y_1^2 + Z_1^4)(Y_1^2 + Z_1^2)]$ $\cdot [r_1 (X_1^2 + Z_1^2)]$	$[r_2^3 (X_1 X_2 Y_1 Y_2 + (Z_1 Z_2)^2)(Y_1 Y_2 + Z_1 Z_2)]$ $\cdot [r_2 (X_1 X_2 + Z_1 Z_2)]$
$X_3' =$ $(m_4 + m_{11}) \cdot m_{11}$ $+ m_{11}^2$ $+ \alpha \cdot m_9 + Z_3$	$[r_1 (X_1^2 + Z_1^2)]$ $+ r_1^3 (X_1^2 Y_1^2 + Z_1^4)(Y_1^2 + Z_1^2)]$ $\cdot [r_1^3 (X_1^2 Y_1^2 + Z_1^4)(Y_1^2 + Z_1^2)]$ $+ (r_1^3 (X_1^2 Y_1^2 + Z_1^4)(Y_1^2 + Z_1^2))^2$ $+ [\alpha] \cdot [r_1^4 X_1^2 Z_1^2 (Y_1^2 + Z_1^2)^2] + Z_3$	$[r_2 (X_1 + Z_1)(X_2 + Z_2)$ $+ r_2^3 (X_1 X_2 Y_1 Y_2 + (Z_1 Z_2)^2)(Y_1 Y_2 + Z_1 Z_2)]$ $\cdot [r_2^3 (X_1 X_2 Y_1 Y_2 + (Z_1 Z_2)^2)(Y_1 Y_2 + Z_1 Z_2)]$ $+ (r_2^3 (X_1 X_2 Y_1 Y_2 + (Z_1 Z_2)^2)(Y_1 Y_2 + Z_1 Z_2))^2$ $+ [\alpha] \cdot [r_2^4 X_1 X_2 Z_1 Z_2 (Y_1 Y_2 + Z_1 Z_2)^2] + Z_3$
$Y_3' =$ $(m_5 + m_{12}) \cdot m_{12}$ $+ m_{12}^2$ $+ \beta \cdot m_{10} + Z_3$	$[r_1 (Y_1^2 + Z_1^2)]$ $+ r_1^3 (X_1^2 Y_1^2 + Z_1^4)(X_1^2 + Z_1^2)]$ $\cdot [r_1^3 (X_1^2 Y_1^2 + Z_1^4)(X_1^2 + Z_1^2)]$ $+ (r_1^3 (X_1^2 Y_1^2 + Z_1^4)(X_1^2 + Z_1^2))^2$ $+ [\beta] \cdot [r_1^4 Y_1^2 Z_1^2 (X_1^2 + Z_1^2)^2] + Z_3$	$[r_2 (Y_1 + Z_1)(Y_2 + Z_2)$ $+ r_2^3 (X_1 X_2 Y_1 Y_2 + (Z_1 Z_2)^2)(X_1 X_2 + Z_1 Z_2)]$ $\cdot [r_2^3 (X_1 X_2 Y_1 Y_2 + (Z_1 Z_2)^2)(X_1 X_2 + Z_1 Z_2)]$ $+ (r_2^3 (X_1 X_2 Y_1 Y_2 + (Z_1 Z_2)^2)(X_1 X_2 + Z_1 Z_2))^2$ $+ [\beta] \cdot [r_2^4 Y_1 Y_2 Z_1 Z_2 (X_1 X_2 + Z_1 Z_2)^2] + Z_3$

Type 1 vulnerability: As shown in Table 2, if $P_1 = P_2(k' = 0)$, then the output of the proposed unified point addition operation is $X_3' = r_1^4 X_3, Y_3' = r_1^4 Y_3, Z_3' = r_1^4 Z_3$, where $(X_3 : Y_3 : Z_3)$ is the output of Table 1. Since $(X_3', Y_3', Z_3') \in (X_3, Y_3, Z_3)$, then $(X_3', Y_3', Z_3') = (X_3 : Y_3 : Z_3)$. In addition, if $P_1 = P_2$, then m_1, m_2, m_3, m_4, and m_5 can be computed as follows:

$$m_1 = (X_1)(r_1X_1), \quad m_2 = (Y_1)(r_1Y_1), \quad m_3 = (Z_1)(r_1Z_1),$$
$$m_4 = (X_1 + Z_1)(r_1X_1 + r_1Z_1), \quad m_5 = (Y_1 + Z_1)(r_1Y_1 + r_1Z_1).$$

For m_1, although $P_1 = P_2$, the operands X_1 and r_1X_1 are different. Similarly, the operands of the field multiplications for m_2, m_3, m_4, and m_5 are different. Also, there is no other field multiplication vulnerable to Type 1. Thus, the proposed algorithm is secure against the Type 1 vulnerability for the binary Huff curve.

Type 2 vulnerability: Although wRPC is applied to the proposed unified point addition operation, $m_4 = [(X_1 + Z_1)] \cdot [(r_1X_1 + r_1Z_1)] = r_1(X_1X_1 + Z_1Z_1)$ and $m_1 + m_3 = r_1X_1X_1 + r_1Z_1Z_1$ when $P_1 = P_2(k' = 0)$. Thus, $m_4 = m_1 + m_3$. Similarly, $m_5m_8 = m_{11}$. Thus, if we compute $[m_{11}] \cdot [(m_1 + m_3)], [m_4] \cdot [m_{11}]$, and $[m_5m_8] \cdot [(m_1 + m_3)]$ as the previous unified point addition operation, their operands are the same when $P_1 = P_2$. On the other hand, they are different when $P_1 \neq P_2$. Likewise, the operands of $[m_8] \cdot [(m_2 + m_3)]$ and $[m_5] \cdot [m_8]$ are the same when $P_1 = P_2$. They become targets to a Type 2 vulnerability. The reason for this vulnerability is that the same intermediate results occur in the previous unified point addition operation. They are used as inputs to more than one multiplication without modification for $P_1 = P_2$. Thus, we used the proposed method to mask the operands of vulnerable multiplications. Considering $[m_4] \cdot [m_{11}]$ and $[m_{11}] \cdot [(m_1 + m_3)]$ in Table 1, since the operand m_{11} is identically used in two multiplications, they do not affect the vulnerability. Thus, we only have to mask the other operand m_4 (or $m_1 + m_3$). Specifically, we computed $[m_4] \cdot [m_{11}]$ as $[(m_4 + M)] \cdot [m_{11}] + M \cdot m_{11}$ so that an adversary cannot distinguish between $P_1 = P_2$ and $P_1 \neq P_2$ using a Type 2 vulnerability. However, we additional cost is incurred for $M \cdot m_{11}$. To reduce this additional cost, we computed $[m_4] \cdot [m_{11}]$ as $[(m_4 + m_{11})] \cdot [m_{11}] + m_{11}^2$ to use the advantage of zero computational cost for squaring in a binary field, which is almost free. Similarly, we applied masking to $[m_5] \cdot [m_{12}]$ as $[(m_5 + m_{12})] \cdot [m_{12}] + m_{12}^2$. In addition, for $[m_8] \cdot [(m_2 + m_3)]$ and $[m_5] \cdot [m_8]$, we modified the computation of Y_3 as $m_{12} = m_8(m_1 + m_3)$ and $Y_3 = (m_5 + m_{12})m_{12} + m_{12}^2 + \beta m_{10} + Z_3$ without performing $[m_5] \cdot [m_8]$. Based on the proposed algorithm, Type 1 and Type 2 vulnerabilities no longer exist (Table 2).

6. Comparisons

We compared the proposed method with the previously presented unified point addition operations with respect to computational cost. Also, we compared the proposed method with the previously unified point addition formulae to which we applied the blinding operands of field multiplication. In this work, as the side-channel atomic algorithms, we considered (i) the proposed method, (ii) the unified point additions in [8,9], and (iii) the application of the blinding operands of a field multiplication [10] on the unified point addition method in [8,9]. We analyzed two aspects, that is, security against SCAs and computational cost. Table 3 shows the security against SCAs. The unified point additions described in [8,9] using the blinding operands in [10] are secure against ROSETTA and HCCA.

Table 3. The security against side-channel attacks (SCAs) of algorithms.

Algorithm	SPA	ROSETTA	HCCA
[8]	insecure	insecure	insecure
[9]	secure	insecure	insecure
[8] using [10]	secure	secure	secure
[9] using [10]	secure	secure	secure
proposed method	secure	secure	secure

The computational costs of [8,9] are the same. Also, the computational cost of the proposed unified point addition method is the same as that of the previous one. Thus, the computational costs of the algorithms are affected by the additional cost of wRPC and [10]. Let $w = 32$ and let n be the bit size of a finite field. Also, let $t = \lceil n/32 \rceil$. We consider that n has one of the bit sizes of the standard binary curve in FIPS 186-3 [20] (233, 283, 409, and 571). The computational cost of an iteration of the algorithms is shown in Table 4.

Table 4. The computational cost of the algorithms of the binary Huff curve.

n	Algorithm	M	Additional Cost	Total Cost	Ratio
	[8,9]	64	-	1088	1.000
233	[8,9] using [10]	81	-	1377	1.266
	proposed method	64	48	1136	1.044
	[8,9]	81	-	1377	1.000
283	[8,9] using [10]	100	-	1700	1.235
	proposed method	81	54	1431	1.039
	[8,9]	169	-	2873	1.000
409	[8,9] using [10]	196	-	3332	1.160
	proposed	169	78	2951	1.027
	[8,9]	324	-	5508	1.000
571	[8,9] using [10]	361	-	6137	1.114
	proposed method	324	108	5616	1.020

In Table 4, M is the number of w-bit multiplications of a field multiplication. Namely, $M = t^2$ in [8,9] and in the proposed method. Also, $M = t^2 + 2t + 1$ in [8,9] with [10]. The additional cost is the number of w-bit multiplications of wRPC in the proposed method. Namely, (additional cost) $= 2 * (3 * t)$ for the proposed method. The total cost is the number of w-bit multiplications of an iteration of the side-channel atomic algorithm using unified point additions. Namely, (total cost) $= 17 * M +$ (additional cost). The ratio is the overhead of the algorithm when the original algorithm [8,9] is assumed as 1. This shows that the proposed algorithm is about 0.2~4.4% slower than [8,9]. However, the methods from [8,9] are not secure against ROSETTA and HCCA. The proposed method is about 8.5~17.5% faster than the previous methods from [8,9] using [10], which are secure against ROSETTA and HCCA. In addition, the previous methods ([8,9] using [10]) also require random number generation for r_1 and r_2 in each field multiplication.

7. Conclusions

In this paper, we present two vulnerabilities of unified point addition on the binary Huff curve; these vulnerabilities are exploitable by ROSETTA and HCCA. In particular, we found these vulnerabilities of unified point addition on the binary Huff curve as presented in [9]. As countermeasures, we propose wRPC and present a new unified point addition method for the binary Huff curve. Additionally, we show the proposed unified point addition method and wRPC applied to the side-channel atomic algorithm. The proposed method is secure against ROSETTA and HCCA. In addition, the proposed unified point addition method has no additional cost compared to the previous one. However, wRPC does incur additional cost. Depending on the size of the base field of an elliptic curve, the proposed method is about 0.2~4.4% slower than the original one. However, it is about 8.5~17.5% faster than unified point additions using blinding operands as a countermeasure. Additionally, we present our analyses of the vulnerabilities of unified point addition on other elliptic curves, such as Weierstraß, Hessian, Edwards, Jacobi intersections, Jacobi quartic, and binary Edwards elliptic curves in the Appendix A.

Author Contributions: S.M.C. and S.J. performed the experiments, analyzed the data, and wrote the paper. H.K. analyzed the data and verified the paper.

Funding: This research was supported by Basic Science Research Program through the National Research Foundation of Korea (NRF) funded by the Ministry of Science and ICT (NRF-2017R1C1B2004583).

Conflicts of Interest: The authors declare no conflict of interest.

Appendix A

We applied Type 1 and Type 2 vulnerabilities to unified point additions on other elliptic curves. As a result, we found that most unified point additions on these elliptic curves (such as Weierstraß, Hessian, Edwards, Jacobi intersections, Jacobi quartic, and binary Edwards elliptic curves) have these vulnerabilities. Table A1 shows the vulnerability of each unified point addition. In the case of Hessian, Edwards, Jacobi intersections, and Jacobi quartic curves, it is enough to apply wRPC to unified point additions to ensure security against Type 1 and Type 2 vulnerabilities. However, in the case of Weierstraß and binary Edwards elliptic curves, we need to modify the unified point addition formula. In this section, we explain the vulnerabilities of unified point addition and its countermeasure for Weierstraß, Hessian, Edwards, Jacobi intersections, Jacobi quartic, and binary Edwards elliptic curves.

Table A1. The vulnerabilities of the elliptic curve forms and it countermeasures.

Curve	Type 1	Type 2	Countermeasures
Weierstraß	insecure	insecure	wRPC The modified unified point addition
Hessian	insecure	insecure	wRPC
Edwards	insecure	secure	wRPC
Jacobi intersections	secure	insecure	wRPC
Jacobi quartic	insecure	insecure	wRPC
binary Edwards	insecure	insecure	wRPC The modified unified point addition

Appendix A.1 Weierstraß Elliptic Curve

A Weierstraß elliptic curve has the parameters a and b that satisfy the following equations:

$$y^2 = x^3 + ax + b \tag{A1}$$

The projective coordinates have the assumption $a = -3$ and represent x, y as X, Y, Z to satisfy the following equations:

$$x = X/Z \quad \text{and} \quad y = Y/Z$$

The equivalence class containing (X, Y, Z) is

$$(X : Y : Z) = (rX, rY, rZ) : r \in \mathbb{F}. \tag{A2}$$

We describe a projective form of the unified point addition method (add-2007-bl) given in [21]. Let $P_1 = (X_1 : Y_1 : Z_1)$ and $P_2 = (X_2 : Y_2 : Z_2)$; then, we can get $P_1 + P_2 = (X_3 : Y_3 : Z_3)$ by the unified point addition formula for the Weierstraß elliptic curve:

$$\begin{cases} X_3 = 2FW \\ Y_3 = R(G - 2W) - 2L^2 \\ Z_3 = 4F^3, \end{cases} \tag{A3}$$

where

$$U_1 = X_1 Z_2, \quad U_2 = X_2 Z_1, \quad S_1 = Y_1 Z_2, \quad S_2 = Y_2 Z_1,$$
$$Z = Z_1 Z_2, \quad T = U_1 + U_2, \quad M = S_1 + S_2,$$
$$R = T^2 - U_1 U_2 + aZ^2, \quad F = ZM, \quad L = MF,$$

$G = (T + L)^2 - T^2 - L^2$, and $W = 2R^2 - G$.

This formula requires 11 field multiplications and 6 field squarings. We found both Type 1 and Type 2 vulnerabilities during the computations of $P_1 + P_2$ for $P_1 \neq P_2$ and $P_1 = P_2$.

Type 1 vulnerability: Let us consider the computation $Z = [Z_1] \cdot [Z_2]$. In this formula, it is computed as $[Z_1] \cdot [Z_1]$ for $P_1 = P_2$, whereas it is computed as $[Z_1] \cdot [Z_2]$ for $P_1 \neq P_2$. Similarly, for $U_1 \cdot U_2$ in R, this is computed as $[X_1 Z_1] \cdot [X_1 Z_1]$ for $P_1 = P_2$. Thus, we can distinguish between $P_1 = P_2$ and $P_1 \neq P_2$ using ROSETTA.

Type 2 vulnerability: Let us consider the computations $U_1 = [X_1] \cdot [Z_2]$ and $U_2 = [X_2] \cdot [Z_1]$. If $P_1 = P_2$, then $[X_1] \cdot [Z_1]$ is computed twice. Namely, the operands of $[X_1] \cdot [Z_2]$ and $[X_2] \cdot [Z_1]$ for U_1 and U_2 are the same for $P_1 = P_2$ but different for $P_1 \neq P_2$. Similarly, considering $S_1 = [Y_1] \cdot [Z_2]$ and $S_2 = [Y_2] \cdot [Z_1]$, the multiplications for S_1 and S_2 have the same operands for $P_1 = P_2$ but different operands for $P_1 \neq P_2$. Therefore, we can distinguish between $P_1 = P_2$ and $P_1 \neq P_2$ using HCCA.

Applying wRPC to unified point addition on the Weierstraß elliptic curve, the two inputs are expressed as follows:

$$
\begin{cases}
P_1 = (X_1 : Y_1 : Z_1), P_2 = (r_1 X_1 : r_1 Y_1 : r_1 Z_1) & \text{If } k' = 0, \\
P_1 = (X_1 : Y_1 : Z_1), P_2 = (r_2 X_2 : r_2 Y_2 : r_2 Z_2) & \text{If } k' = 1.
\end{cases}
\tag{A4}
$$

where $r \neq 1$. Although wRPC is applied to unified point addition, $U_1 \cdot U_2$ in R is computed as $[r X_1 Z_1] \cdot [r X_1 Z_1]$ for $P_1 = P_2$. Thus, we need to modify $U_1 \cdot U_2$ in R. We modified R as follows:

$$
\begin{aligned}
R &= T^2 - U_1 U_2 + a Z^2 \\
&= (U_1 + U_2)^2 - (U_1 + U_2) U_2 + U_2^2 + a Z^2 \\
&= (U_1 + U_2)((U_1 + U_2) - U_2) + U_2^2 + a Z^2 \\
&= T U_1 + U_2^2 + a Z^2
\end{aligned}
$$

After applying the above modification to unified point addition, 11 field multiplications and 6 field squarings were required, which are exactly the same as those required by the original one. After applying wRPC to the modified unified point addition formula, Type 1 and Type 2 vulnerabilities no longer exist (Table A2).

Table A2. The proposed unified point addition method on the Weierstraß elliptic curve by applying wRPC.

Out	$P_1 = P_2$	$P_1 \neq P_2$
U_1	$[X_1] \cdot [r_1 Z_1]$	$[X_1] \cdot [r_2 Z_2]$
U_2	$[r_1 X_1] \cdot [Z_1]$	$[r_2 X_2] \cdot [Z_1]$
S_1	$[Y_1] \cdot [r_1 Z_1]$	$[Y_1] \cdot [r_2 Z_2]$
S_2	$[r_1 Y_1] \cdot [Z_1]$	$[r_2 Y_2] \cdot [Z_1]$
Z	$[Z_1] \cdot [r_1 Z_1]$	$[Z_1] \cdot [r_2 Z_2]$
$T = U_1 + U_2$	$r_1 X_1 Z_1 + r_1 X_1 Z_1$	$r_2 X_1 Z_2 + r_2 X_2 Z_1$
$M = S_1 + S_2$	$r_1 Y_1 Z_1 + r_1 Y_1 Z_1$	$r_2 Y_1 Z_2 + r_2 Y_2 Z_1$
$R = T \cdot U_1 + U_2^2$	$[2 r_1 X_1 Z_1] \cdot [r_1 X_1 Z_1] + (r_1 X_1 Z_1)^2$	$[r_2 (X_1 Z_2 + X_2 Z_1)] \cdot [r_2 X_1 Z_2] + (r_2 X_2 Z_1)^2$
$+ a Z^2$	$+ a (r_1 Z_1^2)^2$	$+ a (r_2 Z_1 Z_2)^2$
$\vdots$	$\vdots$	$\vdots$

Appendix A.2 Hessian Elliptic Curve

A Hessian elliptic curve has a parameter d that satisfies the following equation:

$$
x^3 + y^3 + 1 = 3 d x y
\tag{A5}
$$

The projective coordinates represent x, y as X, Y, Z satisfying the following equation:

$$x = X/Z \quad \text{and} \quad y = Y/Z$$

The equivalence class containing (X, Y, Z) is

$$(X : Y : Z) = (\lambda X, \lambda Y, \lambda Z) : \lambda \in \mathbb{F}. \tag{A6}$$

We describe a projective form of the unified point addition formula (add-2009-bkl) given in [21]. Let $P_1 = (X_1 : Y_1 : Z_1)$ and $P_2 = (X_2 : Y_2 : Z_2)$; then, we get $P_1 + P_2 = (X_3 : Y_3 : Z_3)$ with the unified point addition formula for the Hessian elliptic curve:

$$\begin{cases} X_3 = DC - FA \\ Y_3 = BA - DE \\ Z_3 = FE - BC, \end{cases} \tag{A7}$$

where

$$A = Y_1 X_2, \quad B = Y_1 Y_2, \quad C = Z_1 Y_2,$$
$$D = Z_1 Z_2, \quad E = X_1 Z_2, \quad F = X_1 X_2.$$

This formula requires 12 field multiplications. We can identify vulnerabilities of Type 1 and Type 2 during the computations of $P_1 + P_2$ for $P_1 \neq P_2$ and $P_1 = P_2$.

Type 1 vulnerability: Let us consider the computation $B = [Y_1] \cdot [Y_2]$. In this formula, it is computed as $[Y_1] \cdot [Y_1]$ for $P_1 = P_2$, whereas it is computed as $[Y_1] \cdot [Y_2]$ for $P_1 \neq P_2$. Similarly, in $D = [Z_1] \cdot [Z_2]$ and $F = [X_1] \cdot X_2$, these are computed as $[Z_1] \cdot [Z_1]$ and $[X_1] \cdot [X_1]$ for $P_1 = P_2$, respectively. Thus, we can distinguish between $P_1 = P_2$ and $P_1 \neq P_2$ using ROSETTA.

Type 2 vulnerability: Let us consider the computations $A = [Y_1] \cdot [X_2]$ and $C = [Z_1] \cdot [Y_2]$. If $P_1 = P_2$, then $[Y_1] \cdot [X_1]$ and $[Z_1] \cdot [Y_1]$ are computed. Thus, they have the same operand Y_1 when $P_1 = P_2$ but not when $P_1 \neq P_2$. Similarly, considering $C = [Z_1] \cdot [Y_2]$ and $E = [X_1] \cdot [Z_2]$, the multiplications for C and E have the same operand Z_1 for $P_1 = P_2$ and different operands for $P_1 \neq P_2$. Also, the multiplications for A and E have the same operand X_1 for $P_1 = P_2$. Therefore, we can distinguish between $P_1 = P_2$ and $P_1 \neq P_2$ using HCCA.

When applying wRPC to unified point addition on the Hessian elliptic curve, the two inputs are expressed as follows:

$$\begin{cases} P_1 = (X_1 : Y_1 : Z_1), P_2 = (r_1 X_1 : r_1 Y_1 : r_1 Z_1) & \text{If } k' = 0, \\ P_1 = (X_1 : Y_1 : Z_1), P_2 = (r_2 X_2 : r_2 Y_2 : r_2 Z_2) & \text{If } k' = 1. \end{cases} \tag{A8}$$

where $r \neq 1$. It is sufficient to secure against Type 1 and Type 2 vulnerabilities by applying wRPC to unified point addition. The application of wRPC to unified point addition is evaluated in Table A3. Table A3 shows that vulnerabilities of Type 1 and Type 2 no longer exist.

Table A3. Unified point addition for the Hessian elliptic curve form.

Out	$P = Q$	$P \neq Q$
A	$[Y_1] \cdot [r_1 X_1]$	$[Y_1] \cdot [r_2 X_2]$
B	$[Y_1] \cdot [r_1 Y_1]$	$[Y_1] \cdot [r_2 Y_2]$
C	$[Z_1] \cdot [r_1 Y_1]$	$[Z_1] \cdot [r_2 Y_2]$
D	$[Z_1] \cdot [r_1 Z_1]$	$[Z_1] \cdot [r_2 Z_2]$
E	$[X_1] \cdot [r_1 Z_1]$	$[X_1] \cdot [r_2 Z_2]$
F	$[X_1] \cdot [r_1 X_1]$	$[X_1] \cdot [r_2 X_2]$
$\vdots$	$\vdots$	$\vdots$

Appendix A.3 Edwards Elliptic Curve

An Edwards elliptic curve has the parameters c and d that satisfy the following equation:

$$x^2 + y^2 = c^2(1 + dx^2y^2) \tag{A9}$$

The inverted projective coordinates represent x, y as X, Y, Z to satisfy the following equation:

$$x = Z/X \quad \text{and} \quad y = Z/Y$$

The equivalence class containing (X, Y, Z) is

$$(X : Y : Z) = (\lambda X, \lambda Y, \lambda Z) : \lambda \in \mathbb{F}. \tag{A10}$$

We describe a inverted projective form of the unified point addition formula (add-2007-bl) given in [21]. Let $P_1 = (X_1 : Y_1 : Z_1)$ and $P_2 = (X_2 : Y_2 : Z_2)$. Then, we get $P_1 + P_2 = (X_3 : Y_3 : Z_3)$ by the unified point addition formula for the Edwards elliptic curve:

$$\begin{cases} X_3 = c(E + B)H \\ Y_3 = c(E - B)I \\ Z_3 = AHI, \end{cases} \tag{A11}$$

where

$$A = Z_1 Z_2, \quad B = dA^2, \quad C = X_1 X_2, \quad D = Y_1 Y_2,$$
$$E = CD, \quad H = C - D, \quad I = (X_1 + Y_1)(X_2 + Y_2) - C - D.$$

This formula requires 9 field multiplications and 1 field squaring. We can identify vulnerabilities of Type 1 and Type 2 during the computations of $P_1 + P_2$ for $P_1 \neq P_2$ and $P_1 = P_2$.

Type 1 vulnerability: Let us consider the computation $A = [Z_1] \cdot [Z_2]$. In this formula, it is computed as $[Z_1] \cdot [Z_1]$ for $P_1 = P_2$, whereas it is computed as $[Z_1] \cdot [Z_2]$ for $P_1 \neq P_2$. Similarly, in $C = [X_1] \cdot [X_2]$, $D = [Y_1] \cdot [Y_2]$ and $I = [(X_1 + Y_1)] \cdot [(X_2 + Y_2)] - C - D$, and these are computed as $[X_1] \cdot [X_1]$, $[Y_1] \cdot [Y_1]$, and $[(X_1 + Y_1)] \cdot [(X_1 + Y_1)] - C - D$ for $P_1 = P_2$, respectively. Thus, we can distinguish between $P_1 = P_2$ and $P_1 \neq P_2$ using ROSETTA.

Type 2 vulnerability: The vulnerability of Type 2 does not exist.

When applying wRPC to unified point addition for the Edwards elliptic curve, the two inputs are expressed as follows:

$$\begin{cases} P_1 = (X_1 : Y_1 : Z_1), P_2 = (r_1 X_1 : r_1 Y_1 : r_1 Z_1) & \text{If } k' = 0, \\ P_1 = (X_1 : Y_1 : Z_1), P_2 = (r_2 X_2 : r_2 Y_2 : r_2 Z_2) & \text{If } k' = 1. \end{cases} \tag{A12}$$

where $r \neq 1$. It is sufficient to secure against a Type 1 vulnerability by applying wRPC to unified point addition. The application of wRPC to unified point addition is evaluated in Table A4. Table A4 shows that vulnerability of Type 1 no longer exists.

Table A4. Unified point addition for the Edwards elliptic curve.

Out	$P = Q$	$P \neq Q$
A	$[Z_1] \cdot [r_1 Z_1]$	$[Z_1] \cdot [r_2 Z_2]$
B	$d(r_1 Z_1^2)^2$	$d(r_2 Z_1 Z_2)^2$
C	$[X_1] \cdot [r_1 X_1]$	$[X_1] \cdot [r_2 X_2]$
D	$[Y_1] \cdot [r_1 Y_1]$	$[Y_1] \cdot [r_2 Y_2]$
$E = C \cdot D$	$[r_1 X_1^2] \cdot [r_1 Y_1^2]$	$[r_2 X_1 X_2] \cdot [r_2 Y_1 Y_2]$
$H = C - D$	$[r_1 X_1^2] - [r_1 Y_1^2]$	$[r_2 X_1 X_2] - [r_2 Y_1 Y_2]$
$I = (X_1 + Y_1) \cdot (X_2 + Y_2)$	$[(X_1 + Y_1)] \cdot [(r_1 X_1 + r_1 Y_1)]$	$[(X_1 + Y_1)] \cdot [(r_2 X_2 + r_2 Y_2)]$
$-C - D$	$-[r_1 X_1^2] - [r_1 Y_1^2]$	$-[r_2 X_1 X_2] - [r_2 Y_1 Y_2]$
$\vdots$	$\vdots$	$\vdots$

Appendix A.4 Jacobi Intersections Elliptic Curve

An elliptic curve in Jacobi intersection form has the parameter a and coordinate s, c, d that satisfy the following equations:

$$s^2 + c^2 = 1 \qquad as^2 + d^2 = 1 \tag{A13}$$

The projective coordinates represent s, c, d as S, C, D, Z to satisfy the following equations:

$$s = S/Z, \quad c = C/Z \quad \text{and} \quad d = D/Z$$

The equivalence class containing (S, C, D, Z) is

$$(S : C : D : Z) = (\lambda S, \lambda C, \lambda D, \lambda Z) : \lambda \in \mathbb{F}. \tag{A14}$$

We describe a projective form of the unified point addition formula (add-20080225-hwcd) given in [21]. Let $P_1 = (S_1 : C_1 : D_1 : Z_1)$ and $P_2 = (S_2 : C_2 : D_2 : Z_2)$; then, we get $P_1 + P_2 = (S_3 : C_3 : D_3 : Z_3)$ with the unified point addition formula for the Jacobi intersection elliptic curve:

$$\begin{cases} S_3 = (H + F)(E + G) - J - K \\ C_3 = (H + E)(F - G) - J + K \\ D_3 = (B - aA)(C + D) + aJ - K \\ Z_3 = (H + G)^2 - 2K, \end{cases} \tag{A15}$$

where

$$A = S_1 C_1, \quad B = D_1 Z_1, \quad C = S_2 C_2, \quad D = D_2 Z_2,$$
$$E = S_1 D_2, \quad F = C_1 Z_2, \quad G = D_1 S_2, \quad H = Z_1 C_2,$$
$$J = AD, \quad K = BC.$$

This formula requires 13 field multiplications and 1 field squaring. We can identify vulnerabilities of Type 1 and Type 2 during the computations of $P_1 + P_2$ for $P_1 \neq P_2$ and $P_1 = P_2$.

Type 1 vulnerability: The vulnerability of Type 1 does not exist.

Type 2 vulnerability: Let us consider the computations of $A = [S_1] \cdot [C_1]$ and $C = [S_2] \cdot [C_2]$. If $P_1 = P_2$, then $[S_1] \cdot [C_1]$ are computed twice. Namely, the operands of $[S_1] \cdot [C_1]$ and $[S_2] \cdot [C_2]$ for A and B are the same for $P_1 = P_2$ and different for $P_1 \neq P_2$. Similarly, consider multiplications for B and D, E and G, F and H, and J and K. These multiplication pairs have the same operands for $P_1 = P_2$ and different operands for $P_1 \neq P_2$. Also, consider multiplication of $A = [S_1] \cdot [C_1]$ and $G = [D_1] \cdot [S_1]$. If $P_1 = P_2$, then $[S_1] \cdot [C_1]$ and $[D_1] \cdot [S_1]$ are computed. Thus, they have the same operand S_1 when $P_1 = P_2$ but not when $P_1 \neq P_2$. Similarly, the multiplication pairs A and H, B and E, B and F, C and E, C and F, D and G, and D and H have the same operand C_1, D_1, Z_1, S_1, C_1, D_1, and Z_1 for $P_1 = P_2$, respectively. Therefore, we can distinguish between $P_1 = P_2$ and $P_1 \neq P_2$ using HCCA.

Applying wRPC to unified point addition of the Jacobi intersection elliptic curve, the two inputs are expressed as follows:

$$\begin{cases} P_1 = (S_1 : C_1 : D_1 : Z_1), P_2 = (r_1 S_1 : r_1 C_1 : r_1 D_1 : r_1 Z_1) & \text{If } k' = 0, \\ P_1 = (S_1 : C_1 : D_1 : Z_1), P_2 = (r_2 S_2 : r_2 C_2 : r_2 D_2 : r_2 Z_2) & \text{If } k' = 1. \end{cases} \tag{A16}$$

where $r \neq 1$. It is sufficient to secure against a Type 2 vulnerability by applying wRPC to unified point addition. The application of wRPC to unified point addition is evaluated in Table A5. Table A5 shows that vulnerability of Type 2 no longer exists.

Table A5. Unified point addition for the Jacobi intersection elliptic curve form.

Out	$P = Q$	$P \neq Q$
A	$[S_1] \cdot [C_1]$	$[S_1] \cdot [C_1]$
B	$[D_1] \cdot [Z_1]$	$[D_1] \cdot [Z_1]$
C	$[r_1 S_1] \cdot [r_1 C_1]$	$[r_2 S_2] \cdot [r_2 C_2]$
D	$[r_1 D_1] \cdot [r_1 Z_1]$	$[r_2 D_2] \cdot [r_2 Z_2]$
E	$[S_1] \cdot [r_1 D_1]$	$[S_1] \cdot [r_2 D_2]$
F	$[C_1] \cdot [r_1 Z_1]$	$[C_1] \cdot [r_2 Z_2]$
G	$[D_1] \cdot [r_1 S_1]$	$[D_1] \cdot [r_2 S_2]$
H	$[Z_1] \cdot [r_1 C_1]$	$[Z_1] \cdot [r_2 C_2]$
$J = A \cdot D$	$[S_1 C_1] \cdot [r_1^2 D_1 Z_1]$	$[S_1 C_1] \cdot [r_2^2 D_2 Z_2]$
$K = B \cdot C$	$[D_1 Z_1] \cdot [r_1^2 S_1 C_1]$	$[D_1 Z_1] \cdot [r_2^2 S_2 C_2]$
$\vdots$	$\vdots$	$\vdots$

Appendix A.5 Jacobi Quartic Elliptic Curve

An elliptic curve in the Jacobi quartic form has the parameter a and coordinates x, y that satisfy the following equation:

$$y^2 = x^4 + 2ax^2 + 1 \tag{A17}$$

The projective coordinates represent x, y as X, Y, Z to satisfy the following equations:

$$x = X/Z \quad \text{and} \quad y = Y/Z^2$$

The equivalence class containing (X, Y, Z) is

$$(X : Y : Z) = (\lambda X, \lambda^2 Y, \lambda Z) : \lambda \in \mathbb{F}. \tag{A18}$$

We describe a projective form of the unified point addition formula (add-2007-bl) given in [21]. Let $P_1 = (X_1 : Y_1 : Z_1)$ and $P_2 = (X_2 : Y_2 : Z_2)$; then, we get $P_1 + P_2 = (X_3 : Y_3 : Z_3)$ with the unified point addition formula for the Jacobi quartic elliptic curve:

$$\begin{cases} X_3 = E_1 E_2 - I - K \\ Y_3 = F(4K + aG) + (D_1 D_2 - F)G \\ Z_3 = 2(J - H), \end{cases} \tag{A19}$$

where

$$A_2 = X_2^2, \quad C_2 = Z_2^2, \quad D_2 = A_2 + C_2, \quad B_2 = (X_2 + Z_2)^2 - D_2,$$
$$E_2 = B_2 + Y_2, \quad A_1 = X_1^2, \quad C_1 = Z_1^2, \quad D_1 = A_1 + C_1,$$
$$B_1 = (X_1 + Z_1)^2 - D_1, \quad E_1 = B_1 + Y_1, \quad H = A_1 A_2,$$
$$I = B_1 B_2, \quad J = C_1 C_2, \quad K = Y_1 Y_2, \quad F = J + H, \quad F = 2I.$$

This formula requires 8 field multiplications and 6 field squarings. We can identify vulnerabilities of Type 1 and Type 2 during the computations of $P_1 + P_2$ for $P_1 \neq P_2$ and $P_1 = P_2$.

Type 1 vulnerability: Let us consider the computation $B = [Y_1] \cdot [Y_2]$. In this formula, it is computed as $[Y_1] \cdot [Y_1]$ for $P_1 = P_2$, whereas it is computed as $[Y_1] \cdot [Y_2]$ for $P_1 \neq P_2$. Similarly, in $D = [Z_1] \cdot [Z_2]$ and $F = [X_1] \cdot X_2]$, these are computed as $[Z_1] \cdot [Z_1]$ and $[X_1] \cdot [X_1]$ for $P_1 = P_2$, respectively. Thus, we can distinguish between $P_1 = P_2$ and $P_1 \neq P_2$ using ROSETTA.

Type 2 vulnerability: Let us consider the computations $A = [Y_1] \cdot [X_2]$ and $C = [Z_1] \cdot [Y_2]$. If $P_1 = P_2$; then, $[Y_1] \cdot [X_1]$ and $[Z_1] \cdot [Y_1]$ are computed. Thus, they have the same operand Y_1 when $P_1 = P_2$ but not when $P_1 \neq P_2$. Similarly, considering $C = [Z_1] \cdot [Y_2]$ and $E = [X_1] \cdot [Z_2]$, the multiplications for C and E have the same operand Z_1 for $P_1 = P_2$ and different operands for $P_1 \neq P_2$. Also, the multiplications for A and E have the same operand X_1 for $P_1 = P_2$. Therefore, we can distinguish between $P_1 = P_2$ and $P_1 \neq P_2$ using HCCA.

By Algorithm 2, to use unified point addition on the Jacobi quartic elliptic curve, the two inputs of step 8 are expressed as follows:

$$\begin{cases} P_1 = (X_1 : Y_1 : Z_1), P_2 = (r_1 X_1 : r_1^2 Y_1 : r_1 Z_1) & \text{If } k' = 0, \\ P_1 = (X_1 : Y_1 : Z_1), P_2 = (r_2 X_2 : r_2^2 Y_2 : r_2 Z_2) & \text{If } k' = 1. \end{cases} \tag{A20}$$

where $r \neq 1$. It is sufficient to secure against Type 1 and Type 2 vulnerabilities by applying wRPC to unified point addition. The application of wRPC to unified point addition is evaluated in Table A6. Table A6 shows that vulnerabilities of Type 1 and Type 2 no longer exist.

Table A6. Unified point addition for the Jacobi quartic elliptic curve form.

Out	$P = Q$	$P \neq Q$
A	$[Y_1] \cdot [r_1 X_1]$	$[Y_1] \cdot [r_2 X_2]$
B	$[Y_1] \cdot [r_1^2 Y_1]$	$[Y_1] \cdot [r_2^2 Y_2]$
C	$[Z_1] \cdot [r_1^2 Y_1]$	$[Z_1] \cdot [r_2^2 Y_2]$
D	$[Z_1] \cdot [r_1 Z_1]$	$[Z_1] \cdot [r_2 Z_2]$
E	$[X_1] \cdot [r_1 Z_1]$	$[X_1] \cdot [r_2 Z_2]$
F	$[X_1] \cdot [r_1 X_1]$	$[X_1] \cdot [r_2 X_2]$
$\vdots$	$\vdots$	$\vdots$

Appendix A.6 Binary Edwards Elliptic Curve

A binary Edwards elliptic curve has the parameters d_1 and d_2 that satisfy the following equation:

$$d_1(x + y) + d_2(x^2 + y^2) = (x + x^2)(y + y^2) \tag{A21}$$

The projective coordinates represent x, y as X, Y, Z to satisfy the following equation:

$$x = X/Z \quad \text{and} \quad y = Y/Z$$

The equivalence class containing (X, Y, Z) is

$$(X : Y : Z) = (rX, rY, rZ) : r \in \mathbb{F}. \tag{A22}$$

We describe a projective form of the unified point addition formula (add-2008-blr-4) given in [21]. Let $P_1 = (X_1 : Y_1 : Z_1)$ and $P_2 = (X_2 : Y_2 : Z_2)$; then, we can get $P_1 + P_2 = (X_3 : Y_3 : Z_3)$ with unified point addition for the binary Edwards elliptic curve:

$$\begin{cases} X_3 = V + D(A+D)(G+D) \\ Y_3 = V + D(B+D)(H+D) \\ Z_3 = U + (d_2 + d_1)CK^2, \end{cases} \tag{A23}$$

where

$$\begin{aligned} & A = X_1 X_2, \quad B = Y_1 Y_2, \quad C = Z_1 Z_2, \quad D = d_1 C, \quad E = C^2, \quad F = D^2, \\ & G = (X_1 + Z_1)(X_2 + Z_2), \quad H = (Y_1 + Z_1)(Y_2 + Z_2), \quad I = A + G, \\ & J = B + H, \quad K = (X_1 + Y_1)(X_2 + Y_2), \quad U = C(F + d_1 K(K + I + J + C)), \\ & V = U + DF + K(d_2(d_1 E + GH + AB) + (d_2 + d_1)IJ). \end{aligned}$$

This formula requires 18 field multiplications. We found both Type 1 and Type 2 vulnerabilities during the computations of $P_1 + P_2$ for $P_1 \neq P_2$ and $P_1 = P_2$.

Type 1 vulnerability: Let us consider the computation $A = [X_1] \cdot [X_2]$. In this formula, it is computed as $[X_1] \cdot [X_1]$ for $P_1 = P_2$, whereas it is computed as $[X_1] \cdot [X_2]$ for $P_1 \neq P_2$. Similarly, for $B = [Y_1] \cdot [Y_2], C = [Z_1] \cdot [Z_2], G = [(X_1 + Z_1)] \cdot [(X_2 + Z_2)], H = [(Y_1 + Z_1)] \cdot [(Y_2 + Z_2)]$, and $K = [(X_1 + Y_1)] \cdot [(X_2 + Y_2)]$, these are computed as $B = [Y_1] \cdot [Y_1], C = [Z_1] \cdot [Z_1], G = [(X_1 + Z_1)] \cdot [(X_1 + Z_1)], H = [(Y_1 + Z_1)] \cdot [(Y_1 + Z_1)]$, and $K = [(X_1 + Y_1)] \cdot [(X_1 + Y_1)]$ for $P_1 = P_2$. Also, if $P_1 = P_2$, I and J compute as follows:

$$I = A + G = X_1 X_1 + (X_1 + Z_1)(X_1 + Z_1) = X_1^2 + X_1^2 + Z_1^2 = Z_1^2 \text{ and}$$

$$J = B + H = Y_1 Y_1 + (Y_1 + Z_1)(Y_1 + Z_1) = Y_1^2 + Y_1^2 + Z_1^2 = Z_1^2.$$

Thus, if $P_1 = P_2$, $[I] \cdot [J] = [Z_1^2] \cdot [Z_1^2]$. An adversary can distinguish between $P_1 = P_2$ and $P_1 \neq P_2$ using ROSETTA.

Type 2 vulnerability: Let us consider the computations $U = [C] \cdot [(F + d_1 K(K + I + J + C))]$, $[(d_2 + d_1)] \cdot [I] \cdot [J]$ in V and $[(d_2 + d_1)] \cdot [C] \cdot [K^2]$ in Z^3. If $P_1 = P_2$, since $C = I = J$, both operations have at least one same operand. Therefore, they can be distinguished using HCCA.

By Algorithm 2, to use unified point addition on the binary Edwards elliptic curve, the two inputs of step 8 are expressed as follows:

$$\begin{cases} P_1 = (X_1 : Y_1 : Z_1), P_2 = (r_1 X_1 : r_1 Y_1 : r_1 Z_1) & \text{If } k' = 0, \\ P_1 = (X_1 : Y_1 : Z_1), P_2 = (r_2 X_2 : r_2 Y_2 : r_2 Z_2) & \text{If } k' = 1. \end{cases} \tag{A24}$$

where $r \neq 1$. Although wRPC is applied to unified point addition, $C = I = J$ for $P_1 = P_2$. Thus, we need to modify the unified point addition formula. The collision pairs exposed by HCCA are $(U = [C] \cdot [(F + d_1 K(K + I + J + C))]$ and $[(d_2 + d_1)] \cdot [I] \cdot [J]$ in V or $([(d_2 + d_1)] \cdot [C] \cdot [K^2]$ in Z^3 and $[(d_2 + d_1)] \cdot [I] \cdot [J]$ in V). Since both collision pairs contain the operation $[(d_2 + d_1)] \cdot [I] \cdot [J]$, we only have to mask its operands. We modified $[(d_2 + d_1)] \cdot [I] \cdot [J]$ in V as follows:

$$\begin{aligned} (d_2 + d_1) \cdot I \cdot J & = ((d_2 + d_1) \cdot (I + d_2 + d_1) + (d_2 + d_1)^2) \cdot J \\ & = ((d_2 + d_1) \cdot (I + d_2 + d_1) + (d_2 + d_1)^2) \cdot (J + (d_2 + d_1)I) \\ & \quad + ((d_2 + d_1)I)^2. \end{aligned}$$

To use the advantage of the free computational cost of squaring in a binary field, we configured the masking of $d_2 + d_1$ and $(d_2 + d_1)I$ by squaring. The proposed unified point addition method for the binary Edwards elliptic curve is as follows:

$$\begin{cases} X_3 = V + D(A+D)(G+D) \\ Y_3 = V + D(B+D)(H+D) \\ Z_3 = U + (d_2+d_1)CK^2, \end{cases} \tag{A25}$$

where

$$A = X_1 X_2, \quad B = Y_1 Y_2, \quad C = Z_1 Z_2, \quad D = d_1 C, \quad E = C^2, \quad F = D^2,$$
$$G = (X_1 + Z_1)(X_2 + Z_2), \quad H = (Y_1 + Z_1)(Y_2 + Z_2), \quad I = A + G,$$
$$J = B + H, \quad L = (d_2 + d_1)(I + d_2 + d_1) + (d_2 + d_1)^2 \quad K = (X_1 + Y_1)(X_2 + Y_2),$$
$$U = C(F + d_1 K(K + I + J + C)),$$
$$V = U + DF + K(d_2(d_1 E + GH + AB) + L(J + L) + L^2).$$

After applying the above modification to the unified point addition, 18 field multiplications were required, which was exactly the same as in the original one. After applying *w*RPC to the modified unified point addition method, Type 1 and Type 2 vulnerabilities no longer exist (Table A7).

Table A7. The proposed unified point addition method on the binary Edwards elliptic curve.

Out	$P_1 = P_2 (k' = 0)$	$P_1 \neq P_2 (k' = 1)$
A	$[X_1] \cdot [r_1 X_1]$	$[X_1] \cdot [r_2 X_2]$
B	$[Y_1] \cdot [r_1 Y_1]$	$[Y_1] \cdot [r_2 Y_2]$
C	$[Z_1] \cdot [r_1 Z_1]$	$[Z_1] \cdot [r_2 Z_2]$
$D = d_1 \cdot C$	$[d_1] \cdot [r_1 Z_1^2]$	$[d_1] \cdot [r_2 Z_1 Z_2]$
$E = C^2$	$(r_1 Z_1^2)^2$	$(r_2 Z_1 Z_2)^2$
$F = D^2$	$(r_1 d_1 Z_1^2)^2$	$(r_2 d_1 Z_1 Z_2)^2$
G	$[(X_1 + Z_1)] \cdot [(r_1 X_1 + r_1 Z_1)]$	$[(X_1 + Z_1)] \cdot [(r_2 X_2 + r_2 Z_2)]$
H	$[(Y_1 + Z_1)] \cdot [(r_1 Y_1 + r_1 Z_1)]$	$[(Y_1 + Z_1)] \cdot [(r_2 Y_2 + r_2 Z_2)]$
$I = A + G$	$r_1 X_1^2 + (r_1 X_1^2 + r_1 Z_1^2)$	$r_2 X_1 X_2 + (X_1 + Z_1)(r_2 X_2 + r_2 Z_2)$
$J = B + H$	$r_1 Y_1^2 + (r_1 Y_1^2 + r_1 Z_1^2)$	$r_2 Y_1 Y_2 + (Y_1 + Z_1)(r_2 Y_2 + r_2 Z_2)$
$L = (d_2 + d_1)$	$[(d_2 + d_1)]$	$[(d_2 + d_1)]$
$\cdot (I + d_2 + d_1)$	$\cdot [(r_1 Z_1^2 + d_2 + d_1)]$	$\cdot [(r_2 X_1 X_2 + r_2(X_1 + Z_1)(X_2 + Z_2) + d_2 + d_1)]$
$+ (d_2 + d_1)^2$	$+ (d_2 + d_1)^2$	$+ (d_2 + d_1)^2$
K	$[(X_1 + Y_1)] \cdot [(r_1 X_1 + r_1 Y_1)]$	$[(X_1 + Y_1)] \cdot [(r_2 X_2 + r_2 Y_2)]$
		$[r_2 Z_1 Z_2] \cdot [((r_2 d_1 Z_1 Z_2)^2 + [d_1]$
$U = C \cdot (F + d_1$	$[r_1 Z_1^2] \cdot [((r_1 d_1 Z_1^2)^2 + [d_1]$	$\cdot [r_2 (X_1 + Y_1)(X_2 + Y_2)]$
$\cdot K \cdot (K + I$	$\cdot [(r_1 X_1^2 + r_1 Y_1^2)] \cdot [(r_1 X_1^2 + r_1 Y_1^2$	$\cdot [(r_2(X_1 + Y_1)(X_2 + Y_2) + r_2 X_1 X_2$
$+ J + C))$	$+ r_1 Z_1^2 + r_1 Z_1^2 + r_1 Z_1^2)])]$	$+ r_2(X_1 + Z_1)(X_2 + Z_2) + r_2 Y_1 Y_2$
		$+ r_2(Y_1 + Z_1)(Y_2 + Z_2) + r_2 Z_1 Z_2)])]$
$\vdots$	$\vdots$	$\vdots$

References

1. Kocher, P.; Jaffe, J.; Jun, B. Differential power analysis. In Proceedings of the Annual International Cryptology Conference, Santa Barbara, CA, USA, 15–19 August 1999; Springer: Berlin/Heidelberg, Germany, 1999; pp. 388–397.
2. Coron, J.S. Resistance against differential power analysis for elliptic curve cryptosystems. In Proceedings of the International Workshop on Cryptographic Hardware and Embedded Systems, Worcester, MA, USA, 12–13 August 1999; Springer: Berlin/Heidelberg, Germany, 1999; pp. 292–302.
3. Izu, T.; Takagi, T. A fast parallel elliptic curve multiplication resistant against side channel attacks. In Proceedings of the International Workshop on Public Key Cryptography, Paris, France, 12–14 February 2002; Springer: Berlin/Heidelberg, Germany, 2002; pp. 280–296.

4. Chevallier-Mames, B.; Ciet, M.; Joye, M. Low-cost solutions for preventing simple side-channel analysis: Side-channel atomicity. *IEEE Trans. Comput.* **2004**, *53*, 760–768. [CrossRef]

5. Brier, E.; Joye, M. Weierstraß elliptic curves and side-channel attacks. In Proceedings of the International Workshop on Public Key Cryptography, Paris, France, 12–14 February 2002; Springer: Berlin/Heidelberg, Germany, 2002; pp. 335–345.

6. Bauer, A.; Jaulmes, E.; Prouff, E.; Reinhard, J.R.; Wild, J. Horizontal collision correlation attack on elliptic curves. *Cryptogr. Commun.* **2015**, *7*, 91–119. [CrossRef]

7. Clavier, C.; Feix, B.; Gagnerot, G.; Giraud, C.; Roussellet, M.; Verneuil, V. ROSETTA for single trace analysis. In Proceedings of the International Conference on Cryptology in India, Kolkata, India, 9–12 December 2012; Springer: Berlin/Heidelberg, Germany, 2012; pp. 140–155.

8. Devigne, J.; Joye, M. Binary huff curves. In Proceedings of the Cryptographers' Track at the RSA Conference, San Francisco, CA, USA, 14–18 February 2011; Springer: Berlin/Heidelberg, Germany, 2011; pp. 340–355.

9. Ghosh, S.; Kumar, A.; Das, A.; Verbauwhede, I. On the implementation of unified arithmetic on binary huff curves. In Proceedings of the International Workshop on Cryptographic Hardware and Embedded Systems, Santa Barbara, CA, USA, 20–23 August 2013; Springer: Berlin/Heidelberg, Germany, 2013; pp. 349–364.

10. Clavier, C.; Feix, B.; Gagnerot, G.; Roussellet, M.; Verneuil, V. Horizontal correlation analysis on exponentiation. In Proceedings of the International Conference on Information and Communications Security, Barcelona, Spain, 15–17 December 2010; Springer: Berlin/Heidelberg, Germany, 2010; pp. 46–61.

11. Joye, M.; Tibouchi, M.; Vergnaud, D. Huff's model for elliptic curves. In Proceedings of the International Algorithmic Number Theory Symposium, Nancy, France, 19–23 July 2010; Springer: Berlin/Heidelberg, Germany, 2010; pp. 234–250.

12. O'Flynn, C.; Chen, Z.D. Chipwhisperer: An open-source platform for hardware embedded security research. In Proceedings of the International Workshop on Constructive Side-Channel Analysis and Secure Design, Paris, France, 13–15 April 2014; Springer: Berlin/Heidelberg, Germany, 2014; pp. 243–260.

13. Gierlichs, B.; Lemke-Rust, K.; Paar, C. Templates vs. stochastic methods. In Proceedings of the International Workshop on Cryptographic Hardware and Embedded Systems, Yokohama, Japan, 10–13 October 2006; Springer: Berlin/Heidelberg, Germany, 2006; pp. 15–29.

14. Welch, B.L. The generalization ofstudent's' problem when several different population variances are involved. *Biometrika* **1947**, *34*, 28–35. [PubMed]

15. Hospodar, G.; Gierlichs, B.; De Mulder, E.; Verbauwhede, I.; Vandewalle, J. Machine learning in side-channel analysis: a first study. *J. Cryptogr. Eng.* **2011**, *1*, 293. [CrossRef]

16. Choudary, O.; Kuhn, M.G. Efficient template attacks. In Proceedings of the International Conference on Smart Card Research and Advanced Applications, Berlin, Germany, 27–29 November 2013; Springer: Berlin/Heidelberg, Germany, 2013; pp. 253–270.

17. Durvaux, F.; Standaert, F.X. From improved leakage detection to the detection of points of interests in leakage traces. In Proceedings of the Annual International Conference on the Theory and Applications of Cryptographic Techniques, Vienna, Austria, 8–12 May 2016; Springer: Berlin/Heidelberg, Germany, 2016; pp. 240–262.

18. Hankerson, D.; Menezes, A.J.; Vanstone, S. *Guide to Elliptic Curve Cryptography*; Springer Science & Business Media: Berlin/Heidelberg, Germany, 2006.

19. Fan, J.; Guo, X.; De Mulder, E.; Schaumont, P.; Preneel, B.; Verbauwhede, I. State-of-the-art of secure ECC implementations: a survey on known side-channel attacks and countermeasures. In Proceedings of the 2010 IEEE International Symposium on Hardware-Oriented Security and Trust (HOST), Anaheim, CA, USA, 13–14 June 2010; pp. 76–87.

20. Locke, G.; Gallagher, P. Fips pub 186-3: Digital signature standard (dss). *Fed. Inf. Process. Stand. Publ.* **2009**, *3*, 186-3.

21. Bernstein, D.J. Explicit-Formulas Database. Available online: http://www.hyperelliptic.org/EFD (accessed on 22 October 2018).

applied
sciences

MDPI

Article

Side Channel Leakages Against Financial IC Card of the Republic of Korea [†]

Yoo-Seung Won [‡], Jonghyeok Lee [‡] and Dong-Guk Han [*,‡]

Department of Financial Information Security, Kookmin University, 77 Jeongneung-ro, Seongbuk-gu,
Seoul 02727, Korea; mathwys87@kookmin.ac.kr (Y.-S.W.); n_seeu@kookmin.ac.kr (J.L.)
* Correspondence: christa@kookmin.ac.kr
† This paper is an extended version of paper published in the International Conference on Information
  Security and Cryptology, ICISC held in Seoul, Korea, 29 November–1 December 2017.
‡ These authors contributed equally to this work.

Received: 15 September 2018; Accepted: 7 November 2018; Published: 15 November 2018

Abstract: Integrated circuit (IC) chip cards are commonly used in payment system applications since they can provide security and convenience simultaneously. More precisely, Europay, MasterCard, and VISA (EMV) are widely known to be well equipped with security frameworks that can defend against malicious attacks. On the other hand, there are other payment system applications at the national level. In the case of the Republic of Korea, standards for financial IC card specifications are established by the Korea Financial Telecommunications and Clearings Institute. Furthermore, security features defending against timing analysis, power analysis, electromagnetic analysis, and TEMPEST are required. This paper identifies side channel leakages in the financial IC cards of the Republic of Korea, although there may be side channel countermeasures. Side channel leakages in the financial IC cards of the Republic of Korea are identified for the first time since the side channel countermeasures were included in the standards. The countermeasure that is applied to the IC card from a black box perspective is estimated to measure security features against power analysis. Then, in order to investigate whether an underlying countermeasure is applied, first-order and second-order power analyses are performed on the main target, e.g., a S-box of the block cipher SEED that is employed in the financial system. Furthermore, the latest proposal in ICISC 2017 is examined to apply block cipher SEED to the financial IC card protocol. As a result, it is possible to identify some side channel leakages while expanding the lemma of the paper accepted in ICISC 2017. Algebraic logic is also constructed to recover the master key from some round keys. Finally, it is found that only 20,000 traces are required to find the master key.

Keywords: side channel analysis; financial IC card; first-order analysis; second-order analysis

1. Introduction

Security plays an important role in payment system applications and is directly connected to customer's credibility. An integrated circuit (IC) chip is usually chosen to provide stable security as it offers high performance, data storage, and application processing. Europay, MasterCard and VISA (EMV) are examples. For personal identification number transaction security, physical and logical security requirements are required simultaneously for full payment security. To satisfy physical security requirements, some countermeasures [1–13] against side channel analysis [14–19] (This article is an extended version of the paper [19] accepted in Information Security and Cryptology (ICISC) 2017). The security of the financial IC card protocol will also be evaluated after expanding upon a previous suggestion.) or fault injection should be employed. Alternatively, there are some payment system applications that are only applied in their domestic markets. In the Republic of Korea, a specific payment system [20] is employed when using a credit or debit card. Specifically, the authentication protocol below (Figure 1), which is called the financial IC card protocol in this paper, is carried out when withdrawing cash from an automated teller machine (ATM) in the domestic market.

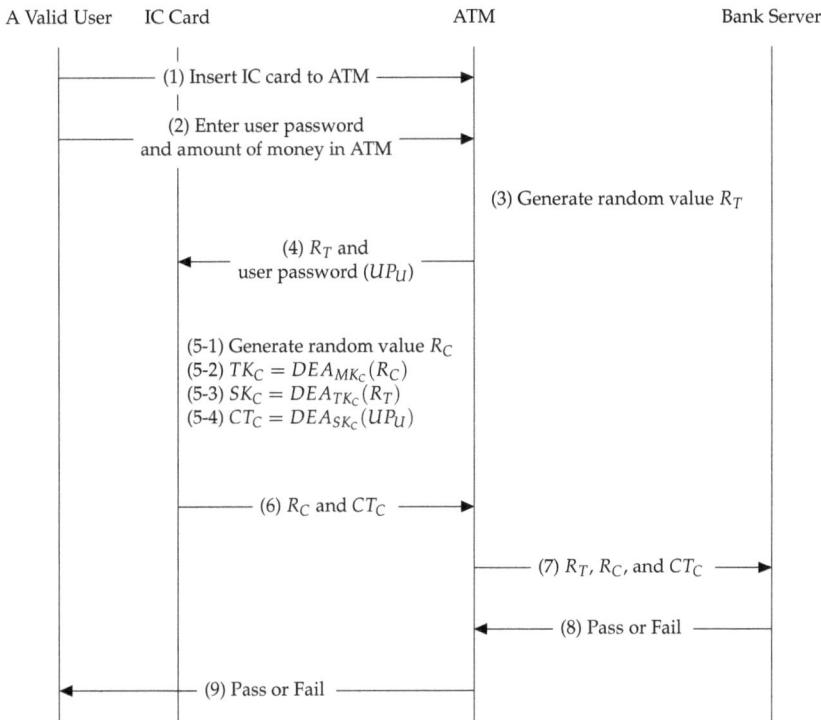

Figure 1. Financial IC (Integrated circuit) card protocol. ATM: automated teller machine.

Here, Data Encryption Algorithm (DEA) indicates block cipher SEED [21]; UP_U indicates user password; $C = DEA_K(P)$ means that the encryption function DEA takes a key K and plaintext P as input and produces a ciphertext C, and all inputs and outputs in the encryption function consist of 128 bits except for UP_U. However, UP_U is encoded to 128 bits using a padding scheme [20]. Moreover, the subscripts U, C, and T for all variables indicate user, IC chip, and ATM, respectively.

(1) A valid user inserts an IC card into an ATM in the baking system of the Republic of Korea.
(2) A valid user enters the user password and amount of desired cash through the ATM interface.
(3) The ATM generates a random number, R_T, which is composed of 128 bits.
(4) The ATM transfers R_T and the user password to the IC card.
(5-1) The IC chip generates a random value, R_C, which consists of 128 bits, for the financial IC card protocol.
(5-2) In the IC chip, R_C is encrypted to TK_C under a fixed key, MK_C, stored in the secure memory.
(5-3) In the same way, R_T is encrypted to R_T under non-fixed key TK_C generated from (5-1).
(5-4) Finally, UP_U is encrypted to CT_C under non-fixed key SK_C generated from (5-2).
(6) The IC chip passes R_C and CT_C to the ATM.
(7) The bank server receives the R_T, R_C, and CT_C from the ATM.

(8)   The bank server possesses keys MK_B and UP_B, which are identical to MK_C and UP_U, respectively. Therefore, by performing a scheme identical to (5-1)$\sim$(5-4), it can determine the truth of this protocol.

(9)   The ATM informs the user whether a cash withdrawal is possible or not.

For a detailed description of this protocol, refer to [20]. In particular, the security features for the financial IC card protocol are clearly stated in Chapter I.8. This document states that there should be resilience to power analysis, timing analysis, electromagnetic leak, fault injection, TEMPEST, etc. This is due to security threats, such as the risk of IC chip duplication. There is currently only one study [22] on the side channel leakage in the financial IC card protocol. Moreover, that study only found a single byte of round key against block cipher SEED without countermeasure, because a countermeasure was not mandatory at the time.

Our Contributions. This paper examines side channel leakage in the financial IC card of the Republic of Korea. As of 2008, financial IC chips must be resilient to state-of-the-art power analysis. However, side channel leakages have not been considered in the financial IC card protocol since 2008. Three main contributions are made by this paper. The first is that it offers the only challenge to the security evaluation of the Korean financial IC card protocol since 2008. Even though the first round key is retrieved via side channel analysis, the master key cannot be recovered in reasonable time. Therefore, the recovery logic of the master key is established. For this, a black box model (the black box model means that an adversary only knows the public information while performing the general protocol, in particular, it makes sense even if a fake ATM is employed, because a fake ATM never damages the financial IC card protocol) is fundamentally employed, i.e., it assumes that the adversary does not have knowledge of the detailed countermeasures. Therefore, underlying schemes such as first-order analysis and second-order analysis can be performed.

The second contribution is that the resilience for the state-of-the-art attack technique [19] is investigated to evaluate the security of financial IC card protocol. Furthermore, we demonstrate that the correct key can be revealed by expanding upon the suggestion of [19]. That is, novel combinations are found in this paper, differing from the original suggestion [19]. Specifically, novel combinations induce that four bits of eight bits are related to second-order leakage, whereas the original combination suggested in [19] derives only two bits of eight bits.

Finally, this study recovered the round key as well as the master key for the financial IC card protocol for the first time, although the countermeasure for state-of-the-art attack techniques should be mandatory. Consequently, the master key dedicated in the IC chip can be sufficiently recovered using our suggested method in reasonable time.

Extension. As written in the footnote of this title, this paper is an extended version of our paper [19] presented at ICISC 2017. In our paper [19], we demonstrated that the diffusion layer of block cipher SEED is vulnerable to our suggestion, although first-order Boolean masking and partial shuffling countermeasures are applied. Furthermore, we found a two-bit correlation between the diffusion layer and confusion layer outputs in terms of second-order leakage. We also implemented our method on simulated trace and a well-leaked board in terms of power analysis (the board is used in academic articles because the power source is well-leaked and the platform is easy to use; detailed information can be found at http://wiki.newae.com.). However, in this paper, we show two differences compared to the previous paper [19]. One is that we extend our previously suggested equation, as shown in the present Appendix A. Therefore, we show four-bit correlation between the diffusion layer and confusion layer outputs rather than two-bit correlation for some special diffusion outputs (refer to Cases 2 and 5 in Appendix A). The other is that we apply our suggestion to the latest smart card. Additionally, the adversary assumption is even a black-box model. Therefore, we propose an attack scheme of financial IC card in terms of side channel analysis as well as the recovery mechanism of the master key in this paper.

2. Preliminaries

In this section, existing knowledge regarding details of the financial IC card, a countermeasure for block cipher SEED, and the side channel analysis for block cipher SEED's countermeasure are presented.

2.1. Known and Unknown Information on Financial IC Card Protocol

If an invalid user acquires a certain IC card, some information can be revealed without significant effort. That is, a fake ATM can be constructed by the invalid user, ignoring the bank server. In other words, a fake ATM transfers R_T and UP_U^* values to the IC chip (* indicates invalid value). Moreover, a valid UP_U value is not required, as our goal is to recover the master key, not a valid CT_C value. The following summarizes the known or unknown information when a financial IC card protocol is performed by an invalid user:

- Known information: R_T, UP_U^*, R_C, and CT_C^*
- Unknown information: MK_C, TK_C, and SK_C^*

The important point here is that R_C cannot be controlled by an invalid user. Due to this fact, a chosen plaintext attack cannot be applied to the financial IC card protocol. Furthermore, a non-profiled attack cannot be performed on second and third DEA operations, because TK_C and SK_C^* are changed whenever this protocol is conducted. In other words, assuming a non-profiled attack, the main target is naturally the first DEA operation. More precisely, the first round of the foremost DEA operation is definitely the main target, since TK_C is unknown information.

2.2. Countermeasure for Block Cipher SEED

The countermeasure of block cipher SEED [21] is analogous to that of the block cipher Advanced Encryption Standard (AES) [23] in terms of S-box operation. However, other countermeasures are required because block cipher SEED is constructed with *AND* and *Addition* operations. A few countermeasures to protect the *AND* operation are suggested in [2]. Moreover, many suggestions [24–32] have been proposed with respect to the *Addition* operation.

When compared with the higher-order masking scheme in software implementation, hiding schemes, such as shuffling and dummy methods, are adequately applied to reasonable countermeasures. Furthermore, the dummy operation can sometimes be employed because the number of S-boxes is four in the G-function of block cipher SEED.

In software implementation, it is sometimes considered reasonable to combine first-order masking, shuffling, and dummy operation schemes [11]. In particular, the application of this combination is relevant to the DPA contest [33].

2.3. Side Channel Analysis for Countermeasure of Block Cipher SEED

A Boolean masking scheme is normally employed, since most block ciphers can be comfortably applied in this case. However, the $(d + 1)$-th order attack theoretically allows the leakage of the master key, even though the d-th order Boolean masking scheme is adopted. Thus, to improve security strength, the Boolean masking scheme is sometimes combined with a hiding countermeasure, such as shuffling and dummy operation. As previously stated, first-order Boolean masking, shuffling, and dummy schemes can be simultaneously selected as a realistic countermeasure, considering the time performance and read-only memory/random access memory (ROM/RAM) size. Some attacks have been proposed to evaluate the security of these countermeasures [13,19,33]. In general, a single signal is shuffled to p signals when a shuffling countermeasure is applied. Moreover, more p^2 traces are required to retrieve the secret key in comparison to the number required in non-shuffled traces. In [13], more p traces were only required when performing windowing attacks. That is, the shuffling complexity was decreased to $\sqrt{p}$ from p. However, there are some barriers of realistic attack in order to perform a windowing attack. After performing the windowing scheme that derives the constant size from the original trace, the points of interest should be overlapped in each trace. Therefore, sometimes numerous traces may have to be compared in terms of theoretical complexity.

3. Evaluation Methodology for the Financial IC Card Protocol

In this section, the process of evaluating the security of the financial IC card protocol is explained. The ultimate aim of this study was to retrieve the entire master key, not just a part of it. First, the evaluation system that was constructed as a fake ATM is described. Then, the analysis methodology under the black box model is explained in order to recover the whole master key.

3.1. Construction of Side Channel Analysis for Financial IC Card Protocol

As previously stated in Section 2.1, a fake ATM can be constructed to communicate with an IC chip card. The related experimental setup can be provided via the SCARF system (Version 4.0.0.17153 (64 bit), Electronics and Telecommunications Research Institute (ETRI), Daejeon metropolitan city, Republic of Korea) [34].

As shown in Figure 2, the CEB board (this board is provided by SCARF system), which plays the role of a fake ATM in the financial IC card protocol, is controlled by the SCARF system. As previously described, R_T and UP_U were generated from script language in the SCARF system. That is, these values can take the desired form, between fixed or random values.

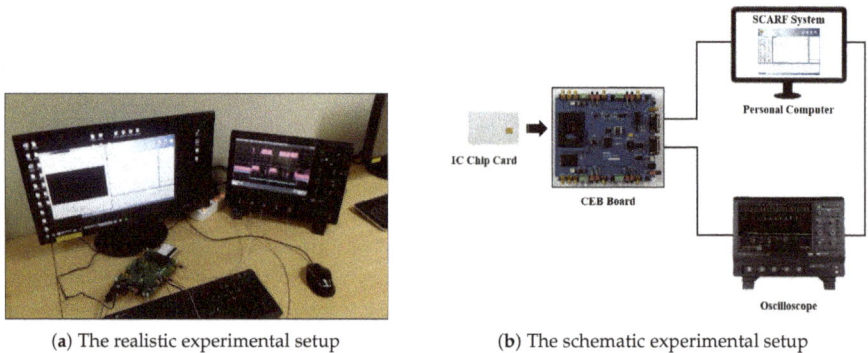

(a) The realistic experimental setup   (b) The schematic experimental setup

Figure 2. Experimental setup for security evaluation of financial IC card protocol.

3.2. Security Evaluation Methodology of Side Channel Analysis Under Black Box Model

Before describing the security evaluation methodology, the intermediate variables were enumerated, depending on the side channel attack schemes. Basically, the intermediate variables in the first round of the foremost DEA should be the main targets, due to the constraint of public information. In other words, the second and third DEAs cannot be considered major objectives under a non-profiling attack, owing to protocol features. Moreover, the last round of the foremost DEA cannot be an available intermediate variable, because TK_C is an unknown information in the usual protocol.

The master key can also be leaked. For this, two round keys should be recovered through side channel analysis, and a brute force attack is required.

3.2.1. Enumeration For Intermediate Variables

To evaluate the security against side channel analysis, the underlying scheme is performed as first and second-order analyses. Thus, the intermediate variables should be enumerated.

1. First-order correlation power analysis

    (a)   Each of the four S-boxes in the G-function

2. Second-order correlation power analysis

    (a)   Combination between two of the four S-boxes in the G-function
    (b)   Combination between two out of the four outputs of the G-function

First-order correlation power analysis is a fundamental attack when the countermeasure is not applied. Sometimes the first-order leakage may be exposed in spite of applying the countermeasure, due to unintended phenomena, such as optimization in the computer language compiler and human error. Therefore, item 1a should be required. Additionally, the four S-boxes in the G-function should be chosen as intermediate variables to recover the single round key.

To conduct second-order correlation power analysis, pre-processing logic must be employed [18]. In addition, two intermediate variables should be required to release the leakage of the round key. In particular, the identical masking value is usually applied to each of the S-boxes to reduce the cost of pre-computed tables when assuming the general first-order Boolean masking scheme. For this reason, item 2a is needed to disclose the masking information. Moreover, the novel leakage in connection with block cipher SEED was recently suggested in [19]. According to this proposal, second-order leakage can be revealed with low attack-complexity, despite first-order Boolean masking and restricted shuffling countermeasures being employed, although a main target is the output of G-function. In other words, first-order Boolean masking and shuffling countermeasures can usually be utilized as reasonable countermeasures in software implementation due to certain merits, such as time performance, memory size, and security. However, in terms of block cipher SEED, side channel leakage can occur via 2b. In particular, authors in [19] only explained that one combination of total six cases was showed.

3.2.2. Security Evaluation Methodology

In this section, the evaluation methodology used to perform side channel analysis under the black box model is described. The black box model assumes that the detailed countermeasure applied to block cipher SEED in the IC chip cannot be known. The first part of the approach is to observe public information such as R_T or CT_C^*, not the intermediate variable related to the secret key. This approach allows for the acquisition of two features. The first is that the operation position of block cipher SEED can be found in the whole protocol. The second is the establishment of the best solution for pre-processing information, such as compression method, compression rate, the number of traces, etc.

Before analyzing the enumeration defined in Section 3.2.1, it should be identified whether the public information can be revealed or not. The enumeration has the potential to reveal a secret key if the public information can be disclosed in traces. Moreover, the best solution for the pre-processing scheme can be found.

- Applied pre-processing scheme: none, Correlation value when performing correlation power analysis: 0.03
- Applied pre-processing scheme: average compression 30, Correlation value when performing correlation power analysis: 0.15
- Applied pre-processing scheme: average compression 50, Correlation value when performing correlation power analysis: 0.05

Here, the average compression N indicates that N points are compressed to 1 point by calculating the average.

The best solution is the average compression 30 scheme, because its correlation value is the highest among the three cases. Subsequently, side channel analysis can be performed using the stored pre-processing information.

3.2.3. Recovering the Master Key via Some Round Keys

The brute force attack for recovery of the master key after retrieving two round keys via side channel analysis is introduced here. Yoo et al. [35] explains how to recover the master key when 1 round and 16 round keys are already known. In the current study, this suggestion is not sufficient, since only 1 and 2 round keys are retrieved via side channel analysis. Therefore, it is necessary to properly apply the suggestion presented in [35] to the current context. Moreover, the recovery scheme of the master key is represented at the appropriate algorithmic level.

Note that notation of all variables used in Algorithm 1 is given in [21,35]. The size of all variables is 32 bits, with the exception of the variable MK. $X^{(n)}$ indicates the n-th significant byte of 32-bit X. $X^{(n-m)}$ means from the n-th significant byte to the m-th significant byte. Additionally, G^{-1} is the inverse of the G-function.

Algorithm 1 Recovery of the master key under two round keys K_1, K_2

Input: 1 round key $K_{1,0}$ and $K_{1,1}$, and 2 round key $K_{2,0}$ and $K_{2,1}$
Output: Candidates for a master key MK

    ▷ Calculating the inverse of G-function and constant value KC_0, KC_1
1:   $TK_{1,0} \leftarrow G^{-1}[K_{1,0}] + KC_0$
2:   $TK_{1,1} \leftarrow G^{-1}[K_{1,1}] - KC_0$
3:   $TK_{2,0} \leftarrow G^{-1}[K_{2,0}] + KC_1$
4:   $TK_{2,1} \leftarrow G^{-1}[K_{2,1}] - KC_1$

    ▷ Guessing a part of A and B for master key
5: **for** $TC = 0$ to $2^{32} - 1$ **do**
6:     $TA0 \leftarrow TK_{1,0} - TC$
7:     $TA1 \leftarrow TK_{2,0} - TC$
8:     **if** $TA0^{(3-1)} = TA1^{(2-0)}$ **then**
9:       Storing a pair $(TA0, TA1)$
10:    **end if**
11: **end for**
12: **for** $TD = 0$ to $2^{32} - 1$ **do**
13:     $TB0 \leftarrow TK_{1,1} + TD$
14:     $TB1 \leftarrow TK_{2,1} + TD$
15:     **if** $TB0^{(3-1)} = TB1^{(2-0)}$ **then**
16:       Storing a pair $(TB0, TB1)$
17:    **end if**
18: **end for**

    ▷ Filtering temporary keys
19: **if** $TA0^{(0)} = TB1^{(3)}$ **then**
20:    **if** $TB0^{(0)} = TA1^{(3)}$ **then**
21:       $MK \leftarrow (TA0)||(TB0)||(TK_{1,0} - TA0)||(TK_{1,1} + TB0)$
22:       Storing candidate MK
23:    **end if**
24: **end if**
25: **return** Candidate MK

The procedures are divided into three steps in Algorithm 1. In addition, the recovery algorithm can be performed in reverse order of generation for the round key. In particular, the first step is directly related to its reverse order, allowing it to be easily computed. Subsequently, the second step is that a part of A and B for the master key can be guessed. Moreover, $TA0^{(3-1)}$ and $TB0^{(3-1)}$ are identical to $TA1^{(2-0)}$ and $TB1^{(2-0)}$, respectively, because the 2 round key is derived from the rotation of 1 round key. There are 256 candidates until the second step, since $TA0$ and $TB0$ depend on the size of a single byte. In the third step, a part of $TA0$ is equal to part of $TB1$, due to the rotation in the generation of a round key. Finally, the number of candidates for the master key is less than 256.

In order to acquire only one master key, a correct single pair $(R_C, R_T, UP_U^*, CT_C^*)$ in the financial IC card protocol is required. Then, the master key can be determined by proceeding with an exhaustive attack.

4. Experimental Results

The security strength against side channel analysis was evaluated to protect against the leakage of private information. Based on Section 3, the security can be evaluated for a specific IC card. For this, a recent IC card issued in the Republic of Korea was used. The expiration date of this specific IC card was 2023, because the issuance date was 2018. In other words, the master key dedicated to the IC chip should not be able to be revealed by any attacks, according to ChapterI.8 of [20].

As previously stated, the public information in the collected traces was first identified, and then the pre-processing information was stored. Afterwards, security evaluation was conducted against the novel side channel analysis for the defined enumeration in Section 3.2.1. Finally, the master key was recovered via the brute force attack defined in Algorithm 1.

4.1. Identifying the Public Information

This section elaborates upon the security evaluation methodology defined in Figure 3, except for the defined enumeration. For this, the experimental setup described in Figure 2 was utilized.

In Figure 4, three identical areas should be identified, due to the *DEA* algorithm being computed three times in the financial IC card protocol. Therefore, when performing correlation power analysis, the plaintext or ciphertext information was searched through until the occurrence of a peak. Due to time and memory limitations, not all points can be collected simultaneously. Thus, a proportion of all points should be measured after estimating the plaintext/cipher operation. Finally, the *DEA* algorithm operation in the red box of Figure 4 was found.

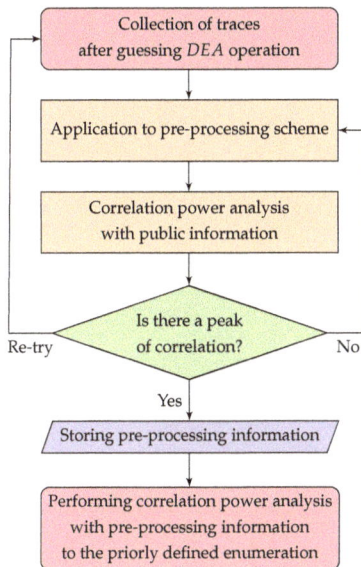

Figure 3. Security evaluation methodology for financial IC card protocol.

In particular, block cipher SEED was composed of 16 rounds. However, 16 identical rounds cannot be found in Figure 5. It can be estimated that the area of dimensions 0.9×10^4 to 1.5×10^4 is divided into 12 identical areas. There are also two identical areas in the front and rear of the middle part. Therefore, it can be estimated that some countermeasures are applied to 1, 2, 15, and 16 rounds. Additionally, the plaintext information was revealed in the front part in Figure 6. The detailed description is given in the next section.

In order to analyze the plaintext operation, the front part of the *DEA* algorithm operation was re-collected. Additionally, as shown in Figure 3, the plaintext operation was analyzed by repeating trial-and-error to obtain the best solution.

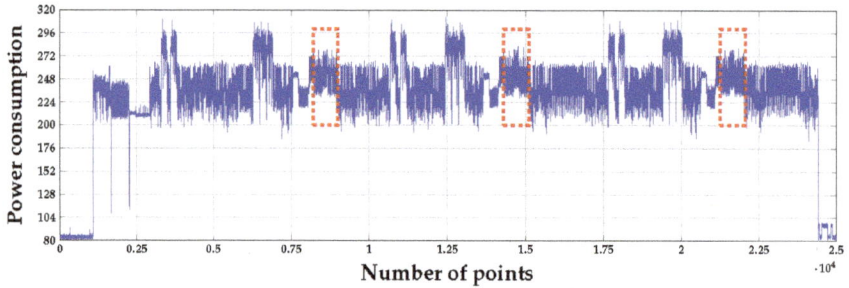

Figure 4. Measured trace when performing the financial IC card protocol.

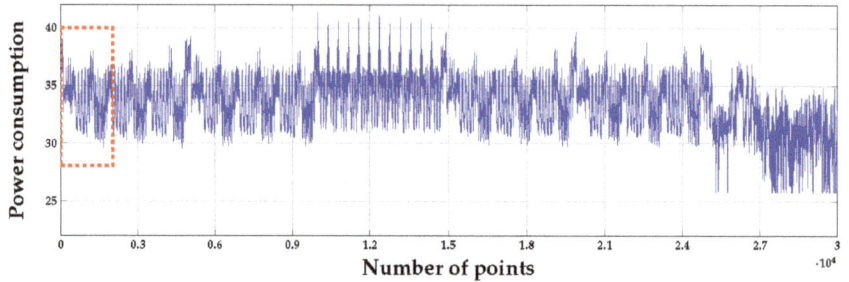

Figure 5. A trace for the *DEA* algorithm.

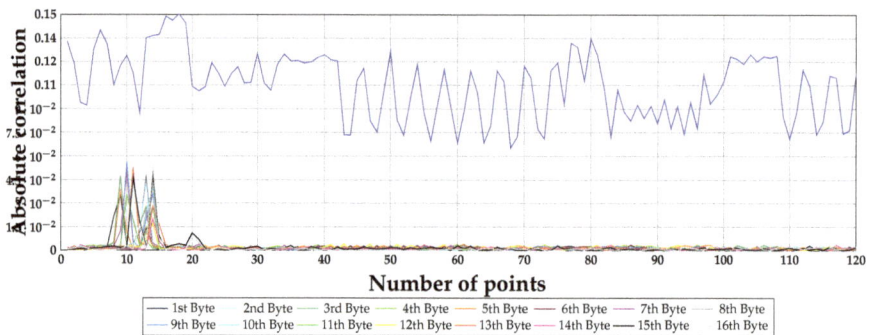

Figure 6. Result of correlation power analysis with plaintext information.

The best solution is as follows.

- Compression scheme: sum of squares
- Compression Ratio : 800

The sum of squares scheme indicates that N points were squared and summed up to a single point, where N implies the compression ratio. All 16 bytes of plaintext were analyzed at the front part of the foremost DEA operation. The alignment based on correlation was performed as a default pre-processing scheme with extensive trial-and-error. Finally, the best solution was obtained. Afterwards, the best solution was utilized to perform side channel analysis for a main target.

4.2. Security Evaluation for Financial IC Card Protocol

As shown in Figure 3, the security for a financial IC card, including state-of-the-art attack schemes described in Section 3.2.1, was evaluated. Before performing the side channel analysis defined in Section 3.2.1, the applied countermeasure was estimated in block cipher SEED by investigating some power traces.

As discussed in Section 4.1, it can be estimated that some countermeasures may be applied to only 1, 2, 15, and 16 rounds in the DEA operation of Figure 5, since the middle part can be divided into 12 rounds. Additionally, 1, 2, 15 and 16 rounds were thicker than the middle part. The first round can be split into three parts, due to the identical operation being repeated three times in 1 round. Therefore, the first part of 1 round can refer to the red box in Figure 5.

Five different behaviors were also confirmed in all the collected traces. As seen in Figure 7, the different points are the middle part of the five behaviors. The middle part of the first behavior is composed of four negative peaks. By the same logic, the middle part of the fifth behavior consists of eight negative peaks. The second, third, and fourth behaviors also have same features. Naturally, it was predicted that these behaviors may be connected to S-box operation, since the G-function of the first round has four S-box operations. It seems that the random dummy operation may be applied to that if this computation is a real S-box operation. Two possible criteria for obtaining some leakages in order to perform side channel analysis are the back and forth of S-box operations. In other words, there are two alignment points. The result of the alignment to the back of S-box operations is shown in Figure 7.

The aim of alignment to the front of the S-box operations is to analyze the plaintext and S-box operations. Furthermore, the purpose of the other alignment is to analyze the output of G-function and S-box operations. This is directly related to the suggestion of [19]. To sum up, the notations are defined as below, according to the following alignment criteria:

- Alignment 1: front of S-box operations.
- Alignment 2: rear of S-box operations (refer to Figure 7).

4.2.1. Result of Performing Side Channel Analysis Defined in Enumeration

This section describes the side channel analysis performed using the defined scheme in Section 3.2.1. As a result, we cannot identify any secret information for Alignment 1. Therefore, it is represented to all results for Alignment 2 in this section. None of the first-order leakage on the collect traces was revealed, and the second-order leakage of the combination between each of the S-boxes was also not exposed. However, by performing item 2b, two round keys can be retrieved as shown in Table 1.

Figure 8 represents the result for the first G-function of 1 round based on all combinations in Appendix A. Additionally, the x-axis and y-axis indicate the absolute correlation and key candidates, respectively. In general, the key candidate with the highest value is considered to be the correct key.

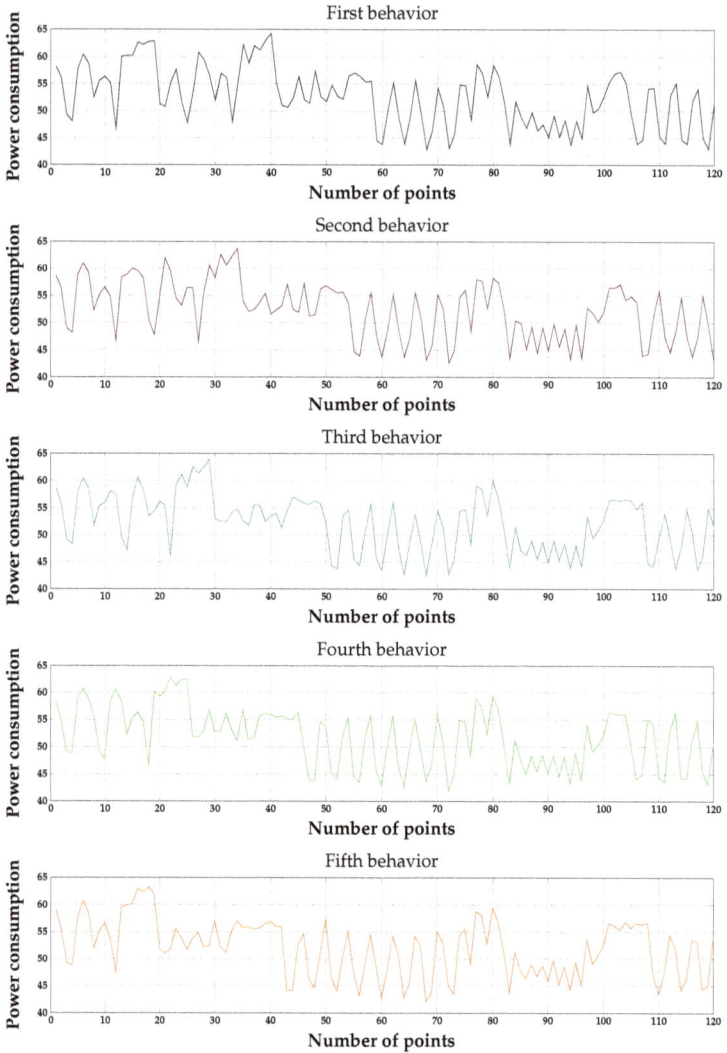

Figure 7. Five behaviors when operating DEA algorithm.

When considering all combinations, there are 20 results. As shown in Figure 8, the result is only represented if the correct key is retrieved. As described in Appendix A, the performance is superior to that of the previous proposal [19], owing to the fact that results Figure 8a–d have higher correlation. In order to analyze all combinations, 1,000,000 traces were collected. However, only 20,000 traces were required to recover the correct key in results Figure 8a–d.

By analyzing the first G-function of 1 round, the xored key $K_{1,0} \oplus K_{1,1}$ can be acquired. Therefore, it is necessary to analyze the second G-function of 1 round, the first G-function of 2 round, and the second G-function of 2 round to obtain two round keys. All results are analogous to Figure 8.

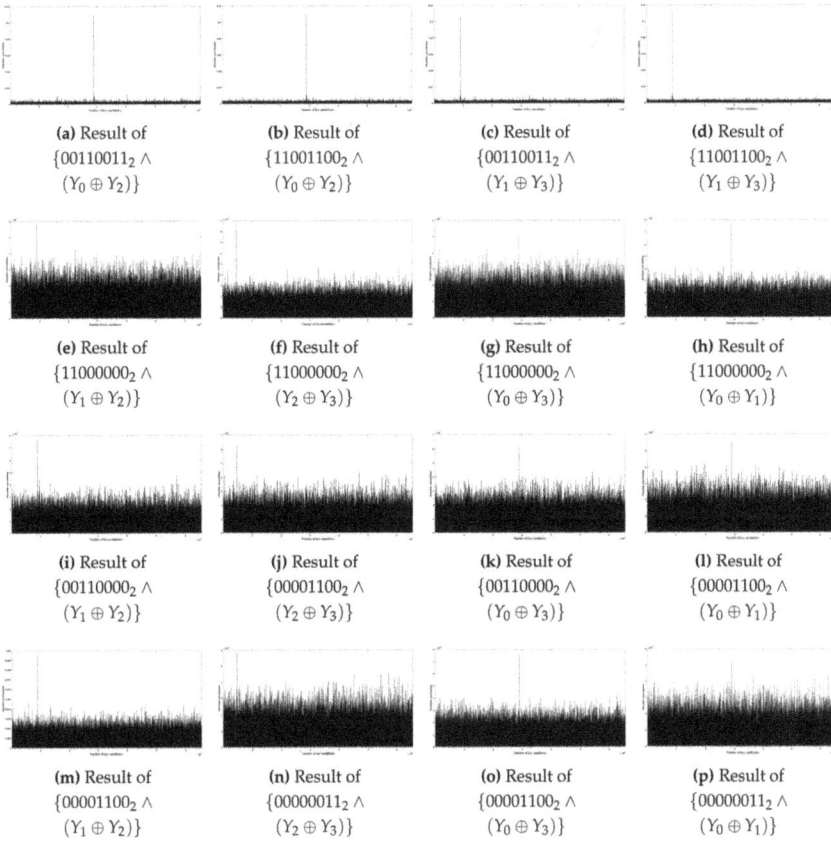

(a) Result of $\{00110011_2 \wedge (Y_0 \oplus Y_2)\}$

(b) Result of $\{11001100_2 \wedge (Y_0 \oplus Y_2)\}$

(c) Result of $\{00110011_2 \wedge (Y_1 \oplus Y_3)\}$

(d) Result of $\{11001100_2 \wedge (Y_1 \oplus Y_3)\}$

(e) Result of $\{11000000_2 \wedge (Y_1 \oplus Y_2)\}$

(f) Result of $\{11000000_2 \wedge (Y_2 \oplus Y_3)\}$

(g) Result of $\{11000000_2 \wedge (Y_0 \oplus Y_3)\}$

(h) Result of $\{11000000_2 \wedge (Y_0 \oplus Y_1)\}$

(i) Result of $\{00110000_2 \wedge (Y_1 \oplus Y_2)\}$

(j) Result of $\{00001100_2 \wedge (Y_2 \oplus Y_3)\}$

(k) Result of $\{00110000_2 \wedge (Y_0 \oplus Y_3)\}$

(l) Result of $\{00001100_2 \wedge (Y_0 \oplus Y_1)\}$

(m) Result of $\{00001100_2 \wedge (Y_1 \oplus Y_2)\}$

(n) Result of $\{00000011_2 \wedge (Y_2 \oplus Y_3)\}$

(o) Result of $\{00001100_2 \wedge (Y_0 \oplus Y_3)\}$

(p) Result of $\{00000011_2 \wedge (Y_0 \oplus Y_1)\}$

Figure 8. When performing combinations defined in item 2b, the result is only represented if the correct key is revealed.

Table 1. Analyzed two round keys.

Round Key	Value
$K_{1,0}$	0x$D555B112$
$K_{1,1}$	0x$A476A39F$
$K_{2,0}$	0x15043F29
$K_{2,1}$	0x1886$A9FA$

4.3. Recovering the Master Key via Two Round Keys

Two round keys were already retrieved to recover the master key. At first, the G-function in 1 round, the $K_{1,0} \oplus K_{1,1}$, is used for the G-function. Therefore, it was necessary to perform side channel analysis for the second G-function in order to recover each $K_{1,0}$ and $K_{1,1}$. By the same logic, the second round key can be recovered.

To calculate the master key, a correct single pair $(R_C, R_T, UP_C^*, CT*_C)$ should be acquired in the financial IC card protocol. Afterwards, Algorithm 1 is operated to recover a correct master key.

As mentioned earlier, R_T and UP_C^* were controlled by the adversary, utilizing a fake ATM. Therefore, we set to the fixed zero value. The result of performing Algorithm 1 is as follows.

As shown in Figure 9, the number of filtering temporary keys is 16. Finally, the analyzed master key (0xE73A67FFE48EF10DD542EC7BE57A9FBE) could be recovered by utilizing a correct pair defined in Table 2.

Table 2. A correct single pair.

A Correct Single Pair	Value
R_C	0x41A4E5052FC75CAB15C2715E8897C11E
R_T	0x0000000000000000000000000000000000
UP_C^*	0x0000000000000000000000000000000000
CT_C^*	0x9B0A4C535916A73D7E328F5EBAF691F9

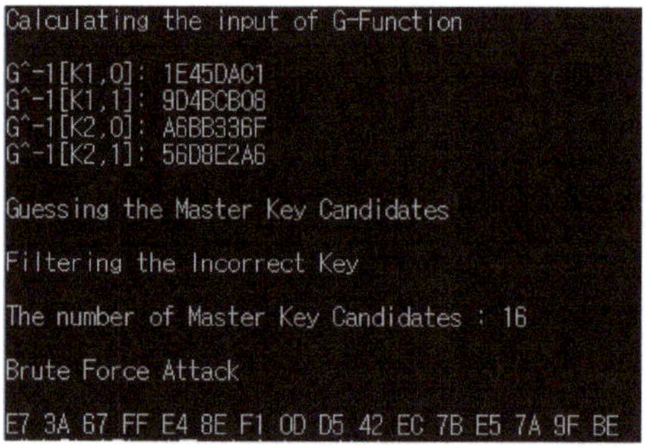

Figure 9. Analyzed master key

5. Conclusions

A specific payment system is one that is only applied to its domestic market. In the Republic of Korea, the specific payment system is employed when using a credit or debit card. A side channel countermeasure should be required when using the financial IC card to prevent state-of-the-art attacks. This study demonstrated that the correct key can be revealed by expanding upon the suggestion of [19]. Consequently, the master key can be completely recovered by utilizing state-of-the-art side channel attacks and brute force attacks. Note that powerful countermeasures should be applied to financial IC card protocol, considering the expiration date and technological advances of side channel attacks.

Instant countermeasure to our attack. To defend our suggestion, each of the diffusion layer outputs should be covered with different masking values. Furthermore, this computation should be done in shuffling countermeasure and with acceptable overhead. In other words, if the diffusion layer consists of AND and XOR operations, the countermeasure should be carefully designed.

Author Contributions: This article is an extended version of [19]. Moreover, the authors evaluated the security for the financial IC card [20], expanding upon the previous paper [19]. Additionally, these authors contributed equally to this work.

Funding: This work was supported by Institute for Information & communications Technology Promotion(IITP) grant funded by the Korea government(MSIT) (No.2017-0-00520, Development of SCR-Friendly Symmetric Key Cryptosystem and Its Application Modes).

Conflicts of Interest: The authors declare no conflict of interest.

Appendix A. All Combinations for Novel Leakage of the Block Cipher SEED

Won et al. [19] showed novel leakage of the block cipher SEED in the output of the G-function. From the perspective of second-order analysis, the combination between the outputs of the G-function allows low complexity in the partial bits of all bits. In other words, powerful countermeasures can sometimes be applied to only confusion layers, because the diffusion layer entail a high complexity to perform side channel analysis. That is, a low-level countermeasure can be employed in the diffusion layer. Assuming this concept, novel side channel leakage for the output of the diffusion layer has been demonstrated. Lemma 1 was established in [19] to apply the combination of second-order leakage in the output of G-function. This Appendix expands the suggestion of [19] to all cases. Note that all notations are the same as in [19].

Case 1 $\quad Z_0 \oplus Z_1$ (Refer to [19] for a detailed description)

$$
\begin{aligned}
Z_0 \oplus Z_1 = &\{11000000_2 \wedge (Y_2 \oplus Y_3)\} \\
&\oplus \{00110000_2 \wedge (Y_1 \oplus Y_2)\} \\
&\oplus \{00001100_2 \wedge (Y_0 \oplus Y_1)\} \\
&\oplus \{00000011_2 \wedge (Y_3 \oplus Y_0)\}
\end{aligned}
$$

Case 2 $\quad Z_0 \oplus Z_2$

$$
\begin{aligned}
Z_0 \oplus Z_2 = &\{(M_0 \wedge Y_0) \oplus (M_1 \wedge Y_1) \oplus (M_2 \wedge Y_2) \oplus (M_3 \wedge Y_3)\} \\
&\oplus \{(M_2 \wedge Y_0) \oplus (M_3 \wedge Y_1) \oplus (M_0 \wedge Y_2) \oplus (M_1 \wedge Y_3)\} \\
= &\{(M_0 \oplus M_2) \wedge Y_0\} \oplus \{(M_1 \oplus M_3) \wedge Y_1\} \\
&\oplus \{(M_2 \oplus M_0) \wedge Y_2\} \oplus \{(M_3 \oplus M_1) \wedge Y_3\} \\
= &(00110011_2 \wedge Y_0) \oplus (11001100_2 \wedge Y_1) \\
&\oplus (00110011_2 \wedge Y_2) \oplus (11001100_2 \wedge Y_3) \\
= &\{11001100_2 \wedge (Y_1 \oplus Y_3)\} \\
&\oplus \{00110011_2 \wedge (Y_0 \oplus Y_2)\}
\end{aligned}
$$

Case 3 $\quad Z_0 \oplus Z_3$

$$
\begin{aligned}
Z_0 \oplus Z_3 = &\{(M_0 \wedge Y_0) \oplus (M_1 \wedge Y_1) \oplus (M_2 \wedge Y_2) \oplus (M_3 \wedge Y_3)\} \\
&\oplus \{(M_3 \wedge Y_0) \oplus (M_0 \wedge Y_1) \oplus (M_1 \wedge Y_2) \oplus (M_2 \wedge Y_3)\} \\
= &\{(M_0 \oplus M_3) \wedge Y_0\} \oplus \{(M_1 \oplus M_0) \wedge Y_1\} \\
&\oplus \{(M_2 \oplus M_1) \wedge Y_2\} \oplus \{(M_3 \oplus M_2) \wedge Y_3\} \\
= &(11000011_2 \wedge Y_0) \oplus (00001111_2 \wedge Y_1) \\
&\oplus (00111100_2 \wedge Y_2) \oplus (11110000_2 \wedge Y_3) \\
= &\{11000000_2 \wedge (Y_3 \oplus Y_0)\} \\
&\oplus \{00110000_2 \wedge (Y_2 \oplus Y_3)\} \\
&\oplus \{00001100_2 \wedge (Y_1 \oplus Y_2)\} \\
&\oplus \{00000011_2 \wedge (Y_0 \oplus Y_1)\}
\end{aligned}
$$

Case 4 $Z_1 \oplus Z_2$

$$
\begin{aligned}
Z_1 \oplus Z_2 =& \{(M_1 \wedge Y_0) \oplus (M_2 \wedge Y_1) \oplus (M_3 \wedge Y_2) \oplus (M_0 \wedge Y_3)\} \\
& \oplus \{(M_2 \wedge Y_0) \oplus (M_3 \wedge Y_1) \oplus (M_0 \wedge Y_2) \oplus (M_1 \wedge Y_3)\} \\
=& \{(M_1 \oplus M_2) \wedge Y_0\} \oplus \{(M_2 \oplus M_3) \wedge Y_1\} \\
& \oplus \{(M_3 \oplus M_0) \wedge Y_2\} \oplus \{(M_0 \oplus M_1) \wedge Y_3\} \\
=& (00111100_2 \wedge Y_0) \oplus (11110000_2 \wedge Y_1) \\
& \oplus (11000011_2 \wedge Y_2) \oplus (00001111_2 \wedge Y_3) \\
=& \{11000000_2 \wedge (Y_1 \oplus Y_2)\} \\
& \oplus \{00110000_2 \wedge (Y_0 \oplus Y_1)\} \\
& \oplus \{00001100_2 \wedge (Y_3 \oplus Y_0)\} \\
& \oplus \{00000011_2 \wedge (Y_2 \oplus Y_3)\}
\end{aligned}
$$

Case 5 $Z_1 \oplus Z_3$

$$
\begin{aligned}
Z_1 \oplus Z_3 =& \{(M_1 \wedge Y_0) \oplus (M_2 \wedge Y_1) \oplus (M_3 \wedge Y_2) \oplus (M_0 \wedge Y_3)\} \\
& \oplus \{(M_3 \wedge Y_0) \oplus (M_0 \wedge Y_1) \oplus (M_1 \wedge Y_2) \oplus (M_2 \wedge Y_3)\} \\
=& \{(M_1 \oplus M_3) \wedge Y_0\} \oplus \{(M_2 \oplus M_0) \wedge Y_1\} \\
& \oplus \{(M_3 \oplus M_1) \wedge Y_2\} \oplus \{(M_0 \oplus M_2) \wedge Y_3\} \\
=& (11001100_2 \wedge Y_0) \oplus (00110011_2 \wedge Y_1) \\
& \oplus (11001100_2 \wedge Y_2) \oplus (00110011_2 \wedge Y_3) \\
=& \{11001100_2 \wedge (Y_0 \oplus Y_2)\} \\
& \oplus \{00110011_2 \wedge (Y_1 \oplus Y_3)\}
\end{aligned}
$$

Case 6 $Z_2 \oplus Z_3$

$$
\begin{aligned}
Z_2 \oplus Z_3 =& \{(M_2 \wedge Y_0) \oplus (M_3 \wedge Y_1) \oplus (M_0 \wedge Y_2) \oplus (M_1 \wedge Y_3)\} \\
& \oplus \{(M_3 \wedge Y_0) \oplus (M_0 \wedge Y_1) \oplus (M_1 \wedge Y_2) \oplus (M_2 \wedge Y_3)\} \\
=& \{(M_2 \oplus M_3) \wedge Y_0\} \oplus \{(M_3 \oplus M_0) \wedge Y_1\} \\
& \oplus \{(M_0 \oplus M_1) \wedge Y_2\} \oplus \{(M_1 \oplus M_2) \wedge Y_3\} \\
=& (11110000_2 \wedge Y_0) \oplus (11000011_2 \wedge Y_1) \\
& \oplus (00001111_2 \wedge Y_2) \oplus (00111100_2 \wedge Y_3) \\
=& \{11000000_2 \wedge (Y_0 \oplus Y_1)\} \\
& \oplus \{00110000_2 \wedge (Y_3 \oplus Y_0)\} \\
& \oplus \{00001100_2 \wedge (Y_2 \oplus Y_3)\} \\
& \oplus \{00000011_2 \wedge (Y_1 \oplus Y_2)\}
\end{aligned}
$$

Case 1, which is suggested in [19], is significantly analogous to Cases 3, 4, and 6. However, Cases 2 and 5 are distinct from the proposal in [19]. In other words, these combinations can only be leaked information about $(Y_1 \oplus Y_3)$ and $(Y_0 \oplus Y_2)$. Although Cases 2 and 5 are only associated with two combinations, more correlation bits are leaked for these cases due to four bits of eight bits being related to second-order leakage. Therefore, side channel analysis related to Cases 2 and 5 can first be performed to discover some leakages.

References

1. Herbst, C.; Oswald, E.; Mangard, S. An AES Smart Card Implementation Resistant to Power Analysis Attacks. In Proceedings of the International Conference on Applied Cryptography & Network Security (ACNS), Singapore, 6–9 June 2006; pp. 239–252.
2. Kim, H.; Cho, Y.I.; Choi, D.; Han, D.G.; Hong, S. Efficient masked implementation for SEED based on combined masking. *ETRI J.* **2011**, *33*, 267–274. [CrossRef]
3. Bonnecaze, A.; Liardet, P.; Venelli, A. AES Side-Channel Countermeasure using Random Tower Field Constructions. *Des. Codes Cryptogr.* **2013**, *69*, 331–349. [CrossRef]
4. Coron, J.-S. Higher Order Masking of Look-up Tables. In Proceedings of the Annual International Conference on the Theory and Applications of Cryptographic Techniques (EUROCRYPT), Cophenhagen, Denmark, 11–15 May 2014; pp. 441–458.
5. Coron, J.-S.; Prouff, E.; Rivain, M.; Roche, T. Higher-Order Side Channel Security and Mask Refreshing. In Proceedings of the International Workshop on Fast Software Encryption (FSE), Singapore, 11–13 March 2013; pp. 410–424.
6. Fumaroli, G.; Martinelli, A.; Prouff, E.; Rivain, M. Affine Masking against Higher-Order Side Channel Analysis. In Proceedings of the International Workshop on Selected Areas in Cryptography (SAC), Waterloo, ON, Canada, 12–13 August 2010; pp. 262–280.
7. Goudarzi, D.; Rivain, M. How Fast Can Higher-Order Masking Be in Software? In Proceedings of the Annual International Conference on the Theory and Applications of Cryptographic Techniques (EUROCRYPT), Paris, France, 30 April–4 May 2017; pp. 567–597.
8. Grosso, V.; Standaert, F.-X.; Prouff, E. Low Entropy Masking Schemes, Revisited. In Proceedings of the Smart Card Research and Advanced Application Conference (CARDIS), Berlin, Germany, 27–29 November 2013; pp. 33–43.
9. Kim, H.; Hong, S.; Lim, J. A Fast and Provably Secure Higher-Order Masking of AES S-box. In Proceedings of the International Conference on Cryptographic Hardware and Embedded Systems (CHES), Nara, Japan, 25–27 September 2011; pp. 95–107.
10. Prouff, E.; Rivain, M. A Generic Method for Secure SBox Implementation. In Proceedings of the International Workshop on Information Security Applications (WISA), Juju Island, Korea, 27–29 August 2007; pp. 227–244.
11. Rivain, M.; Prouff, E.; Doget, J. Higher-Order Masking and Shuffling for Software Implementations of Block Ciphers. In Proceedings of the International Conference on Cryptographic Hardware and Embedded Systems (CHES), Lausanne, Switzerland, 6–9 September 2009; pp. 171–188.
12. Tillich, S.; Herbst, C. Attacking State-of-the-Art Software Countermeasures—A Case Study for AES. In Proceedings of the International Conference on Cryptographic Hardware and Embedded Systems (CHES), Washington, DC, USA, 10–13 August 2008; pp. 228–243.
13. Tillich, S.; Herbst, C.; Mangard, S. Protecting AES software implementations on 32-bit processors against power analysis. In Proceedings of the International Conference on Applied Cryptography & Network Security (ACNS), Zhuhai, China, 5–8 June 2007; pp. 141–157.
14. Brier, E.; Clavier, C.; Olivier, F. Correlation Power Analysis with a Leakage Model. In Proceedings of the International Conference on Cryptographic Hardware and Embedded Systems (CHES), Cambridge, MA, USA, 11–13 August 2004; pp. 16–29.
15. Balasch, J.; Gierlichs, B.; Grosso, V.; Reparaz, O.; Standaert, F.-X. On the Cost of Lazy Engineering for Masked Software Implementations. In Proceedings of the Smart Card Research and Advanced Application Conference (CARDIS), Paris, France, 5–7 November 2014; pp. 64–81.
16. Kocher, P.; Jaffe, J.; Jun, B. Differential Power Analysis. In Proceedings of the International Cryptology Conference (CRYPTO), Santa Barbara, CA, USA, 15–19 August 1999; pp. 388–397.
17. Pan, J.; den Hartog, J.I.; Lu, J. You Cannot Hide behind the Mask: Power Analysis on a Provably Secure S-Box Implementation. In Proceedings of the International Workshop on Information Security Applications (WISA), Jeju Ireland, Korea, 25–27 August 2009; pp. 178–192.
18. Prouff, E.; Rivain, M.; Bevan, R. Statistical Analysis of Second Order Differential Power Analysis. *IEEE Trans. Comput.* **2009**, *58*, 799–814. [CrossRef]

19. Won, Y.-S.; Park, A.; Han, D.-G. Novel Leakage against Realistic Masking and Shuffling Countermeasures—Case study on PRINCE and SEED. In Proceedings of the International Conference on Information Security and Cryptology (ICISC), Seoul, Korea, 29 November–1 December 2017; pp. 139–154.

20. The Bank of Korea. *CFIP.ST.FINIC-02-2012: Standards for Financial IC Cards Specification Part 2–Open Platform*; The Bank of Korea: Seoul, Korea, 2012.

21. Korea Information Security Agency (KISA). *TTAS KO-12.0004: Block Cipher Algorithm SEED*; KISA: Seoul, Korea, 1999.

22. Kim, C.; Park, I. Investigation of Side Channel Analysis Attacks on Financial IC Cards. In Proceedings of the Korea Institute of Information Security & Cryptology (KIISC), Seoul, Korea, 29 November–1 December 2017; pp. 31–39.

23. Federal Information Processing Standards Publication 197, Advanced Encryption Standard (AES). Available online: https://nvlpubs.nist.gov/nistpubs/FIPS/NIST.FIPS.197.pdf (accessed on 10 November 2018).

24. Goubin, L. A Sound Method for Switching between Boolean and Arithmetic Masking. In Proceedings of the International Conference on Cryptographic Hardware and Embedded Systems (CHES), Paris, France, 13–16 May 2001; pp. 3–15.

25. Coron, J.-S.; Tchulkine, A. A New Algorithm for Switching from Arithmetic to Boolean Masking. In Proceedings of the International Conference on Cryptographic Hardware and Embedded Systems (CHES), Cologne, Germany, 7–10 September 2003; pp. 89–97.

26. Neiße, O.; Pulkus, J. Switching Blindings with a View Towards IDEA. In Proceedings of the International Conference on Cryptographic Hardware and Embedded Systems (CHES), Cambridge, MA, USA, 11–13 August 2004; pp. 230–239.

27. Rivain, M.; Dottax, E.; Prouff, E. Block Ciphers Implementations Provably Secure Against Second Order Side Channel Analysis. In Proceedings of the International Workshop on Fast Software Encryption (FSE), Lausanne, Switzerland, 10–13 February 2008; pp. 127–143.

28. Debaize, B. Efficient and Provably Secure Methods for Switching from Arithmetic to Boolean Masking. In Proceedings of the International Conference on Cryptographic Hardware and Embedded Systems (CHES), Leuven, Belgium, 9–12 September 2012; pp. 107–121.

29. Coron, J.-S.; Großschädl, J.; Vadnala, P.K. Secure Conversion between Boolean and Arithmetic Masking of Any Order. In Proceedings of the International Conference on Cryptographic Hardware and Embedded Systems (CHES), Busan, Korea, 23–26 September 2014; pp. 188–205.

30. Karroumi, M.; Richard, B.; Joye, M. Addition with Blinded Operands. In Proceedings of the International Workshop on Constructive Side-Channel Analysis and Secure Design, Paris, France, 14–15 April 2014; pp. 41–55.

31. Vadnala, P.K.; Großschädl, J. Faster Mask Conversion with Lookup Tables. In Proceedings of the International Workshop on Constructive Side-Channel Analysis and Secure Design, Berlin, Germany, 13–14 April 2015; pp. 207–221.

32. Won, Y.-S.; Han, D.-G. Efficient Conversion Method from Arithmetic to Boolean Masking in Constrained Devices. In Proceedings of the International Workshop on Constructive Side-Channel Analysis and Secure Design, Paris, France, 13–14 April 2017; pp. 120–137.

33. Available online: http://www.dpacontest.org/home (accessed on 10 November 2018).

34. SCARF Project. Available online: http://www.k-scarf.or.kr/ (accessed on 10 November 2018).

35. Yoo, H.; Kim, C.; Ha, J.; Moon, S.; Park, I. Side Channel Cryptanalysis on SEED. In Proceedings of the International Workshop on Information Security Applications (WISA), Jeju Island, Korea, 23–25 August 2004; pp. 411–424.

Article

Key Bit-Dependent Side-Channel Attacks on Protected Binary Scalar Multiplication †

Bo-Yeon Sim [1,‡], Junki Kang [2,‡] and Dong-Guk Han [1,*]

[1] Department of Mathematics, Kookmin University, 77 Jeongneung-ro, Seongbuk-gu, Seol 02707, Korea; qjdusls@kookmin.ac.kr

[2] The Affiliated Institute of ETRI, 1559 Yuseong-daero, Yuseong-gu, Daejeon 34044, Korea; kang.junki@gmail.com

* Correspondence: christa@kookmin.ac.kr

† This paper is an extended version of our paper published in ISPEC 2017, Melbourne, Australia, 13–15 December 2017.

‡ These authors contributed equally to this work.

Received: 15 September 2018; Accepted: 22 October 2018; Published: 6 November 2018

Abstract: Binary scalar multiplication, which is the main operation of elliptic curve cryptography, is vulnerable to side-channel analysis. It is especially vulnerable to side-channel analysis using power consumption and electromagnetic emission patterns. Thus, various countermeasures have been reported. However, they focused on eliminating patterns of conditional branches, statistical characteristics according to intermediate values, or data inter-relationships. Even though secret scalar bits are directly loaded during the check phase, countermeasures for this phase have not been considered. Therefore, in this paper, we show that there is side-channel leakage associated with secret scalar bit values. We experimented with hardware and software implementations, and experiments were focused on the Montgomery–López–Dahab ladder algorithm protected by scalar randomization in hardware implementations. We show that we could extract secret key bits with a 100% success rate using a single trace. Moreover, our attack did not require sophisticated preprocessing and could defeat existing countermeasures using a single trace. We focused on the key bit identification functions of mbedTLS and OpenSSL in software implementations. The success rate was over 94%, so brute-force attacks could still be able to recover the whole secret scalar bits. We propose a countermeasure and demonstrate experimentally that it can be effectively applied.

Keywords: side-channel analysis; elliptic curve cryptography; single-trace attack; key bit-dependent attack; countermeasure

---

1. Introduction

The blockchain and fast identity online (FIDO), which are emerging as key technologies to lead the Fourth Industrial Revolution, authenticate users by using an elliptic-curve digital signature algorithm (ECDSA). However, scalar multiplication, which is the core operation of ECDSA, is vulnerable to side-channel analysis (SCA). SCAs were first proposed by Paul Kocher in 1996 [1]; they use the leakage consumed while cryptographic algorithms are performed on embedded systems. Various side-channel attacks against elliptic-curve cryptography (ECC) have been researched [2–16]. Among them, power analysis using power patterns consumed during algorithm operations is known as the most powerful. Electromagnetic analysis using emitted electromagnetic patterns is similar to power analysis, but there is a difference in useable side-channel information. Therefore, in this paper, we focus on power analysis.

As SCAs become more powerful, various countermeasures to resist them have been studied [17–22]. However, only countermeasures to eliminate patterns of data-dependent conditional branches, statistical characteristic according to intermediate values, or data inter-relationships have been studied. No countermeasure has been taken into account for the secure design of the key bit identification phase even though secret scalar bits are directly loaded during that phase. Since the secret scalar bit value is extracted and stored in the variable, the secret scalar can be exposed if the vulnerability is discovered.

Our Contributions. In this paper, we analyzed the power consumption (we also considered information leakage via electromagnetic emanation throughout this paper.) properties of the key bit identification phase and experimentally showed that attacks based on these properties can recover secret scalar bits. Our proposed attacks require only a single power consumption or electromagnetic trace. They also do not require any knowledge of in–out values; thus, they can defeat any combination of existing countermeasures. Two implementations (i.e., hardware and software) were targeted, and we could recover secret scalar bits by applying SPA-VI (SPA based on visual inspection) and a k-means clustering algorithm. Among various scalar multiplication algorithms, we focused on binary scalar multiplication algorithms. The first set of experiments is based on hardware implementation of the Montgomery–López–Dahab ladder algorithm protected by scalar randomization. Experimental results show that the secret scalar bits can be recovered with a 100% success rate using only single power consumption or electromagnetic trace. In the second set of experiments, on software implementation, we targeted algorithms composed using the key bit identification functions of mbedTLS and OpenSSL. Here, secret scalar bits could be recovered with over 94% success rate. If we attacked the power consumption trace using the leakage associated with referenced register addresses, the success rate was 100%. We propose two kinds of countermeasures, one each for hardware and software implementations. Their effectiveness is experimentally demonstrated.

Extension. This paper is an extended version of our paper published in ISPEC 2017 [23]. In that paper, we showed key bit-dependent attack results using only a single power consumption trace. However, in this paper, we show new key bit-dependent attack results using a single electromagnetic trace and a low-pass filter. Thus, we show four experimental results using a power-consumption trace, a power-consumption trace passed through a low-pass filter, electromagnetic trace, and electromagnetic trace passed through a low-pass filter. Measuring electromagnetic traces is not an easy task because it very much depends on the angle and position of the probe. Moreover, in the case of hardware implementation, our latest results using electromagnetic traces have a higher success rate than previous results.

Organization. The rest of this paper is organized as follows. In Section 2, we describe SCAs in scalar multiplication algorithms. In Section 3, we regulate the leakage properties of the attack targets; in Section 4, we establish the attack framework. Experimental results are described in Section 5. We discuss countermeasures in Section 6, and conclusions are presented in Section 7.

2. Conventional SCAs on Scalar Multiplication

2.1. Simple-Power Analysis

Simple-power analysis (SPA) is a method of directly analyzing a secret scalar using only one trace or a few traces collected during cryptographic operations [9]. Because cryptographic algorithms have different power-consumption patterns according to the instructions of the processor, the secret scalar or instantaneous command could be analyzed from these patterns. For instance, in the case of a binary scalar multiplication algorithm that performs a point-doubling operation at all times, and performs a point addition operation only when the secret key bit value is 1, the secret key can be found if the point-doubling and point-addition operations have different power-consumption patterns. That is, as per Figure 1a, this irregular sequence of instructions according to the secret scalar bit (i.e., the data-dependent conditional branch) leads to a serious security problem.

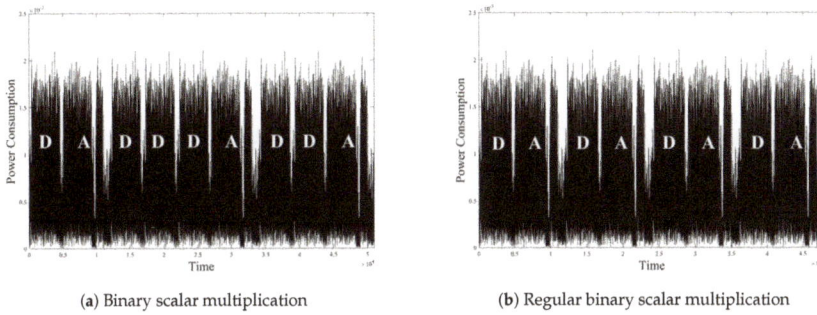

(a) Binary scalar multiplication

(b) Regular binary scalar multiplication

Figure 1. Power-consumption trace of binary scalar multiplication. (**a**) Binary scalar multiplication; (**b**) regular binary scalar multiplication.

2.2. Differential Power Analysis (DPA)

DPA is a statistical analysis method that analyzes multiple power-consumption traces to find the secret scalar [9]. Typically, DPA is based on the fact that power consumption depends on data values being manipulated. To perform DPA, input or output values of cryptographic algorithms have to be known. Similarly, there is an address-bit DPA based on the fact that power consumption depends on the address value of the register that loads or stores data during the operation. Thus, even if an SPA countermeasure [19,21,22], which has a regular power-consumption sequence, as shown in Figure 1b, is applied, it is vulnerable to DPA. To cope with this, randomization techniques that eliminate association between all possible intermediate values and power consumption are generally used [17,18,20].

2.3. Sophisticated Power Analysis

SPA-and DPA-resistant countermeasures can be defeated by sophisticated attacks, such as a template attack (TA) [6,10,12] or collision attack (CA) [7,11,13]. A TA characterizes power-consumption traces by a multivariate normal distribution to build templates, and matches power-consumption leakage to the templates to find a secret scalar value. A CA is a kind of higher-order DPA and is an attack based on the inter-relationships among intermediate data (i.e., collisions of two intermediate values). So far, no theoretically perfect countermeasures against TAs and CAs have been presented. However, there is a disadvantage, in that they require precise preprocessing, such as decapsulation, localization, and a multiprobe to obtain a power-consumption trace having a high signal-to-noise ratio [6,7,11,13]. Decapsulation in particular requires to physically modify the target devices, and numerous traces are required to build templates.

To thwart previous attacks, various countermeasures to eliminate patterns of data-dependent conditional branches, statistical characteristic according to intermediate values, or data inter-relationships have been studied. However, no countermeasure has been taken into account for the secure design of the key bit identification phase, although secret scalar bits are directly loaded during that phase. Since the secret scalar bit value is extracted and stored in the variable, the secret scalar can be exposed if the vulnerability is discovered. Thus, in this paper, we verify that this vulnerability is sufficient to find a secret scalar.

3. Materials

3.1. Key Bit Identification Phase

Elliptic-curve scalar multiplication is a method for computing dP, where d is a secret scalar and P is a point on an elliptic curve. It is an elementary operation of ECC, so it has been used in numerous

PKCs. It basically consists of iterative operations determined according to the i-th bit d_i value of the secret scalar d, where d is a λ-bit scalar, so $d = (d_{\lambda-1}, d_{\lambda-2}, \cdots, d_1, d_0)_2$ and $0 \le i < \lambda$ [19,21,22,24]. For instance, in the algorithms shown in Figure 2, while performing Steps 2 to 5, addresses of registers $R_x(x = 0 \text{ or } 1)$ to be referenced are determined by the d_i value.

Thus, at the beginning of the i-th iterative operation, the i-th secret scalar bit value d_i is extracted from a λ-bit scalar string and stored in a variable. This phase exists in almost all elliptic-curve scalar multiplication algorithms because they are composed of iterative operations based on the value of d_i. At this phase, secret scalar bits, d_i, are extracted at the beginning of each iterative operations. We define this step as the key bit identification phase.

3.1.1. Key Bit-Dependent Properties

Binary scalar multiplication consists of iterative operations determined according to the i-th bit d_i value of secret scalar d (Figure 2). Therefore, there exists a key bit identification phase in which the i-th scalar bit value is extracted from a λ-bit scalar string $d = (d_{\lambda-1}, d_{\lambda-2}, \cdots, d_1, d_0)_2$ and stored in a d_i variable at the beginning of each i-th iteration. Thus, power consumption associated with the d_i value occurs. We can categorize these properties according to hamming distance (HD) and hamming weight (HW), mainly used as power-consumption models as follows.

Left to Right	Right to Left
Input : P is a point on an elliptic curve, a λ-bit scalar $d = (d_{\lambda-1}, \cdots, d_0)_2$ **Output** : $Q = dP$	**Input** : P is a point on an elliptic curve, a λ-bit scalar $d = (d_{\lambda-1}, \cdots, d_0)_2$ **Output** : $Q = dP$
1: $R_0 \leftarrow \infty$, $R_1 \leftarrow P$ 2: **for** $i = \lambda - 1$ down to 0 **do** 3:　　$R_{1-d_i} \leftarrow R_{d_i} + R_{1-d_i}$ 4:　　$R_{d_i} \leftarrow 2R_{d_i}$ 5: **end for** 6: **Return** R_0	1: $R_0 \leftarrow \infty$, $R_1 \leftarrow P$, $R_2 \leftarrow P$ 2: **for** $i = 0$ up to $\lambda - 1$ **do** 3:　　$R_{1-d_i} \leftarrow R_{1-d_i} + R_2$ 4:　　$R_2 \leftarrow R_0 + R_1$ 5: **end for** 6: **Return** R_0

Figure 2. Examples of simple-power analysis (SPA)-resistant regular algorithms for binary scalar multiplication.

Property 1. *In hardware implementations, power consumption in the key bit identification phase is simultaneously affected by the hamming distance between two consecutive bits d_{i+1} and d_i, i.e., $d_{i+1} \oplus d_i$ ($0 \le i < \lambda - 1$). Thus, if two consecutive bits are the same, i.e., $d_{i+1} = d_i$, power consumption related to $d_{i+1} \oplus d_i = 0$ occurs. Otherwise, power consumption related to $d_{i+1} \oplus d_i = 1$ occurs.*

Property 2. *In software implementations, power consumption in the key bit identification phase is affected by the hamming weight of d_i ($0 \le i \le \lambda - 1$). Thus, if the value of i-th secret bit is 0, i.e., $d_i = 0$, then power consumption is related to 0. Otherwise, power consumption related to 1 occurs.*

3.1.2. Key Bit-Dependent Properties of SPA-Resistant Regular Algorithms

The binary scalar multiplication algorithm (Reference [25], Algorithm 3.26 and 3.27) can be easily broken by SPA. Therefore, various SPA-resistant regular algorithms, as shown in Figure 2, have been used. In regular algorithms, the referred register addresses $Reg\,Addr_{d_i}$ differ depending on the d_i value, and these influence power consumption. Since hardware and software operating structures are different from each other, the effect on power consumption especially differs then.

In hardware implementations, operations are executed in parallel. Thus, at the same time as the secret scalar bits d_i are extracted at the beginning of each iterative operation, register addresses $RegAddr_{d_i}$ to be referenced are also determined. In accordance with this characteristic, power consumption when the secret scalar bit d_i is determined is also influenced by the HD between the register addresses used in two successive loops. In software implementations, differing from hardware implementations, operations are executed sequentially. Hence, register addresses $RegAddr_{d_i}$ to be referenced do not affect power consumption at the same time as the secret scalar value d_i. In the following, we describe additional power-consumption properties of SPA-resistant regular algorithms. Note that $RegAddr_0$ is different from $RegAddr_1$.

Property 3. *In hardware implementations, power consumption in the key bit identification phase is simultaneously affected by:*

(a) *the hamming distance between two consecutive bits d_{i+1} and d_i, i.e., $d_{i+1} \oplus d_i$*

(b) *the hamming distance between referred register addresses $RegAddr_{d_{i+1}}$ and $RegAddr_{d_i}$ determined by d_{i+1} and d_i, i.e., $RegAddr_{d_{i+1}} \oplus RegAddr_{d_i}$*

Thus, if two consecutive bits are the same, i.e., $d_{i+1} = d_i$; power consumption related to $d_{i+1} \oplus d_i = 0$ and $RegAddr_{d_{i+1}} \oplus RegAddr_{d_i} = 0$ occurs at the same time. Otherwise, power consumption related to $d_{i+1} \oplus d_i = 1$ and $RegAddr_{d_{i+1}} \oplus RegAddr_{d_i} \neq 0$ occurs at the same time $(0 \leq i < \lambda - 1)$.

Property 4. *In software implementations, power consumption is affected by:*

(a) *the hamming weight of i-th secret bit value d_i*

(b) *the hamming weight of referred register address $RegAddr_{d_i}$ determined by value of i-th secret bit d_i*

Thus, if the i-th secret bit value is 0, i.e., $d_i = 0$, then power consumption related to 0 and $RegAddr_0$ occurs. Otherwise, power consumption related to 1 and $RegAddr_1$ occurs $(0 \leq i \leq \lambda - 1)$.

We can classify power-consumption traces into two groups, G_0 and G_1, using the properties. G_0 includes power-consumption traces when leakage is zero, and G_1 includes traces when leakage is nonzero. Once the traces are classified into two groups, we can recover the respective bit d_i, since the most significant bit is always 1. We define a study exploiting Property 1 and 2 as *Case Study 1* (Figures 3a,c, 4a,c and 5a,c,e). Then, we define a study exploiting Property 3 and 4 as *Case Study 2* (Figures 3b,d, 4b,d and 5b,d,f).

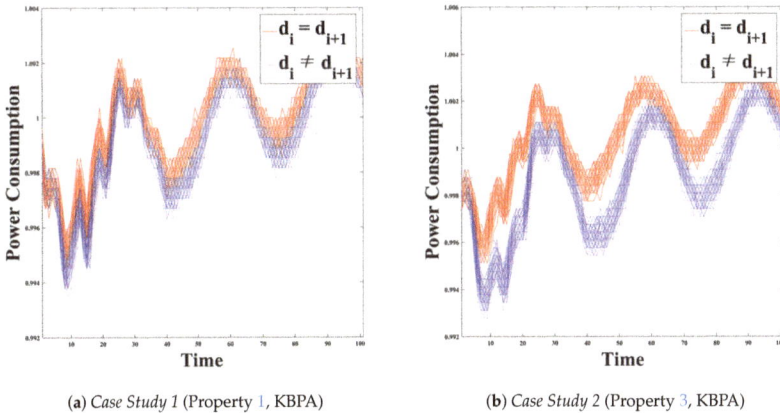

(**a**) *Case Study 1 (Property 1, KBPA)*

(**b**) *Case Study 2 (Property 3, KBPA)*

Figure 3. *Cont.*

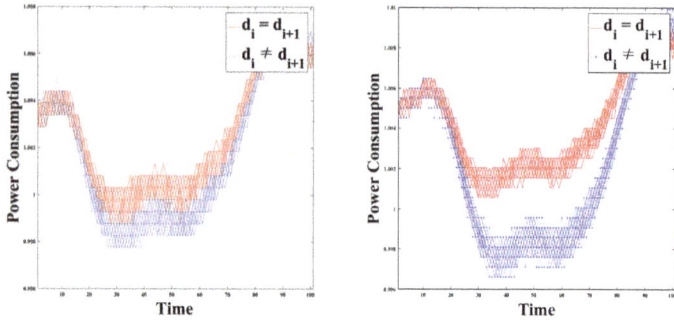

(**c**) *Case Study 1* (Property 1, KBPA, BLP 100-75) (**d**) *Case Study 2* (Property 3, KBPA, BLP-100-75)

Figure 3. Classification result of Points of Interest (PoIs) (Hardware Implementation, Power Consumption). (**a**) *Case Study 1* (Property 1, KBPA); (**b**) *Case Study 2* (Property 3, KBPA); (**c**) *Case Study 1* (Property 1, KBPA, BLP 100-75); (**d**) *Case Study 2* (Property 3, KBPA, BLP-100-75).

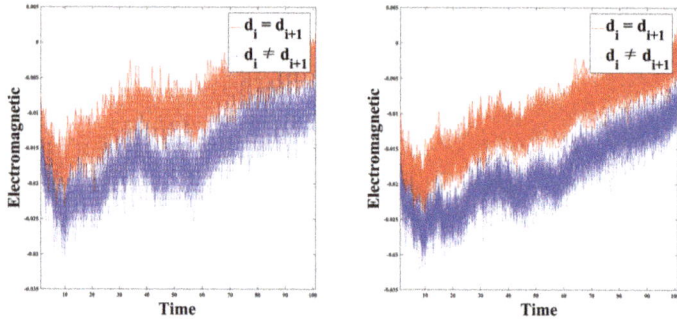

(**a**) *Case Study 1* (Property 1, KBEA, LF-R 400) (**b**) *Case Study 2* (Property 3, KBEA, LF-R 400)

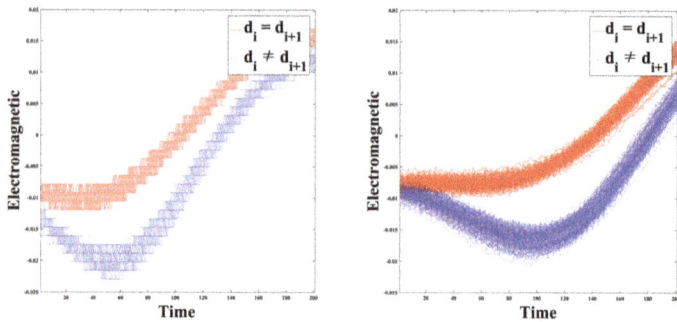

(**c**) *Case Study 1* (Property 1, KBEA, LF-R 400, BLP (**d**) *Case Study 2* (Property 3, KBEA, LF-R 400, BLP
15-75+)   10.7-75+)

Figure 4. Classification result of PoIs (Hardware Implementation, Electromagnetic). (**a**) *Case Study 1* (Property 1, KBEA, LF-R 400); (**b**) *Case Study 2* (Property 3, KBEA, LF-R 400); (**c**) *Case Study 1* (Property 1, KBEA, LF-R 400, BLP 15-75+); (**d**) *Case Study 2* (Property 3, KBEA, LF-R 400, BLP 10.7-75+).

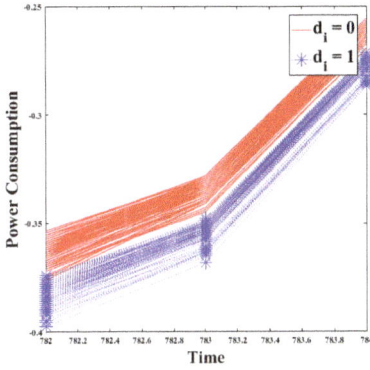

(**a**) *Case Study 1* (Property 2, KBPA)

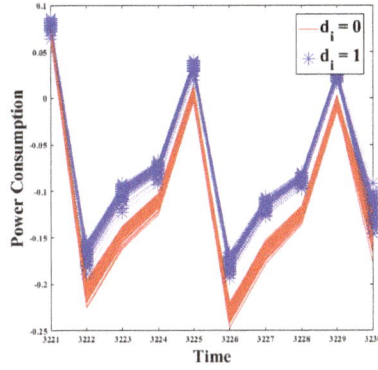

(**b**) *Case Study 2* (Property 4, KBPA)

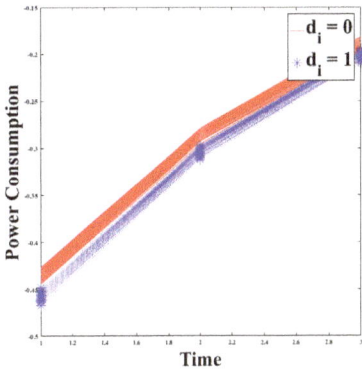

(**c**) *Case Study 1* (Property 2, KBPA, BLP 70+)

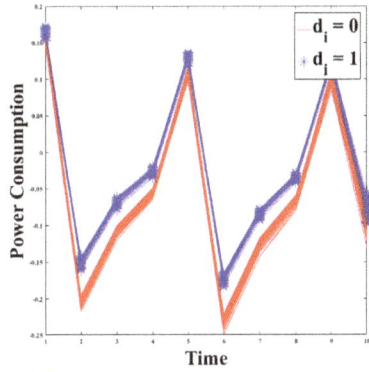

(**d**) *Case Study 2* (Property 4, KBPA, BLP 70+)

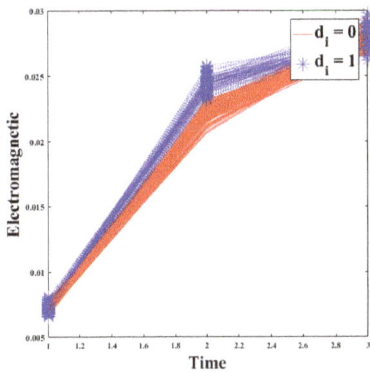

(**e**) *Case Study 1* (Property 2, KBEA, LF-U 5, BLP 10.7+)

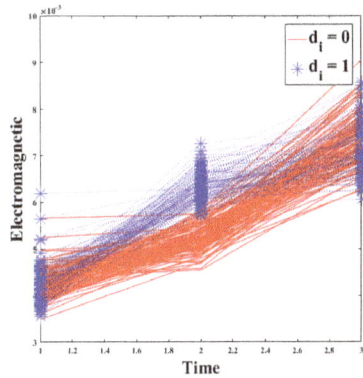

(**f**) *Case Study 2* (Property 4, KBEA, LF-U 5, BLP 10.7+)

Figure 5. Classification result of PoIs (Software Implementation, mbedTLS). (**a**) *Case Study 1* (Property 2, KBPA); (**b**) *Case Study 2* (Property 4, KBPA); (**c**) *Case Study 1* (Property 2, KBPA, BLP 70+); (**d**) *Case Study 2* (Property 4, KBPA, BLP 70+); (**e**) *Case Study 1* (Property 2, KBEA, LF-U 5, BLP 10.7+); (**f**) *Case Study 2* (Property 4, KBEA, LF-U 5, BLP 10.7+).

4. Methods

4.1. Key Bit-Dependent Attack Framework

In this paper, we consider binary scalar multiplication algorithms that are resistant against SPA and DPA. In particular, we targeted algorithms based on regular algorithms protected by intermediate data randomization. Therefore, we suppose that an attacker is obliged to use a single trace rather than numerous traces. In addition, we assumed that the attacker could distinguish the iterative structure in the traces of regular algorithms. We categorized the attack framework in four steps as follows. Note that we did not consider side-channel atomicity algorithms that are SPA-resistant since it is impossible to distinguish the starting point of iterative loop operations.

- Preprocessing
  The attacker can divide trace T into λ subtraces, O_i, corresponding to each iteration ($0 \leq i \leq \lambda - 1$). As shown in Figure 6, trace T is described as a series of λ sub-races as

$$T = \{O_{\lambda-1} \, || O_{\lambda-2} \, || \cdots || O_0\}$$

  since λ iterative operations are performed when the secret scalar is λ-bit, we divide trace T into λ subtraces and align them.

- Select Points of Interest (PoIs)
  If the attacker can use the same device as the target and acquire a trace with a known key, it is easy to find PoIs. The attacker can calculate the sum of squared pairwise t-differences (SOST) [26] of the subtraces classified based on the properties described in Sections 3.1.1 and 3.1.2. Then, the PoIs are the points that have high SOST values. SOST is calculated as follows:

$$SOST = \left(\frac{m_{G_1} - m_{G_2}}{\sqrt{\frac{\sigma_{G_1}^2}{n_{G_1}} + \frac{\sigma_{G_2}^2}{n_{G_2}}}} \right)^2 \tag{1}$$

  where m denotes the mean, σ is standard deviation, and n is the number of elements. If it is not possible to use the same device, the attacker must know how the target algorithm is implemented to find PoIs. Moreover, the key bit identification phase section should be recognized in the trace. In general, since the d_i value must be decided in advance before each loop operation, the target phase is positioned near the beginning of each subtrace O_i. We represent p_i as PoIs of each subtrace O_i ($0 \leq i \leq \lambda - 1$).

- Classify into Two Groups and Extract Secret Scalar Bits
  The attacker can separate p_i into two groups, G_0 and G_1, applying SPA-VI or a clustering algorithm (e.g., k-means, fuzzy k-means, or EM algorithm [27,28]). Because the most significant bit is always 1, the attacker can configure $d_{\lambda-1}$ as 1 and find the respective scalar bit d_i based on the power model and properties described in Sections 3.1.1 and 3.1.2. For instance, when power consumption complies with the HD model, the attacker can recover secret scalar bits d_i as follows. It is possible to assume that the group that contains $p_{\lambda-1}$ indicates that leakage is nonzero, if d_i is at first initialized as zero. Consequently, if p_i is contained in the same group that contains $p_{\lambda-1}$, d_i is one; otherwise, d_i is zero ($0 \leq i < \lambda - 1$). Similarly, when power consumption complies with the HW model, the group that includes $p_{\lambda-1}$ indicates that leakage is non-zero, and the other group indicate that leakage is zero. Consequently, if p_i is contained in the same group that contains $p_{\lambda-1}$, then d_i is one; otherwise, d_i is zero ($0 \leq i < \lambda - 1$).

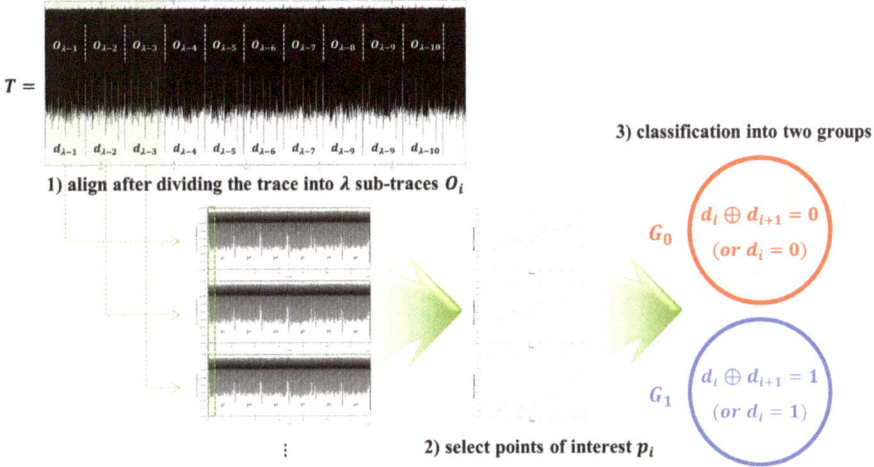

Figure 6. Key bit-dependent attack framework.

4.2. Experiment Environments

The first experimental platform is VHDL implementation on a SASEBO-GII FPGA board, as shown in Figure 7. We measured traces using a Teledyne Lecroy HDO6104A oscilloscope at a sampling rate of 2.5 GS/s. Electromagnetic traces were recorded using a Langer LF-R 400. Additionally, we used Mini Circuit BLP (low-pass filter) to increase the signal-to-noise ratio. The second experimental platform was software implementation on an Atmel AVR XMEGA 128D4 microcontroller equipped with a CW-Lite XMEGA target board, as shown in Figure 7. We measured power-consumption traces using the CW-Lite main board at a sampling rate of 29.5 MS/s. Electromagnetic traces were recorded using a Teledyne Lecroy HDO6104A oscilloscope at a sampling rate of 2.5 GS/s, using a Langer LF-U 5 and Mini Circuit BLP(low-pass filter) to increase the signal-to-noise ratio.

Figure 7. Key bit-dependent attack experiment environments.

5. Experimental Results

In this section, we demonstrate that a key bit-dependent attack could extract secret scalar bit using a single trace.

5.1. Key Bit-Dependent Power/Electromagnetic Attack on Hardware Implementation

Our target binary scalar multiplication algorithm was the Montgomery–López–Dahab ladder algorithm [24] protected by scalar randomization [18]. Therefore, the attacker is restricted to using a single trace. To attack algorithms operating on the first experimental platform, we focused on Properties 1 and 3, described in Sections 3.1.1 and 3.1.2, respectively. However, in hardware implementations, operations are executed in parallel. Thus, at the same time as secret scalar bits d_i are extracted at the beginning of each iterative operation, the addresses of registers $R_x(x = 0 \text{ or } 1)$ to be referenced are also determined. Thus, there is no SPA-resistant regular algorithm that only satisfies Property 1. Our target was an SPA-resistant regular algorithm. Hence, we modified the code as shown in Figure 8a to identify how much information was present according to Property 1. The code as shown in Figure 8b is a general implementation that satisfies Property 3.

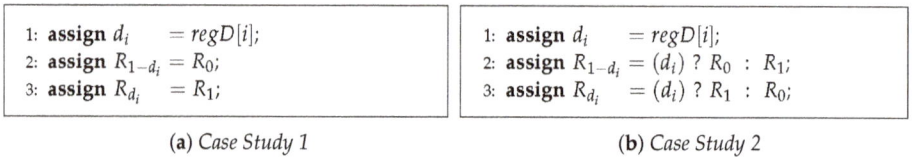

1: **assign** d_i $= regD[i];$ 2: **assign** $R_{1-d_i} = R_0;$ 3: **assign** R_{d_i} $= R_1;$	1: **assign** d_i $= regD[i];$ 2: **assign** $R_{1-d_i} = (d_i)\ ?\ R_0\ :\ R_1;$ 3: **assign** R_{d_i} $= (d_i)\ ?\ R_1\ :\ R_0;$
(**a**) *Case Study 1*	(**b**) *Case Study 2*

Figure 8. Hardware implementation: *Case Study 1* (**a**) and *Case Study 2* (**b**).

- Preprocessing
  Operations for the most significant bit $d_{\lambda-1}$ do not exist in the Montgomery–López–Dahab ladder algorithm, as shown in Algorithm A1 and Appendix A. In accordance, trace T is composed of λ-1 subtraces for a λ-bit scalar, so we divided trace T into $\lambda - 1$ subtraces O_i, and aligned them ($0 \leq i \leq \lambda - 2$). Figure 9 (top) shows one of the subtraces, consisting of six finite-field multiplications, captured from the first experimental platform.

- Select Points of Interest
  The key bit identification phase is operated on the second clock cycle of each subtrace of the target algorithm. We also confirmed that points of the second clock cycle of each subtrace are PoIs p_i, since the SOST value is the greatest on the points of the second clock cycle, as shown in Figure 9 (bottom) ($0 \leq i \leq \lambda - 2$). When we calculated the SOST value, we classified PoIs of subtraces p_i into two groups according to Property 1 (or 3).

- Classify into Two Groups and Extract Secret Scalar Bits
  (**1**) When we targeted *Case Study 1* and exploited the power-consumption trace, it was impossible to clearly split them into two groups through SPA-VI. Since two distributions are overlapped as shown in Figure 3a, we could extract secret scalar bits d_i with a 96.75% success rate when we classified p_i into two groups based on the differences from an average trace. We could also extract secret scalar bits d_i with a 96.74% success rate when we applied the k-means clustering algorithm to classify p_i into two groups (i.e., 8 errors). Consequently, a brute-force attack to recover the entire secret scalar could be viable, because the error rate is sufficiently small. Therefore, it was confirmed that the key bit-dependent leakage based on Property 1 was sufficiently large to recover the secret scalar bits.
  (**2**) By using the low-pass filter to increase the signal-to-noise ratio, the success rate slightly improved to 97.17% when we applied the k-means clustering algorithm. It could not be classified into two groups through SPA-VI because distribution overlapped, as shown in Figure 3c.
  (**3, 4**) When we targeted *Case Study 2* and exploited the power-consumption trace, it was possible

to perfectly split it into two groups via SPA-VI, as shown in Figure 3b,d. The success rate of classification was 100%, as shown in Table 1; thus, we could acquire all the secret scalar bits d_i. From this result, we noticed that changing a referring register leaks more significant information than changing the secret scalar bits. Moreover, we demonstrated that we could recover whole secret scalar bits based on Property 3 using only one power-consumption trace. We define attacks such as in Steps (1) to (4) as key bit-dependent power attacks (KBPA).

(5) Figure 4a shows the PoIs chosen from electromagnetic subtraces when we targeted *Case Study 1*. Although it was not easy to clearly divide them into two groups via SPA-VI, we could classify p_i into two groups with a 100% success rate using the differences from an average trace. Accordingly, the classification success rate based on k-means clustering algorithm was also 100%; thus, we could find the entire secret scalar bits d_i based on Property 1.

(6) Moreover, if we could use the low-pass filter to increase the signal-to-noise ratio, we could extract whole secret scalar bits through SPA-VI as shown in Figure 4c.

(7) Figure 4b shows the PoIs chosen from electromagnetic subtraces when we targeted *Case Study 2*. Unlike the result of **(3, 4)**, it was not easy to clearly divide them into two groups via SPA-VI. However, it was possible to divide p_i into two groups based on the differences from an average trace. Thus, the classification success rate based on k-means clustering algorithm was also 100%; therefore, we could find all the secret scalar bits d_i based on Property 3 ($0 \leq i \leq \lambda - 1$).

(8) Moreover, if we could use the low-pass filter to increase the signal-to-noise ratio, we could extract whole secret scalar bits through SPA-VI as shown in Figure 4d.

To sum up, we also showed that we could recover all secret scalar bits using only one electromagnetic trace. Compared to the key bit-dependent power attack, the secret scalar bits could be recovered with a 100% success rate based on Property 1. We define attacks such as in Steps (5) to (8) as key bit-dependent electromagnetic attacks (KBEA).

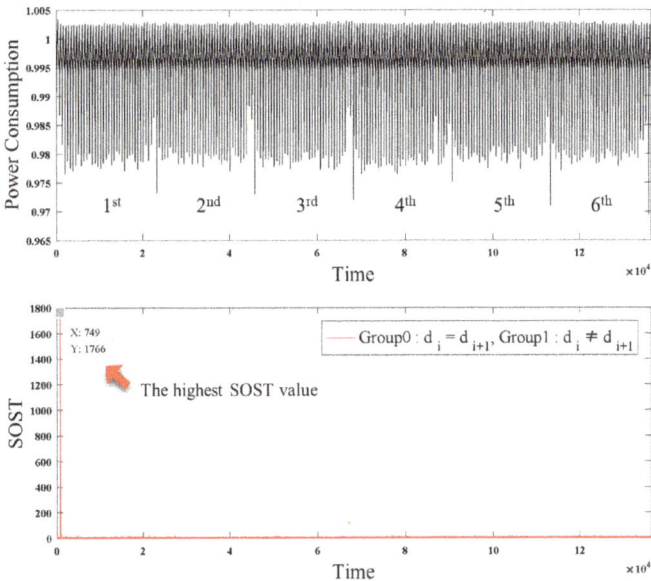

Figure 9. Hardware implementation: one of the subtraces (**top**) and the SOST value between two subgroups (**bottom**).

Table 1. Experimental results: hardware implementations.

| Hardware | Power Consumption (Figure 3) | | | | Electromagnetic (Figure 4) | | | |
| | None | | Low-Pass Filter | | None | | Low-Pass Filter | |
	diff	k-means	diff	k-means	diff	k-means	diff	k-means
Property 1	96.75%	96.74%	96.75%	97.71%	100%	100%	100%	100%
Property 3	100%	100%	100%	100%	100%	100%	100%	100%

diff: clustering based on the difference from an average trace; k-means: clustering using the k-means clustering algorithm.

5.2. Key Bit-Dependent Power/Electromagnetic Attack on Software Implementation

In this section, we focus on the key bit identification function of mbedTLS (polarSSL) as shown in Figure 10, which is an extensively used embedded transmission security TLS/SSL public encryption library. It should be noted that to capture an entire binary scalar multiplication trace using the CW-Lite main board is impossible; thus, we used the modified algorithm shown in Figure 11 based on the function in Figure 10 to identify how much information exists. In Appendix B, we describe the key bit identification function of OpenSSL as shown in Figure A1. We also show the experimental results of when we used it. For attack algorithms operating on the second experimental platform, we focused on Properties 2 and 4 described in Sections 3.1.1 and 3.1.2, respectively.

```
1: int mbedtls_mpi_get_bit (const mbedtls_mpi *X, size_t pos)
2: {
3:      if (X-> n * biL <= pos)
4:          return(0);
5:
6:      return ((X->p[pos / biL] ≫ (pos % biL)) & 0x01);
7: }
```

Figure 10. Key bit identification function of mbedTLS.

```
1: d_i = mbedtls_mpi_get_bit (&d, i);
2: BN_MUL(&R_2, &R_{d_i}, &R_2);
```

Figure 11. Software implementation (mbedTLS): acquisition range.

- Preprocessing
  We uniformly divided trace T into λ subtraces, O_i, and aligned them ($0 \leq i \leq \lambda - 1$). Figure 12 (top) shows one of the subtraces.

- Select Points of Interest
  In software implementation, operations are sequentially executed. Hence, differing from hardware implementations, we targeted two positions. The first came immediately after the & $0x01$ operation was performed, as shown in Figure 10. The second was where the register was referred to. The register addresses to be referenced were determined according to secret scalar bit d_i, so there was information associated with d_i. Thus, we targeted where the register LOAD operation was performed for a long integer operation. Points with high SOST values are located where the key bit identification function is performed (see Figure 12). The second target points were located behind the key bit identification function. Here, we chose points with high SOST values as PoIs. When we calculated SOST values, we classified PoIs of subtraces p_i into two groups according to Property 2 (or 4).

- Classify into Two Groups and Extract Secret Scalar Bits

  (1) When we targeted *Case Study 1* and exploited the power-consumption trace, we could not clearly split it into two groups via SPA-VI, because the two distributions overlapped as shown in Figure 5a, so we applied the k-means clustering algorithm to classify p_i into two groups ($0 \leq i \leq \lambda - 1$). Approximately 97.60% of the secret scalar bits d_i could be extracted, as shown in Table 2. There are misclassified bits, but the number of error bits is sufficiently small. Hence, it is possible to recover whole secret scalar bits with a brute-force attack. Consequently, we confirmed that the key bit-dependent leakage based on Property 2 was sufficiently large to recover the secret scalar bits.

  (2) By using the low-pass filter to increase the signal-to-noise ratio, success rate was slightly improved to 98.24% when we applied the k-means clustering algorithm. It could not be classified into two groups through SPA-VI because the distribution overlapped, as shown in Figure 5c.

  (3, 4) We investigated leakage associated with referred register addresses determined according to d_i in *Case Study 2*. When we exploited the power-consumption trace, subtraces p_i could be divided into two groups through SPA-VI with a 100% success rate, see Figure 5b,d.

  (5) Figure 5e shows the PoIs chosen from electromagnetic subtraces when we targeted *Case Study 1*. They could not be clearly divided into two groups via SPA-VI; thus, the k-means clustering algorithm was needed. Secret scalar bits recovery rate was 94.17%, as shown in Table 2. This was slightly higher (0.17%) than the success rate when we divided p_i into two groups based on the differences from an average trace. **(6)** Unlike the result of **(2)**, PoIs could not be perfectly split into two groups by SPA-VI, as shown in Figure 5f. Thus, we applied the k-means clustering algorithm and we could find approximately 95.96% of the secret scalar bits. This was slightly better than the 93.72% success rate when we divided p_i into two groups based on differences from an average trace. Here, we demonstrated that single-trace KBPA and KBEA can also defeat binary scalar multiplication algorithms that are resistant against SPA and DPA in software implementations.

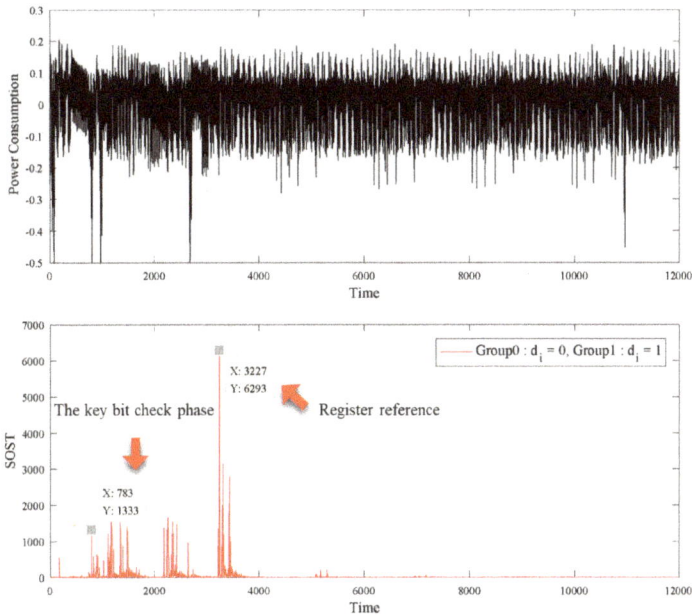

Figure 12. Software implementation (mbedTLS): one of the subtraces (**top**) and the SOST value between two subgroups (**bottom**).

Table 2. Experimental results: software implementations.

| Hardware | Power Consumption | | | | Electromagnetic | |
| | None | | Low-Pass Filter | | Low-Pass Filter | |
	diff	k-means	diff	k-means	diff	k-means
Property 2	97.60%	97.60%	97.31%	98.24%	93.72%	94.17%
Property 4	100%	100%	100%	100%	94.17%	95.96%

diff: clustering based on the difference from average trace; k-means: clustering using the k-means clustering algorithm.

6. Countermeasures

We have shown that single-trace KBPA and KBEA could recover whole secret scalar bits. Here, we discuss countermeasures against KBPA and KBEA. We propose two kinds of countermeasures, one each for hardware and software implementations.

6.1. Countermeasure for Hardware Implementations

For hardware implementations, we suggest random initialization that initializes the d_i variable with random bit before each key bit identification phase, as per Algorithm 1. We verified that the leakage based on Properties 1 and 3 could be efficiently eliminated. The result of the classification of p_i is shown in Figures 13a,b, and 14 (top). The success rate of the attack was approximately 50%, and it was similar to randomly guessing the secret scalar bits with a probability of 1/2.

Algorithm 1: ECC Scalar Multiplication (initialized by random bit)

> **Input** : P is a point on an elliptic curve, a λ-bit scalar $d = (d_{\lambda-1}, d_{\lambda-2}, \cdots, d_0)_2$
> **Output:** $Q = dP$

1: $regD[\lambda - 1 : 0] \leftarrow \{d_{\lambda-1}, d_{\lambda-2}, \cdots, d_0\}$
2: $R_0 \leftarrow \infty, R_1 \leftarrow P$
3: $d_i \leftarrow$ *random bit*
4: **for** $i = n - 1$ down to 0 **do**
5:    $d_i \leftarrow regD[i]$
6:    $R_{1-d_i} \leftarrow R_{d_i} + R_{1-d_i}$
7:    $R_{d_i} \leftarrow 2R_{d_i}$
8:    $d_i \leftarrow$ *random bit*
9: **end for**
10: **Return** R_0

6.2. Countermeasure for Software Implementations

As a countermeasure for software implementations, we propose bit masking to remove the leakage of Properties 2 and 4, as Algorithm 2. This method is a type of address-bit randomization [29,30]. However, there is an important difference, in that bit masking must be performed before loop operation begins, which is shown in Step 2 of Algorithm 2. The result of classification of p_i is shown in Figures 13c,d and 14 (bottom). The success rate of the attack is also approximately 50%, and it is similar to randomly guessing the secret scalar bits with a probability of 1/2.

Algorithm 2: ECC Scalar Multiplication (Masking with random bit)

Input : P is a point on an elliptic curve, a λ-bit scalar $d = (d_{\lambda-1}, d_{\lambda-2}, \cdots, d_0)_2$
Output : $Q = dP$

1: Generate λ-bit random number $r = (r_{\lambda-1}, r_{\lambda-2}, \cdots, r_0)_2$
2: $md \leftarrow d \oplus (d \ll 1) \oplus r$
3: $R_{r_{\lambda-1}} \leftarrow 2P$, $R_{1-r_{\lambda-1}} \leftarrow P$
4: **for** $i = n - 1$ down to 1 **do**
5:     $R_2 \leftarrow 2R_{md_i}$
6:     $R_{1-r_{i-1}} \leftarrow R_0 + R_1$
7:     $R_{r_{i-1}} \leftarrow R_2$
8: **end for**
9: **Return** R_{r_0}

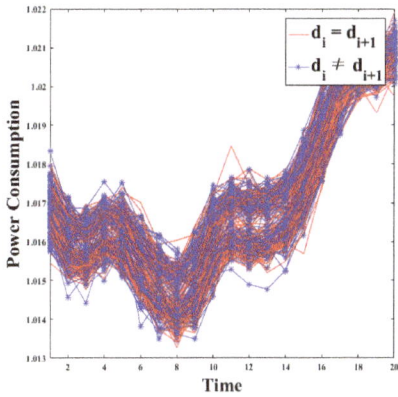

(**a**) Property 1 (hardware implementation)

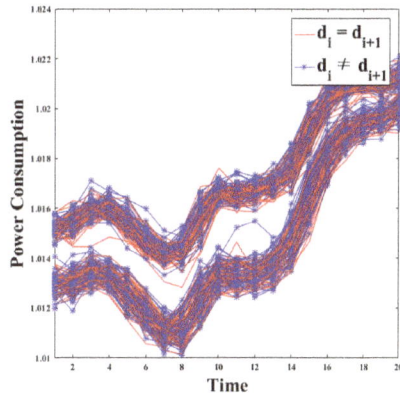

(**b**) Property 3 (hardware implementation)

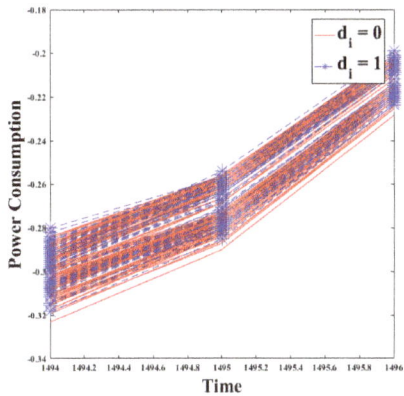

(**c**) Property 2 (software implementation)

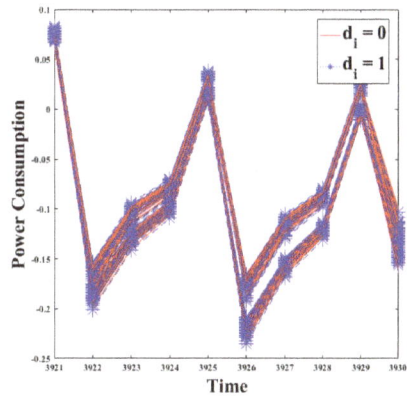

(**d**) Property 4 (software implementation)

Figure 13. Countermeasure (Power). (**a**) Property 1 (hardware implementation); (**b**) Property 3 (hardware implementation); (**c**) Property 2 (software implementation); (**d**) Property 4 (software implementation).

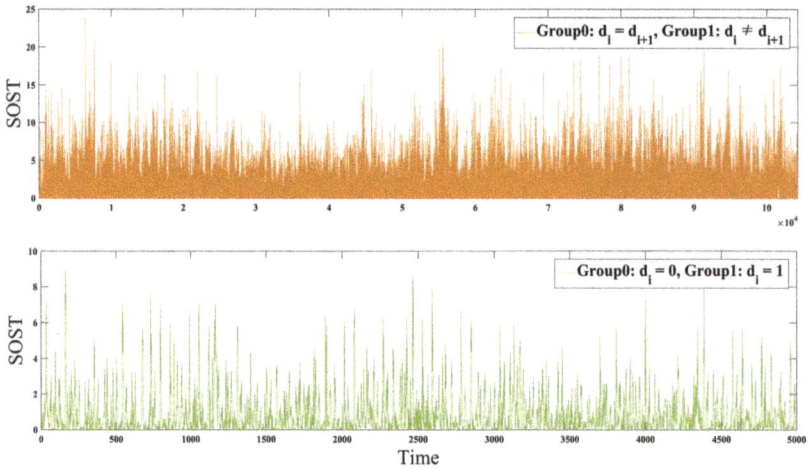

Figure 14. SOST value of countermeasure: hardware implementations (**top**) and software implementations (**bottom**).

7. Conclusions

In this paper, we suggested attacks using the leakage that occurs on the key bit identification phase and demonstrated that such attacks could extract secret scalar bits using a single trace without profiling. The attacks could be done not only by power consumption, but also by electromagnetic trace. Compared with previous attacks that required sophisticated preprocessing and multitraces, this represents a significant advantage. There is no need to apply preprocessing, and we could recover the entire secret scalar bits through SPA-VI. Since the proposed KBPA and KBEA attacks could defeat existing countermeasures, this leads to a very robust attack model. Although we focused on ECC binary scalar multiplication algorithms, our proposed attacks are also applicable to RSA binary modular exponentiation algorithms. We proposed countermeasures and experimentally verified that the leakage was removed.

8. Patents

This section is not mandatory, but may be added if there are patents resulting from the work reported in this manuscript.

Author Contributions: These authors contributed equally to this work.

Funding: This research received no external funding.

Acknowledgments: This work was supported by Institute for Information & communications Technology Promotion (IITP) grant funded by the Korea government (MSIT) (No. 2017-0-00520, Development of SCR-Friendly Symmetric Key Cryptosystem and Its Application Modes).

Conflicts of Interest: The authors declare no conflict of interest.

Appendix A. Montgomery–López–Dahab Ladder Algorithm

Algorithm A1 performs except operations for the most significant bit, $d_{\lambda-1}$. Each iterative operation consists of six patterns, because Steps 7 to 9 (Steps 10 to 11) consist of six finite-field multiplications.

Algorithm A1: Binary scalar multiplication: Montgomery–López–Dahab Ladder

Input : $P = (x, y)$ is a point on elliptic curve, an λ-bit scalar $d = (d_{\lambda-1}, d_{\lambda-2}, \cdots, d_0)_2$
Output : $Q = dP$

1: **if** $d = 0 \ or \ x = 0$ **then**
2:     output $(0, 0)$ and stop
3: **end if**
4: $X_1 \leftarrow x, \ Z_1 \leftarrow 1, \ X_2 \leftarrow x^2 + b, \ Z_2 \leftarrow x^2$
5: **for** $i := \lambda - 2$ down to 0 **do**
6:     $Z_3 \leftarrow (X_1 Z_2 + X_2 Z_1)^2$
7:     **if** $d_i = 1$ **then**
8:        $X_1 \leftarrow xZ_3 + (X_1 Z_2)(X_2 Z_1), \ Z_1 \leftarrow Z_3$
9:        $X_2 \leftarrow X_2^4 + bZ_2^4, \ Z_2 \leftarrow X_2^2 Z_2^2$
10:     **else**
11:        $X_2 \leftarrow xZ_3 + (X_1 Z_2)(X_2 Z_1), \ Z_2 \leftarrow Z_3$
12:        $X_1 \leftarrow X_1^4 + bZ_1^4, \ Z_1 \leftarrow X_1^2 Z_1^2$
13:     **end if**
14: **end for**
15: $A \leftarrow Z_1 Z_2, \ B \leftarrow xZ_2, \ C \leftarrow (xA)^{-1}$
16: $D \leftarrow ((x^2 + y) A + (B + X_2)(xZ_1 + X_1)) C$
17: $x_0 \leftarrow BX_1 C, \ y_0 \leftarrow (x + x_0) + y$
18: **Return** $dP = (x_0, y_0)$

Appendix B. openSSL

The first target points come immediately after the & ()(BN_ULONG)1) operation is performed, as shown in Figure A3. The second is where the register is referred according to the d_i value, and it is located after the key bit identification function is performed. We chose points with a high SOST value as PoIs. **(1)** When we targeted *Case Study 1* and exploited the power-consumption trace, it was impossible to clearly split it into two groups through SPA-VI. Thus, we used the k-means clustering algorithm and recovered secret scalar bits d_i with a 96.25% success rate. **(2)** Figure 4b shows the PoIs chosen from power-consumption subtraces when we targeted *Case Study 2*. They could be clearly divided into two groups through SPA-VI. The success rate was 100%, i.e., we could recover the whole secret scalar bits.

```
1:  int BN_is_bit_set (const BIGNUM *a, int n)
2:  {
3:          int i, j;
4:
5:          bn_check_top(a);
6:          if (n < 0)
7:                  return(0);
8:
9:          i = n / BN_BITS2;
10:         j = n % BN_BITS2;
11:         if (a->top <= i)
12:                 return(0);
13:
14:         return (int)(((a->d[i] >> j) & ((BN_ULONG)1));
15: }
```

Figure A1. Key bit identification function of OpenSSL.

```
1: $d_i$ = BN_is_bit_set (&$d$, i);
2: BN_MUL(&$R_2$, &$R_{d_i}$, &$R_2$);
```

Figure A2. Software implementation (OpenSSL) acquisition range.

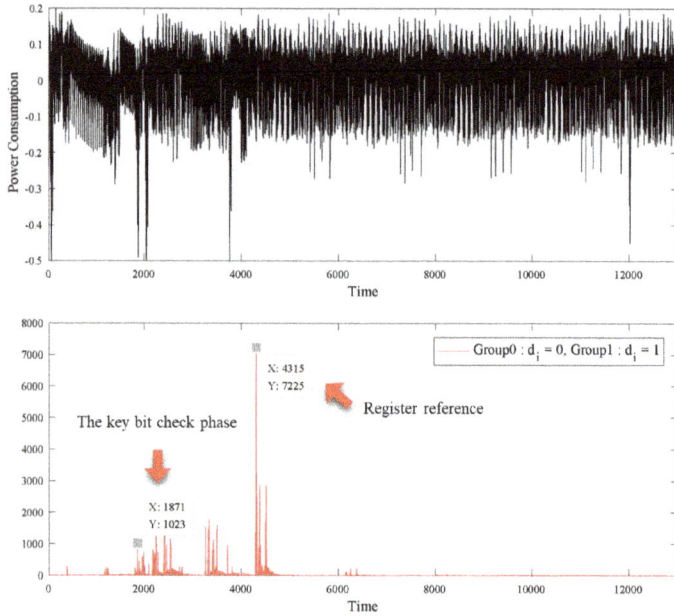

Figure A3. Software implementation (OpenSSL): one of the subtraces (**top**), and the SOST value between two subgroups (**bottom**).

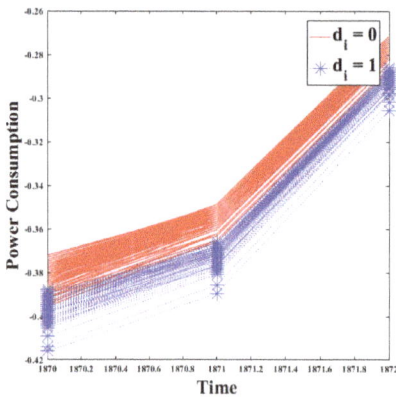

(**a**) *Case Study 1* (Property 2)

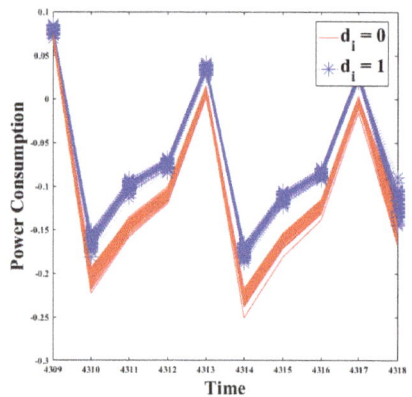

(**b**) *Case Study 2* (Property 4)

Figure A4. Points of Interest in the power-consumption trace (Software Implementation, OpenSSL). (**a**) *Case Study 1* (Property 2); (**b**) *Case Study 2* (Property 4).

References

1. Kocher, P. Timing Attacks on Implementation of Diffie-Hellman, RSA, DSS, and Other Systems. *CRYPTO* **1996**, *9*, 104–113. [CrossRef]
2. Diop, I.; Liardet, P.Y.; Maurine, P. Collision Based Attacks in Practice. *DSD* **2015**, *24*, 367–374. [CrossRef]
3. Clavier, C.; Feix, B.; Gagnerot, G.; Roussellet, M.; Verneuil, V. Horizontal Correlation Analysis on Exponentiation. *ICISC* **2010**, *5*, 46–61. [CrossRef]
4. Diop, I.; Carbone, M.; Ordas, S.; Linge, Y.; Liardet, P.Y.; Maurine, P. Collision for estimating SCA Measurement Quality and Related Applications. *CARDIS* **2015**, *9*, 143–157. [CrossRef]
5. Hanley, N.; Kim, H.S.; Tunstall, M. Exploiting Collisions in Addition Chain-based Exponentiation Algorithms Using a Single Trace. *CT-RSA* **2015**, *23*, 431–448. [CrossRef]
6. Heyszl, J.; Mangard, S.; Heinz, B.; Stumpf, F.; Sigl, G. Localized Electromagnetic Analysis of Cryptographic Implementations. *CT-RSA* **2012**, *15*, 231–244. [CrossRef]
7. Heyszl, J.; Ibing, A.; Mangard, S.; Saints, F.; Sigl, G. Clustering Algorithms for Non-Profiled Single-Execution Attacks on Exponentiations. *CARDIS* **2013**, *6*, 79–93. [CrossRef]
8. Homma, N.; Miyamoto, A.; Aoki, T.; Satoh, A. Comparative Power Analysis of Modular Exponentiation Algorithms. *IEEE* **2010**, *176*, 795–807. [CrossRef]
9. Kocher, P.; Jaffe, J.; Jun, B. Differential Power Analysis. *CRYPTO* **1999**, *6*, 388–397. [CrossRef]
10. Nascimento, E.; Chmielewski, L.; Oswald, D.; Schwabe, P. Attacking embedded ECC implementations through cmov side channels. *SAC* **2016**, *6*, 99–119. [CrossRef]
11. Perin, G.; Imbert, L.; Torres, L.; Maurine, P. Attacking Randomized Exponentiations Using Unsupervised Learning. *COSADE* **2014**, *11*, 144–160. [CrossRef]
12. Perin, G.; Chmielewski, L. A Semi-Parametric Approach for Side-Channel Attacks on Protected RSA Implementations. *CARDIS* **2015**, *3*, 34–53. [CrossRef]
13. Specht, R.; Heyszl, J.; Kleinsteuber, M.; Sigl, G. Improving Non-profiled Attacks on Exponentiations Based on Clustering and Extracting Leakage from Multichannel High-Resolution EM Measurements. *COSADE* **2015**, *1*, 3–19. [CrossRef]
14. Sugawara, T.; Suzuki, D.; Saeki, M. Internal collision attack on RSA under closed EM measurement. *SCIS* **2014**, *1*, 1–8.
15. Sugawara, T.; Suzuki, D.; Saeki, M. Two Operands of Multipliers in Side-Channel Attack. *COSADE* **2015**, *5*, 64–78. [CrossRef]
16. Walter, C.D. Sliding Windows Succumbs to Big Mac Attack. *CHES* **2001**, *24*, 286–299. [CrossRef]
17. Ciet, M.; Joye, M. (Virtually) Free Randomization Techniques for Elliptic Curve Cryptography. *ICISC* **2003**, *32*, 348–359. [CrossRef]
18. Coron, J. Resistance Against Differential Power Analysis for Elliptic Curve Cryptosystems. *CHES* **1999**, *25*, 292–302. [CrossRef]
19. Joye, M.; Yen, S.M. The Montgomery Powering Ladder. *CHES* **2002**, *22*, 291–302. [CrossRef]
20. May, D.; Muller, H.; Smart, N. Random Register Renaming to Foil DPA. *CHES* **2001**, *4*, 28–38. [CrossRef]
21. Montgomery, P. Speeding the Pollard and Elliptic Curve Methods of Factorization. *Math. Comput.* **1987**, *48*, 243–264. [CrossRef]
22. Joye, M. Highly regular right-to-left algorithms for scalar multiplications. *CHES* **2007**, *10*, 135–147. [CrossRef]
23. Sim, B.-Y.; Han, D.-G. Key Bit-Dependent Attack on Protected PKC Using a Single Trace. *ISPEC* **2017**, *168*–185. [CrossRef]
24. Lopez, J.; Dahab, R. Fast Multiplication on Elliptic Curves over $GF(2^m)$ without Precomputation. *CHES* **1999**, *27*, 316–327. [CrossRef]
25. Hankerson, D.; Menezes, A.; Vanstone, S. Elliptic Curve Arithmetic. In *Guide to Elliptic Curve Cryptography*; Springer: New York, NY, USA, 2004; pp. 75–152, ISBN 0-387-95273-X.
26. Gierlichs, B.; Lemke-Rust, K.; Paar, C. Templates vs. Stochastic Methos. A Performance Analysis for Side Channel Cryptanalysis. *CHES* **2006**, *2*, 15–29. [CrossRef]
27. Duda, R.O.; Hart, P.E.; Stork, D.G. *Pattern Classification*, 2nd ed.; Wiley Interscience: Hoboken, NJ, USA, 2001; ISBN 978-0-471-05669-0.
28. Bishop, C. Pattern Recognition and Machine Learning. In *Information Science and Statistics*; Springer: New York, NY, USA, 2007; ISBN 978-1-4939-3843-8.

Appl. Sci. **2018**, *8*, 2168

29. Izumi, M.; Ikegami, J.; Sakiyama, K.; Ohta, K. Improved Countermeasure against Address-bit DPA for ECC Scalar Multiplication. *IEEE* **2010**, 981–984. [CrossRef]
30. Itoh, K.; Izu, T.; Takenaka, M. A Practical Countermeasure against Address-Bit Differential Power Analysis. *CHES* **2003**, *30*, 382–396. [CrossRef]

applied sciences

MDPI

Article

Cache Misses and the Recovery of the Full AES 256 Key

Samira Briongos [1,2,*], Pedro Malagón [1,2,*], Juan-Mariano de Goyeneche [1,2] and Jose M. Moya [1,2]

[1] Integrated Systems Lab, Universidad Politécnica de Madrid, ETSI Telecomunicación, 28040 Madrid, Spain; goyeneche@die.upm.es (J.-M.d.G.); josem@die.upm.es (J.M.M.)

[2] Center for Computational Simulation, Universidad Politécnica de Madrid, Campus de Montegancedo, 28660 Madrid, Spain

* Correspondence: samirabriongos@die.upm.es (S.B.); malagon@die.upm.es (P.M.)

Received: 8 February 2019; Accepted: 27 February 2019; Published: 6 March 2019

Featured Application: This work introduces a new approach to exploit the information gained from cache attacks. Contrary to previous approaches, we focus on cache misses to retrieve the secret AES encryption key more advantageously: we target the OpenSSL AES T-table-based implementation and show that our approach greatly reduces the number of samples and, consequently, the time required to obtain a 128-bit key. Moreover, we demonstrate a practical recovery of a 256-bit AES key.

Abstract: The CPU cache is a hardware element that leaks significant information about the software running on the CPU. Particularly, any application performing sequences of memory access that depend on sensitive information, such as private keys, is susceptible to suffer a cache attack, which would reveal this information. In most cases, side-channel cache attacks do not require any specific permission and just need access to a shared cache. This fact, combined with the spread of cloud computing, where the infrastructure is shared between different customers, has made these attacks quite popular. Traditionally, cache attacks against AES use the information about the victim to access an address. In contrast, we show that using non-access provides much more information and demonstrate that the power of cache attacks has been underestimated during these last years. This novel approach is applicable to existing attacks: *Prime+Probe, Flush+Reload, Flush+Flush* and *Prime+Abort*. In all cases, using cache misses as source of information, we could retrieve the 128-bit AES key with a reduction in the number of samples of between 93% and 98% compared to the traditional approach. Further, this attack was adapted and extended in what we call the encryption-by-decryption cache attack (EBD), to obtain a 256-bit AES key. In the best scenario, our approach obtained the 256 bits of the key of the OpenSSL AES T-table-based implementation using fewer than 10,000 samples, i.e., 135 milliseconds, proving that AES-256 is only about three times more complex to attack than AES-128 via cache attacks. Additionally, the proposed approach was successfully tested in a cross-VM scenario.

Keywords: side-channel cache attacks; cache misses; AES; cloud computing

---

1. Introduction

Cloud computing aims to provide its users with compute resources as they are required, eliminating the need of acquiring and maintaining expensive computing infrastructure. Its low cost, ease of use and on demand access to computing resources have made both government and industry increasingly adopt these technologies. Major companies such as Amazon, Google and Microsoft have become Cloud Service Providers. Cloud providers offer computing capabilities at low prices because of economies of scale: by achieving high utilization of their servers, they can divide costs between more customers. This means that multiple virtual machines (VMs) are co-hosted on a single physical host relying on a virtual machine manager (VMM) to provide logical isolation between them.

However, physical isolation between tenants does not exist, and, as a consequence, shared hardware can create covert channels between different VMs. CPU caches are one of these shared resources, which can be exploited to recover fine grain information from co-resident tenants. Well known attack techniques, such as *Evict+Time*, *Prime+Probe* [1], *Flush+Reload* [2], and *Flush+Flush* [3], or the recently introduced *Prime+Abort* [4], allow inferring instructions or data accessed by a victim program. Thus, when this memory access depends on sensitive information, this information is leaked.

The most impressive demonstrations of the ability of such attacks to extract private information from their victims target both symmetric [1,5,6] and asymmetric [7–9] cryptographic implementations. Intel CPU caches have traditionally been victims of cache attacks, due to their inclusive architecture and replacement policies. However, several researchers [10–12] have also proven that AMD and ARM processors, which have different architectures and replacement policies, are also vulnerable to cache attacks.

Cache attacks require the victim and the attacker to share the cache. In cloud environments this means that, prior to the attack, the attacker must achieve co-residency with the victim. As first shown by Ristenpart et al. [13], co-residency is achievable and detectable in well-known cloud platforms (Amazon EC2). After this work was published, cloud providers obfuscated the techniques they employed. However, recent works [9,14] have shown that it is still possible to determine if two VMs are allocated in the same physical machine. Once a VM is allocated in the same host as a target victim, its owner is ready to gain information about what the neighbors are doing.

Researchers have also become concerned that these attacks represent a serious threat. For this reason, many approaches have been proposed that try to avoid the leakage [15–17] or to detect attacks [18–21]. To the best of our knowledge, no hardware countermeasure is implemented in real cloud environments and no hardware manufacturer has changed the cache architecture. We believe that the main reason for this absence of countermeasures is the performance penalty they introduce. As a result, an attacker wishing to exploit side-channel cache attacks will only have to take into the account the countermeasures based on detection that can be implemented by users.

In this work, we consider the T-table-based implementation of AES. This implementation is known to be vulnerable to cache attacks. However, it is commonly used for comparison, and to demonstrate different attack techniques. In this work, it serves to our purpose of showing the accuracy of the information gained from the non-access to memory and to quantify the improvement that this approach represents compared to the traditionally used approach based on access.

We could improve the results of previously published side channel attacks, decreasing the number of samples required to retrieve the whole key by multiple orders of magnitude. Every known technique (*Flush+Reload*, *Flush+Flush Prime+Probe* and *Prime+Abort*) can benefit from this approach. Moreover, the presented approach is less prone to consider false positives memory accesses. Regarding non-access, a false positive is considered if the data have been loaded into the cache and removed from the cache in a short period of time, which is unlikely to happen. Whereas a false positive in the traditional approach is considered whenever the data are loaded into the cache in any other round of the AES encryption, rather than the target round, or whenever these data are predicted to be used, thus are speculatively loaded into the cache, being both frequent options.

In summary, the main contributions of this work are the following

- We present a non-access attack, a novel approach to exploit information gained from the cache misses.
- We show that our approach improved the performance of previous cache attacks and demonstrate its effectiveness for *Flush+Reload*, *Flush+Flush Prime+Probe* and *Prime+Abort*. We could reduce the amount of encryptions required to derive the key between 93% and 98%. That is, if an access

attack requires 100,000 samples, our approach requires fewer than 3000, performed in a similar and real experimental setup.

- We extend the non-access attack to gain information from more rounds of the algorithm, introducing EBD, a practical attack implementation that provides the full 256-bit AES encryption key using cache side channel attacks.
- We show that the complexity of the attack does not depend on the size of the key but on the number of rounds of each encryption. Passing from 10 rounds (AES-128) to 14 rounds (AES-256) increases the complexity by a factor of around 3.

The remainder of this paper is organized as follows. In Section 2, we provide information on the relevant concepts to understand the attacks. Section 3 describes AES implementations and the attacks we performed to extract the keys. In Section 4, we provide a discussion of our results. Finally, in Section 5, we draw some conclusions.

2. Background and Related Work

To achieve a better understanding of side channel cache-attacks, we give a basic introduction to some required architectural concepts followed by a description of existing cache attacks and their principles.

2.1. Architectural Concepts

We explain the fundamental ideas of three features that have been key to existing cache attacks: CPU cache memories, shared memory and transactional memory.

2.1.1. CPU Caches

CPU caches are small and fast memories located between the CPU and main memory, specially designed to hide main memory access latencies. As the cost of memory is related to the speed, the cache memory is structured in a hierarchy of typically three levels. Level 1 is the smallest, fastest cache memory and similar to level 2 is usually core private. Level 3, the biggest and slowest cache memory, is commonly shared among processing cores. They hold a copy of recently used data, which will likely be requested by the processor in a short period of time. If the data requested by the processor are not present in any level of the cache, they will be loaded from main memory and stored in the cache. If a cache level is completely filled at that point, some of the data currently contained in it will be replaced (evicted) to store the loaded data instead. The selection of the evicted data depends on the processor architecture. Processors, including AMD, Intel and some ARM high-performance processors, typically use a policy related to least recently used (LRU). However, the replacement policy of most ARM processors is based in random replacement or round-robin replacement.

In Intel processors, such as the one used in our experiments, caches are inclusive memories. This means that high level caches, such as level 3 (L3) cache, have a copy of the data of lower level caches. In addition, what is more important and relevant to most attacks, L3 cache is shared among all cores. Consequently, a process being executed in a core, related to L3 cache data, may produce side effects in other core processes. Other processors use exclusive caches, where a memory block is present in only one level of the cache hierarchy.

Caches are usually set associative: they are organized as S sets of W lines (also called ways) each holding B bytes. It is common to name them as *W-way set associative* cache. Given a memory address, the less significant log_2B bits locate the byte on the line, the previous log_2S bits do the same for the set and the remaining high-order bits are used as a tag for each line. The tag will be used to discern whether a line is already loaded or not in the cache. In modern processors, the last level cache (LLC) is divided into slices. Thus, it is necessary to know the slice, and the set in which a datum will be placed. The slice is computed as a hash function of the remaining bits of the address. If the number of cores is a power of two, the hash will be a function of the tag bits. Otherwise, it will use all the bits of the address [22,23] Note that, as main memory is much larger than CPU caches, multiple memory lines map to the same cache set.

2.1.2. Shared Memory

Programs are loaded from hard disk drives to RAM to be executed, using RAM for both instructions and data, because the access to RAM is much faster than to hard disk. RAM memory is a limited resource, so when there is no free memory available, some data and instructions are moved to virtual memory, which is a copy of RAM stored in the hard disk. Using virtual memory in an application severely affects its performance. It is natural that operating systems employ mechanisms such as memory sharing to reduce memory utilization. Whenever there are two different processes using the same library, instead of loading it twice into RAM memory, sharing memory capabilities allow mapping the same physical memory page into the address spaces of each process.

Deduplication is a concrete method of shared memory, which was originally introduced to improve the memory utilization of VMMs and was later applied to non-virtualized environments. The hypervisor or the operating system scans the physical memory and recognizes processes that place the same data in memory; that is, pages with identical content. When several pages happen to include the same content, all the mappings to these identical pages are redirected to one of them, and the other pages are released. However, if any change is performed by any process in the merged pages, memory is duplicated again.

The Linux memory deduplication feature implementation is called KSM (Kernel Same-page Merging) and appeared for the first time in Linux kernel version 2.6.32. KSM is used as a page sharing technique by the Kernel-based Virtual Machine (KVM), which we used as hypervisor within our experiments. KSM scans the user memory for potential pages to be shared, scanning only potential candidates instead of the whole memory continuously [24].

The deduplication optimization saves memory allowing more virtual machines to run on the host machine. To exemplify this statement, we refer to onl [25], who stated that it is possible to run over 50 Windows XP VMx with 1 GB of RAM each on a machine with just 16 GB of RAM. In terms of performance, deduplication is an attractive feature for cloud providers. However, after several demonstrations of side-channel attacks exploiting page sharing, they are advised to disable this feature and no cloud provider is ignoring this advice.

2.1.3. Transactional Memory

When multiple threads are running in parallel and try to access a shared resource, a synchronization mechanism is required to avoid conflicts. Transactional memory is an attempt to simplify concurrent programming, avoiding common problems with mutual exclusion and improving the performance of other locking techniques. Transactional memory enables optimistic execution of the transactional code regions specified by the programmer. The processor executes the specified sections assuming that it is going to be possible to complete them without any interference. The programmer is no longer responsible for identifying the locks and the order in which they are acquired; he only needs to identify the regions that are going to be defined as part of the "transaction".

During the execution of a transaction, the variables and the results of the operations are only visible for that thread; that is, any update performed in these regions is not visible to other threads. If the execution ends successfully, the processor commits all the changes as if they had occurred instantaneously, making them visible to any process. If, on the other hand, there is a conflict with another process, the transaction is unsuccessful and the processor aborts its execution. Consequently, the processor requires a mechanism to undo all the updates, discard all the changes and restore the architectural state to pretend that the execution never happened.

Intel Transactional Synchronization Extensions (TSX) are the Intel's implementation of hardware transactional memory. TSX provides two software interfaces: Hardware Lock Elision (HLE) and Restricted Transactional Memory (RTM).

2.2. Cache Attacks

Cache memories create covert channels that can be exploited to extract sensitive information. When a process tries to access some data, if it is already loaded into the cache (namely, a cache hit), the time required to recover these data is significantly lower than the access time in the case the data have to be retrieved from main memory (cache miss). Therefore, if the execution time of a cryptographic process depends on the previous presence (or not) of the accessed data on the cache memory, this time information can be exploited to gain private information (such as secret keys) from cryptographic processes. Traditionally, cache attacks have been classified as:

- *Time driven attacks:* The information is learned by observing the timing profile for multiple executions of a target cipher.
- *Trace driven attacks:* The information is gained by monitoring the cache directly, considering that the attacker has access to the cache profile when running the target process.
- *Access driven attacks:* The information is retrieved from the sets of the cache accessed during the execution of the target process

The cache memory was first mentioned as a covert channel to extract sensitive information by Hu [26] in 1992. In 1996, Kocher [27] introduced the first theoretical attacks, as did Kelsey [28] in 1998 describing the possibility of performing attacks based on cache hit ratios. In 2002, Page [29] studied a theoretical example of cache attacks for DES, which was also used by Tsunoo [30] to study timing side-channels created due to table lookups. In 2004, Bernstein [5] proposed the first time-driven attack after observing non-constant times when executing cryptographic algorithms. Although the attack he presented was not practical, his correlation attack has been investigated extensively and even retried between VMs recently [31]. Soon after, another attack was proposed by Percival [32], who suggested that an attacker could determine cache ways occupied by other processes, measuring access times to all ways of a cache set. He found that these times are correlated with the number of occupied ways.

In 2006, Osvik et al. [1] proposed two techniques, which have been widely used since then, named *Evict+Time* and *Prime+Probe*. These techniques are intended to allow an attacker to determine the cache sets accessed by a victim process. A significantly more powerful attack that exploits shared memory and the completely fair scheduler (CFS) was proposed by Gullasch [33]. The same principles of the attack were later exploited by Yarom and Falkner [2], who named the attack *Flush+Reload*. The target of this attack was the L3 cache, as it is inclusive and shared among cores. From those, the *Flush+Reload* and the *Prime+Probe* attacks (and their variants) standout over the rest due to their higher resolution.

2.2.1. Evict+Time

This technique consists of three steps: First, an encryption is triggered and its execution time measured. Second, an attacker evicts some line. Third, the encryption time is measured again. By comparing the second time with the first measure, an attacker can decide whether the cache line was accessed, as higher times will be related to the use of the mentioned line.

2.2.2. Flush+Reload

This technique relies on the existence of shared memory: thus, when an attacker flushes the desired lines from the cache, he can be sure that, if the victim process needs to retrieve the flushed line, it will have to load it from main memory. *Flush+Reload* also works in three stages: First, the desired lines from the cache are flushed (reverse engineering may be required to determine the addresses that can leak information). Second, the victim runs its process or a fragment of it. Third, the flushed line is accessed by the attacker, measuring the time required to do it. Depending on the reload time, the attacker decides whether the line was accessed.

This attack uses the `clflush` instruction to remove the target lines from the cache in the initialization stage. It is easy to implement and quite resistant to micro-architectural noise, thus it has become popular. However, its main drawback is that it requires memory deduplication to be enabled. Deduplication is an optimization technique designed to improve memory utilization by merging duplicate memory pages. Consequently, it can only recover information coming from statically allocated data. Shared memory also implies that the attacker and the victim have to be using the same cryptographic library. To eliminate the need of knowing the version of the algorithm attacked, Gruss et al. presented the cache template attack [34], where they enforced the existence of shared memory between the attacking and the attacked processes.

The attack was first introduced by Gullasch et al. [33], and later extended to target the LLC to retrieve cryptographic keys, TLS protocol session messages or keyboard keystrokes across VMs [2,34,35]. It has also demonstrated its power against AES T-table based implementations [36], RSA implementations [2], or ECDSA [37], among others. It can also detect cryptographic libraries [38]. Further, Zhang et al. [39] showed that it is applicable in several commercial PaaS clouds, where it is possible to achieve co-residency with a victim [13].

Relying on the `clflush` instruction and with the same requirements as *Flush+Reload*, Gruss et al. [3] proposed the *Flush+Flush* attack. It was intended to be stealthy and bypass attack monitoring systems [19,20]. The main difference with *Flush+Reload* is that this variant recovers the information by measuring the execution time of the `clflush` instruction instead of the reload time, thus avoiding direct cache accesses. This was the key fact to avoid detection. However, recent works have demonstrated that it is detectable [21,40].

2.2.3. Prime+Probe

Shared memory is not always available through deduplication in virtual environments as most cloud providers turned off this feature after several attack demonstrations. However, *Prime+Probe* [9,41] still works, and, by targeting the L3 cache, an attacker can still extract sensitive information. Since *Prime+Probe* is agnostic to special OS features in the system, it can be applied in virtually every system.

Prime+Probe consist of three steps: First, the attacker fills a cache set with known data. Second, the attacker triggers (or waits for) the victim to perform an encryption. Third, after the encryption has finished, the attacker tries to access the known data place during the first step and measures the time it takes to determine if the previously loaded data have been evicted by the victim process. If so, the attacker discovers which lines the victim process has used.

This attack was first proposed for the L1 data cache by [1] and later expanded to the L1 instruction cache [42]. These approaches require both victim and attacker to share the same core, which diminishes practicality. However, it has been recently shown to be applicable to LLC. Researchers have bypassed several difficulties to target the LLC, such as retrieving its complex address mapping [22,23,43], and recovered cryptographic keys or keyboard typed keystrokes [11,41,44]. Furthermore, the *Prime+Probe* attack was used to retrieve a RSA key in the Amazon EC2 cloud [45].

These attacks highly rely on precise timers to retrieve the desired information about the victim memory access. If a defense system tries to either restrict access to the timers or to generate noise that could hide timing information, the attack is less likely to succeed. Once again, attackers have been able to overcome this difficulty. The *Prime+Abort* attack [4] exploits Intel's implementation of Hardware Transactional Memory (TSX) so it does not require timers to retrieve the information. It first starts a transaction to prime the targeted set, waits and finally it may or may not receive and abort depending on whether the victim has or has not accessed this set. That is, no need for timers.

To summarize, the presented attacks target the cache, selecting one memory location that is expected to be accessed by the victim process. They consist of three stages: initialization (the attacker prepares the cache somehow), waiting (the attacker waits while the victim executes) and recovering (the attacker checks the state of the cache to retrieve information about the victim). They differ in the implementation, requirements and achievable resolution.

3. Attacks on AES

In this section, we describe the fundamentals of AES encryption and decryption algorithms. We explain the insides of the attacked T-table-based OpenSSL AES implementation. We present the non-access cache attack against AES-128, which outperforms previously published cache attacks. Afterwards, we explain how the approach followed in this non-access attack can be extended to perform a practical attack on AES-256. We name it encryption-by-decryption cache attack, as we use the information from the encryption to obtain a decryption round key, which can be transformed into an encryption round key. This way, we are able to obtain information from two different and consecutive encryption rounds.

3.1. AES Fundamentals

The AES algorithm is explained in [46]. It is a symmetric block cipher that operates with data in blocks of 16 bytes. Both encryption and decryption behave similarly. They repeatedly apply a round transformation on the state, denoted S. The number of iteration rounds, N_r, depends on the size of the key: 10 rounds for 128-bits, 12 rounds for 192-bits and 14 rounds for 256-bits. The encryption process is depicted in Figure 1. In each round, the algorithm uses a different round key K^r, which is obtained from the algorithm key using a known and fixed scheme. Once a round key is known, there is a straight forward algorithm to recover the same amount of bits of the encryption key. Consequently, to obtain a 256-bit AES key, we need to obtain at least two consecutive round keys.

AES arranges both round key and the data in form of a 4×4 matrix of byte elements. If we follow the terms used in [46], we can denote $S^r, 0 \leq r \leq N_r$ the input state to round r. Each of its byte elements is then denoted as $S^r_{i,j}, 0 \leq i < 4, 0 \leq j < 4$. i indicates the row of the state and j the columns. We use only one subindex j to refer to a column of the state S_j. Considering a plaintext block $p_0 p_1 p_2 \ldots p_{15}$, the elements of the initial state S^0 of the algorithm are $S^0_{i,j} = p_{i+4j}$.

The normal round transformation in the encryption process consists of four steps denoted: SubBytes, ShiftRows, MixColumns and AddRoundKey, being SubBytes the only non-linear transformation. SubBytes is an S-box S_{RD} applied to the elements of the state. ShiftRows is an element transposition that rotates the rows of the state i positions to the left. MixColumns operates on the state column by column. The operation is a matrix multiplication of a 4×4 known matrix and each of four the original columns. AddRoundKey is the addition (bitwise XOR in GF(2)) of the elements of the state with the corresponding element of the round key.

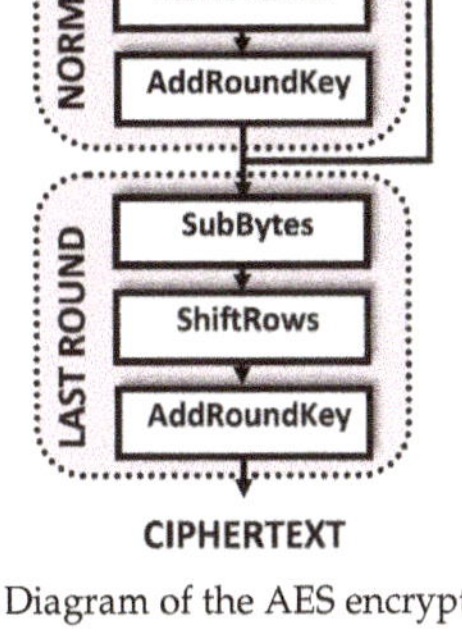

Figure 1. Diagram of the AES encryption process.

All these steps have an inverse step, which are used in the decryption process: InvSubBytes (using S-box S_{RD}^{-1}), InvShiftRows (rotating to the right), InvMixColumns (multiplying by the inverse matrix) and AddRoundKey. The normal round transformation is applied in all but the first and last rounds. The first round applies only the AddRoundKey step. The last round is a normal round without the MixColumns step.

The decryption algorithm can be implemented by applying the inverses of the steps in a reverse order, being InvSubBytes the last step of each round excepting the last one. It is preferable to perform the non-linear step first in typical implementations. In [46], the authors presented the Equivalent Decryption Algorithm, with a sequence of steps equal to the encryption, by switching the order of application of InvSubBytes and InvShiftRows (the order is indifferent) and InvMixColumns and AddRoundKey. The round key of all but the first and last rounds needs to be adapted for this purpose, by applying the InvMixColumns step to it.

The key fact to understand the attack against AES-256 is that the state S^r can be reached in the r round of the plaintext encryption process or in the $N_r + 1 - r$ round of the ciphertext decryption process.

3.2. AES T-Table Implementation

The steps of the round transformation include a non-linear S-box substitution and a matrix multiplication. To reduce the number of operations performed during each round and to improve performance, AES software implementations without special assembly instructions use tables (T-tables) with precalculated values for each of the possible 256 values of an input element. In this work, we evaluate the T-table-based implementation available in OpenSSL version 1.0.1f or newer versions when compiled with the no-asm flag.

In the aforementioned implementation, the state is represented by four 32-bit variables starting with s or t, one for each column of the state. The name of the variable starts with s for odd rounds and with t for even rounds, considering the first round to be 0. The Most Significant Byte (MSB) of the variable represents the element in Row 0, while the Least Significant Byte (LSB) represents the element in Row 3. Encryption and decryption algorithms use four tables, each containing the result of S-box substitution and part of the matrix multiplication. The tables start with Te for encryption and with Td for decryption. Generically, we denote each of the four tables T_i (Te_i and Td_i), where i is the column of the multiplication matrix considered in the table. Table T_i contains the contribution of the element in row i to each row of the resulting column state. That is, each of the 256 positions of the table (2^8) contains a 32-bit value, where each byte is the contribution to each row, aligned to perform the addition of the different contributions. The size of each table is 1024 bytes. The column state is calculated by adding (bitwise XOR in GF(2)) the contributions from the different rows and the round key. Therefore, the four tables are accessed for each output column state. An appropriate selection of the input variable s_j for the different rows is used to perform the ShiftRows step; for the destination variable t_j, variable s_{i+j} is used for row i (index of table T_i). In this example, and for the entire work, the addition in subindexes i and j is done modulus 4. For example, in Round 1, we can obtain t_0 as follows:

$$t_0 = Te_0[s_0 \gg 24] \oplus Te_1[(s_1 \gg 16)\&0xff] \oplus Te_2[(s_2 \gg 8)\&0xff] \oplus Te_3[s_3\&0xff] \oplus rk[4];$$

The last round does not include the matrix multiplication. A table for S_{RD}^{-1} (only InvSubBytes step) is required for decryption algorithm (table Td_4). As there are coefficients with value 1 in the multiplication matrix of step MixColumns, the table is not needed for the encryption algorithm. It is implicitly available in the Te tables. In this case, $Te_{(i+2)mod4}$ is used for row i, because it has coefficient 1 for that row.

3.3. Attacking AES

As it can be inferred from the previous description of the T-table implementation, the accesses to each Te_i depend on the key. Thus, cache hits and misses depend on the key. In Section 3.3.1, we explain how to retrieve the 128 bits key using information about cache misses and the ciphertext.

We show that, by using cache misses, we are not only able to recover an AES 128 bit key using significantly fewer samples than previous approaches, but we can also recover the 256 bits of an AES key. To this end, we need to recover information about two consecutive rounds. As we explain below, we perform this attack in two steps. The first one targets the last round and the second one the penultimate round. A direct cache attack to the penultimate round key is not feasible, so we use the equivalences between encryption and decryption to transform our data and to be able to use cache misses to retrieve the key.

When describing the attacks, we use a nomenclature that is consistent with the terms used when describing AES. The iteration variables, subindexes and superindexes follow this rules:

i is used to refer to the rows of the state or round key. As a subindex, it indicates the row of the element. As an iteration value, it is used for accessing elements related to a row (the T-tables).

j is used to iterate the columns of the state or round key. As a subindex, it indicates the column of an element.

t is the iteration variable for traversing the different encryptions of the experiment.

l is the iteration variable for traversing the values of a Cache line.

r is used as superindex to indicate the round of a variable (state, round key).

The elements used in the algorithm are represented by capital letters.

S^r represents the state of an AES encryption algorithm before round r. The 16 bytes of the state can be accessed using i and j subindexes. The state is different for each encryption performed, and the array $\overline{S^r}$ represents the whole set.

D^r is the equivalent to S for the decryption algorithm.

K^r represents the encryption round key for round r. Each of the 16 bytes can be accessed using i and j subindexes.

Kd^r represents the decryption round key for round r. It can be calculated from the corresponding encryption key K^{N_r+1-r} by applying the InvMixColumns step.

X_i is information on access (1) or non-access (0) to a subset of table Te_i. The subset contains 16 values (64 bytes), which is the cache line length. $\overline{X_i}$ represents the array containing the access information for each encryption performed in the experiment.

CK^r is a set of arrays with 256 positions (the possible values), one for each round key byte. Each position represents the amount of discarding votes received by the candidate for that concrete key byte. The position with the minimum value after the analysis (we call the function argmin) is the correct key byte value. Its elements can be accessed using i and j subindexes.

CKd^r is the equivalent to CK^r for a decryption round key.

3.3.1. Attack on AES-128

We assume that the attacker shares the cache with the victim, which means he can monitor accesses to the cache with line granularity. We also assume that the attacker has access to the ciphertext. The attacker needs to recover information from each of the four T-tables. To do so, he can monitor one line of one table, one line of each T-table or even try to monitor the 16 lines of each of the four tables at the same time (64 lines). The number of samples required to retrieve all the bits of the key will vary in each case as will the effect of the attack observed by the victim [18]. The more lines monitored at a time, the more noticeable the attack will be. While a slight increase in the encryption time can be assumed to have been caused by other programs or virtual machines running on the same CPU, a higher increase in the encryption time would be quite suspicious.

For each of the evaluated techniques, i.e., *Prime+Probe*, *Prime+Abort*, *Flush+Reload* and *Flush+Flush*, we use a similar process to retrieve the desired information. The process followed during the attack involves three steps and is the same in all cases:

1. **Setup:** Prior to the attack, the attacker has to retrieve the necessary information about where the T-tables are located in memory; that is, their virtual addresses or the cache set in which they are

going to be loaded. This way he ensures that the cache is in a known state, ready to perform the attack.

2. **Measurement collection:** In this step, the attacker monitors the desired tables, applying each technique between the requested encryptions and stores the gathered information together with the ciphertext. As explained in Section 2.2, this step can be subdivided into three stages: initialization, waiting and recovering.

3. **Information processing:** Finally, the attacker uses the information recovered in the previous steps and the information about the T-tables (the contents of the monitored line) to retrieve the secret key.

These main steps are equivalent to previous proposals [36,44]. Other proposals are also applicable to our non-access attack, such as the one suggested by Gülmezoglu et al. [47], which aims to detect the bounds of the encryptions instead of triggering them. However, this approach also depends on the accuracy of the detection, which introduces a new variable to be considered when comparing the effectiveness of the attack approaches. In contrast, the aforementioned setup can be generalized for all the considered techniques, allowing a fair comparison.

In the following, we describe each of the steps for each of the considered algorithms and their particularities in our experimental setup. We consider two scenarios. In the first one, the attack is performed from a spy process running in the same OS as the victim. In the second one, the attacker runs in a VM and the victim runs in a different VM and they are both allocated in the same host. Table 1 includes the details of the machine in which the experiments are performed. The steps are equal in the two considered scenarios.

Table 1. Experimental platform details.

Processor	Cores	Frequency	OS	LLC Slices	LLC Size	LLC Ways	VMM	Guest OS
Intel core i5-7600K	4	3.8 GHz	CentOS 7.6	8	6 MB	12 ways	KVM	CentOS 7.6 minimal

Setup

The tasks required to perform *Flush+Reload* and *Flush+Flush* are different from those required for *Prime+Probe* and *Prime+Abort*. Since *Flush+Something* attacks rely on shared memory, an attacker wishing to exploit this feature needs to do some reverse engineering on the library used by the victim to retrieve the target virtual addresses. Note that the offset between the addresses of the beginning of the library and the target symbol (the table) is constant. Thus, once this offset is known, the attacker can easily get the target address by adding this offset to the virtual address where the library is loaded.

In contrast, *Prime+Something* attacks require the physical address of the target; more accurately, it is necessary to know the set and the slice in which the data will be loaded. This information can be extracted from the physical address. Part of the physical address can be directly inferred from the virtual address. Indeed, the bits of the address that points to the elements within a cache page are part of their physical address. For example, if the size of a cache page is 4 KB, the 12 lowest significant bits of the address will keep the same when translating it from virtual to physical. However, both virtual and physical addresses of the data that the victim is processing are unknown to the attacker. To overcome this difficulty, an attacker needs to create its own eviction set (a group of W elements that map exactly to one set) and profile the whole cache looking for accesses being carried out by the victim.

Instead of recovering the complex address function of our processor, we create the eviction sets dynamically using the technique summarized in the Algorithm 1 of the work of Liu et al. [41]. We have also enabled *hugepages* of 2 MB in our system to work with 21 known bits. Since the number of cores of our processor is a power of two, we only need to obtain the mapping of one of the sets for each slice (the function that determines the slice in which the data will be placed only uses the *tag* bits). The remaining sets that will be used during the profiling phase and the attack itself can be constructed using this retrieved mapping.

Measurement collection

The measurement collection process is somehow similar for all the considered techniques. Algorithm 1 summarizes the general method employed to acquire the necessary information about the victim accesses to the tables. This information will be later used to obtain the full key.

Algorithm 1 Generic attack algorithm for cache attacks against T-Table AES implementations.

Input: Address(Te$_0$),Address(Te$_1$),Address(Te$_2$),Address(Te$_3$) ▷ Addresses of the T-Tables
Output: $\overline{X}_0, \overline{X}_1, \overline{X}_2, \overline{X}_3, \overline{S}^{N_r+1}$ ▷ Information about the accesses and ciphertext
 1: **for** $t = 0$ to *number_of_encryptions* **do**
 2:     **for** $m = 0$ to 4 **do** ▷ INITIALIZATION
 3:         **Evict from cache (Te$_m$);** ▷ The attacker prepares the cache for the attack
 4:     **end for**
 5:      ▷ WAITING
 6:     $\overline{S}^{N_r+1}[t]$=**encrypt**(random plaintext); ▷ The victim performs one encryption
 7:     **for** $m = 0$ to 4 **do**
 8:         **Infer victim accesses to((Te$_m$)** ▷ RECOVERING
 9:         **if** hasAccessed((Te$_m$) **then** ▷ The attacker Reloads, Flushes, Probes the target or gets the
      Abort
10:            $\overline{X}_m[t] = 1$; ▷ The attacker decides if the victim has used the data.
11:         **else**
12:            $\overline{X}_m[t] = 0$;
13:         **end if**
14:     **end for**
15: **end for**
16: **return** $\overline{X}_0, \overline{X}_1, \overline{X}_2, \overline{X}_3, \overline{S}^{N_r+1}$;

In the initialization, the attacker has to evict the data of one line of each Te$_i$ table from the cache. That is, the attacker ensures he knows the state of the cache before the victim uses it. Since each cache line holds 64 bytes, and each entry of the table has 4 bytes, it holds 16 of the 256 possible values of the table. Then, in the waiting stage, he triggers an encryption. Once it has finished, in the recovering stage, the attacker checks the state of the cache. If the cache state has changed, this means that it is likely that the victim had used the data of that T-Table. The different approaches considered in this work differ in the way they evict the data from the cache during the initialization and in the way they recover the information about utilization of the evicted table line.

In the initialization stage of *Flush+Something*, the target lines are evicted from the cache using the *clflush* instruction available in Intel processors. In *Prime+Something*, the attacker evicts the target lines by accessing the data of the created eviction sets that map to the same region as the target lines. Intel implements a pseudo LRU eviction policy. As a result, accessing W elements that map to a set implies that any older data in the cache will be replaced with the attacker's data. In *Prime+Abort*, this operation is performed inside a transactional region defined using the Intel TSX extension.

In the recovering stage of *Flush+Reload*, the attacker accesses each of the evicted lines measuring the access times. If the time is below a threshold, it means the line was in the cache, so the algorithm must have accessed it in any of the rounds. *Flush+Flush* decides if the data have been used by measuring the time it takes to flush the line again. The flushing time depends on the presence of the line on the cache; it takes longer to flush data that are located in the cache. The *Prime+Probe* recovers information on access by accessing the data from the eviction set and measuring the access time. As recommended in previous works [41], we access the data within the eviction set in reverse order in the Probe step. That is, the eviction set is read following a zigzag pattern. If the data of the table are used by the encryption process, part of the attacker's data will be evicted; thus, the time it takes to "Probe" the set will be higher. In the *Prime+Abort*, as the eviction set was accessed inside a transaction, any eviction of the attacker's data from the cache causes an abort. The attacker defines a handler for this abort that evaluates its cause. There are different abort causes which are identified with an abort code [48]. This abort code is loaded into the *eax* register and it is read in the handler function to check if the abort was due to an eviction.

Figure 2 shows the distribution of the measured times during the Reload step (Figure 2a) and during the Probe step (Figure 2b). Note that it is easy to establish a threshold to distinguish between accesses and not-accesses to the monitored Te table.

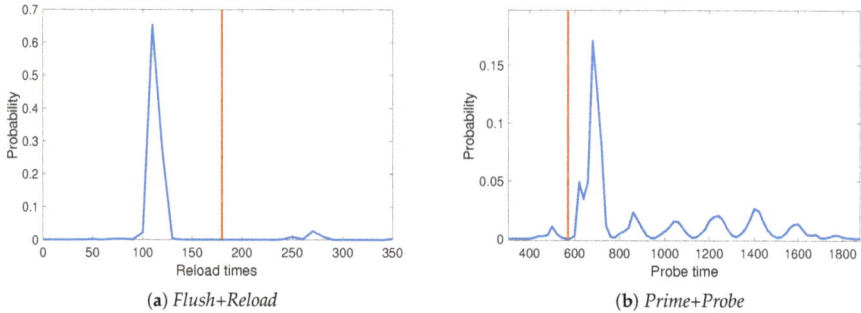

(a) *Flush+Reload*   (b) *Prime+Probe*

Figure 2. Timing information obtained when attacking AES in the Reload and Probe steps in our test machine. The threshold is shown in red.

Information Processing

This final step is common for all the attacks. In this stage, our approach differs from previous attack strategies. Traditional approaches for the considered attacks take into account if the data stored in the monitored cache line has been used by the victim. In contrast, we consider the opposite situation, i.e., the victim has not used that data. Our approach is based on the following observations:

- Whenever the retrieved information shows that the victim has used the considered data, they could have been used during the last round or during any of the remaining ones. Thus, whenever the data are used in a different round, this approach introduces "noise". On the other hand, if a monitored cache line has not been accessed during the encryption, neither has it been accessed in the last round nor in any other round.
- Speculative execution of pieces of code, or prefetching of the data that the processor believes are going to be used in the near future, may involve that the attacker sees in the cache data that have not been used by the victim (false positive). Nonetheless, to obtain a false positive in the non-access approach, the data have to be used during the last round of the encryption and then flushed away by other process during the short period of time after the end of the encryption and before the recovering step.
- Each non-access sample reduces the key search space discarding up to n^4 key values (n equals to 16 table entries stored in the cache line, which are used to obtain 4 output bytes). Thus, we obtain information from each of the samples. In contrast, when considering accesses, only when one candidate value towers over the rest, the retrieved information is useful to retrieve the secret key.
- The information obtained this way is more likely to be applicable to all the rounds of an encryption, or at least to the penultimate round, as it refers to the whole encryption, so we can use this information to retrieve longer keys (192 or 256 bits).

We have seen in Section 3.2 that each table Te is accessed four times during each round. Therefore, the probability of not accessing a concrete cache line within an encryption is provided by Equation (1), where n represents the number of table entries a cache line can hold, and N_r the number of rounds of the algorithm. Particularly, in our case, where the size of a cache line is 64 bytes, n is 16 (each entry of

Te has 32-bits) and the number of rounds (N_r) for AES-128 is 10. Consequently, the probability of not accessing the cache line is 7.5%. This means that one of each fourteen encryptions performed gives us useful information.

$$Pr[no\ access\ Te_i] = \left(1 - \frac{n}{256}\right)^{N_r*4} \tag{1}$$

As stated before, we focus on the last round of the encryption process. This round includes SubBytes, ShiftRows and AddRoundKey operations. Referred to the last round input state and the T-tables, we can express an output element by $S_{i,j}^{N_r+1} = K_{i,j}^{N_r} \oplus Te_{i+2}\left[s_{i,j+i}^{N_r}\right]$. When a cache line has not been used during the encryption, we can discard all the key values which, given an output byte, would have had to access any of the Te-table entries hold in the mentioned cache line. For example, given a ciphertext byte "0F" and a cache line holding "01", "AF", "B4", and "29", if this line remains unused after performing the encryption, we discard the key byte values 0F⊕01, 0F⊕AF, 0F⊕B4 and 0F⊕29. As shown in Section 3.2, the same table is used to obtain all the elements of the row of the output state. This means each Te_i table also discards the $K_{i,j}$ values with the same i value independently of the j index.

A general description of the key recovery algorithm for the last round key is provided in Algorithm 2. Key bytes of the last round, $K_{i,j}^{N_r}$, are recovered from the output state $\overline{S}^{N_r+1}$ (ciphertext) and the information about the accesses to each table Te_i, which is stored in $\overline{X}_i$ and recovered using Algorithm 1. The output state ($\overline{S}^{N_r+1}$) is obtained by arranging the known ciphertext vector as a matrix, as indicated in Section 3.1. The algorithm first initializes an array of 256 elements for each of the 16 bytes of the key ($K_{i,j}$). This array ($CK_{i,j}^{N_r}$) will contain the discarding votes for each key byte candidate. The candidate with less negative votes is the value of the secret key.

Algorithm 2 Recovery algorithm for key byte $K_{i,j}^{N_r}$.

Input: $\overline{X}_0, \overline{X}_1, \overline{X}_2, \overline{X}_3, \overline{S}^{N_r+1}$ ▷ Information about the accesses and ciphertext collected in the previous

   step.
Output: $\mathbf{K}_{i,j}^{N_r}$ ▷ Secret key values
  1: **for** $l = 0$ to 256 **do**
  2:　　$CK_{i,j}^{N_r}[l] = 0;$ ▷ Initialization of the candidates array
  3: **end for**
  4: **for** $t = 0$ to *number_of_encryptions* **do**
  5:　　**if** $\overline{X}_{i+2}[t] == 0$ **then**
  6:　　　　**for** $l = 0$ to n **do** ▷ n stands for the number of entries that a cache line holds
  7:　　　　　　$CK_{i,j}^{N_r}\left[\overline{S}_{i,j}^{N_r+1}[t] \oplus Te_{i+2}[l]\right]$ ++; ▷ Vote for discard key candidate
  8:　　　　**end for**
  9:　　**end if**
 10: **end for**
 11: **return** argmin($CK_{i,j}^{N_r}$); ▷ The candidate with the fewest votes for discard is the **secret key**.

Figure 3 shows the values of one of the 16 $CK_{i,j}^{N_r}$ array, which contains the discarding votes each possible key value has received during the attack for a concrete byte of the key. The x-axis represents each possible key value, and the y-axis the number of times each option has been discarded. The minimum is well distinguishable from the rest of possible key values and is highlighted in red. The index with the minimum score represents the secret key. This approach allows recovering the 128 bits of the key with fewer than 3000 encryptions on average in the best case. We present further results and provide further details in Section 4.

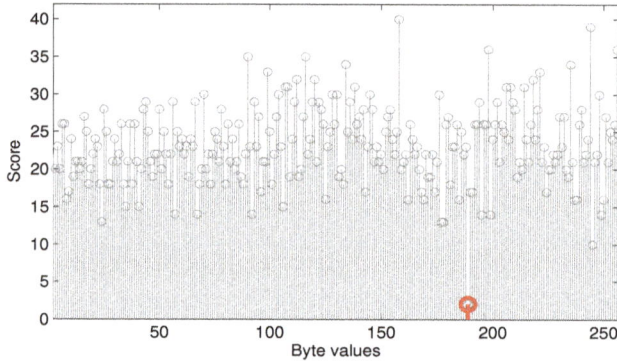

Figure 3. Key byte possible values with their associated discarding scores.

3.3.2. Attack on AES-256

The attack against AES-256 comes as an extension of the previously explained attack on the last round of an encryption. There are no further assumptions than in the previous case. The setup and measurement collection steps are exactly the same for both attacks, and only the information processing phase is different since it includes an extra step. We retrieve 128 bits of the key using Algorithm 2 and use the guessed 128 bits of the key and the same information that has already been used (non-accesses and ciphertext) to derive the remaining 128 unknown bits of the key. We similarly use information about the non-used cache lines to discard key values until we obtain the key. This is possible because this information refers to the whole encryption and, as previously stated, if a cache line has not been accessed within an encryption, it has not been accessed in any of its rounds.

Note that the probability of non-accessing one line when performing an AES-256 encryption, which is a function of the number of rounds, and is given by Equation (1), is 2.69% (14 rounds). Consequently, the attack on AES-256 would require more samples to successfully guess the whole key. In this case, about 1 out of 37 samples carries useful information. This means we need to collect at least the same number of useful samples than in the previous case.

The key to understand the attack is to understand the equivalence between encryption and decryption. Using the decryption algorithm, we transform the data referring to the encryption process in a way that it is possible to easily derive the round key discarding its possible values. For this reason, we call this approach encryption-by-decryption attack (EBD).

If we analyze the AES encryption process, we see that the output of each round is used as an index to access the T-tables in the next round after a ShiftRows operation. Taking into account the valuable information of the unused cache lines, it is possible to figure out which are the state values that are not possible before the penultimate round, S^{Nr-1}. That is, if a cache line has not been accessed, all the state values that would have forced an access to this line can be discarded.

Once we know the last round key, it is straightforward to obtain the input to that round (S^{Nr-1}) only taking into the account the encryption function. However, because of the MixColumns operation which is performed in all the previous rounds (except for the initial one), the certainty about the non-usage of a cache line of one table and the knowledge of the encryption function are not enough to obtain information about the key of these rounds. However, if we use the decryption function and take into the account the equivalence between encryption and decryption, we can transform our data so we can derive information about a state of the decryption function, which is equivalent to the S^{Nr-1} state in the encryption function.

Figure 4 shows with dotted lines the instants where equivalent decryption and encryption algorithms have the same values. According to the figure, we establish a bidirectional relation between an encryption state S^r and a decryption state D^{Nr+1-r} using SubBytes and ShiftRows steps.

The relations are defined in Equations (2a) and (2b). These relations and the knowledge of the values which are not possible in the state S^{N_r-1} give us the values which are not possible in the state D^2. Since the value of what we call intermediate value IV^1 can be easily obtained from the decryption function, our scenario is similar to the previous one ($D^2_{i,j} = IV^1_{i,j} \oplus CKd^1_{i,j}$).

$$D^{N_r+1-r} = \text{ShiftRows}(S_{RD}(S^r)) \tag{2a}$$

$$S^r = S_{RD}(\text{InvShiftRows}(D^{N_r+1-r})) \tag{2b}$$

Algorithm 3 shows how to obtain the correct candidates for the round key of decryption algorithm Round 1 (the second round). The key of the first decryption round is exactly the same as the key used in the last round of the encryption, previously obtained using the non-access cache attack and targeting the last round (Algorithm 2). As we know both input data (ciphertext) and round key, we can perform the first round of the decryption algorithm and obtain the state of round 1, D^1. Applying the Td-tables on the state, we calculate an intermediate value of Round 1, IV^1. At this point, if we perform the AddRoundKey, the next step of the decryption algorithm, the result would be D^2, which can be transformed into the encryption S^{13} state, as explained using Equation (2b).

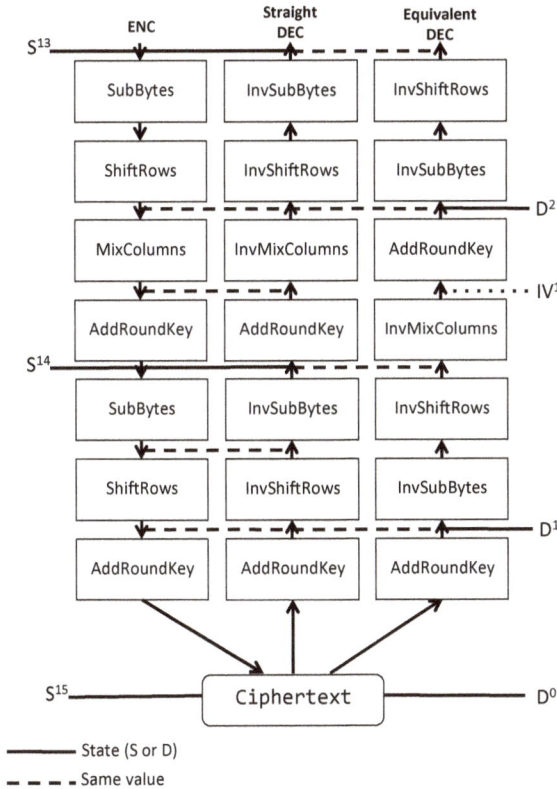

Figure 4. Relation between encryption, straight decryption and equivalent decryption.

In the T-table implementation, the state (including S^{N_r-1} or S^{13} for AES256) is used as index to access the Te tables. Note that not accessing a cache line holding one entry of the table also indicates which indexes of the table are not being used, which is similar to indicating which values of the state can be discarded. Then, we apply Equation (2a) to transform the discarded values for state S^{N_r-1} to discarded values for state D^2. After this transformation, we have information on both sides of the

AddRoundKey in round 1 of the decryption sequence. We can vote to discard possible values of the decryption round key. We vote to those values, which, once the decryption state is transformed into the encryption state, would lead to a memory access on the unused lines.

The least voted value for each byte is the decryption round key. Both the encryption and decryption keys are related and one can be obtained from the other applying the InvMixColumns operation, or an equivalent operation depending on the direction of the conversion. As a result of this information processing, we know two consecutive round keys; that is, the full AES-256 key. This algorithm could be applied recursively for the whole encryption/decryption process, as the cache information is applicable to the whole encryption.

Using this encryption-by-decryption cache attack on AES-256, we obtain the key with fewer than 10,000 samples in the best scenario. We provide further details in Section 4.

Algorithm 3 Recovery algorithm for key byte $Kd^1_{i,j}$.

Input: $\overline{X}_0, \overline{X}_1, \overline{X}_2, \overline{X}_3, \overline{S}^{N_r+1}$ and $\mathbf{K}^{N_r}$ ▷ Information about the accesses, ciphertext and last round key.
Output: $Kd^1_{i,j}$ ▷ Secret decryption key values
 1: $\mathbf{Kd}^0 = \mathbf{K}^{N_r}$
 2: **for** $l = 0$ to 256 **do**
 3:     $CKd^1_{ij}[l] = 0;$ ▷ Initialization of the candidates array
 4: **end for**
 5: **for** $t = 0$ to *number_of_encryptions* **do**
 6:     **if** $\overline{X}_i[t] == 0$ **then**
 7:         $D^0 = \overline{S}^{N_r+1}[t];$
 8:         $D^1 = D^0 \oplus \mathbf{Kd}^0;$
 9:         **for** $j = 0$ to 4 **do**
10:             $\mathbf{IV}^1_j = \mathbf{Td}_0\left[D^1_{0,j}\right] \oplus \mathbf{Td}_1\left[D^1_{1,j-1}\right] \oplus \mathbf{Td}_2\left[D^1_{2,j-2}\right] \oplus \mathbf{Td}_3\left[D^1_{3,j-3}\right]$ ▷ Obtain intermediate
    value
11:         **end for**
12:         **for** $l = 0$ to n **do**   ▷ n stands for the number of entries that a cache line holds
13:             $CKd^1_{i,j} = \left[\mathbf{IV}^1_{ij} \oplus S_{RD}[l]\right] + +;$ ▷ Vote for discard key candidate
14:         **end for**
15:     **end if**
16: **end for**
17: **return argmin**(CKd^1_{ij});   ▷ The candidate with the fewest votes for discard is the **secret key**.

---

4. Discussion of the Results

Our experimental setup included an Intel Core i5-7600K processor (3.80 GHz) with 4 CPUs and two core exclusive L1 caches of 32 KB (one for data and other for instructions), a L2 cache of 256 KB and, finally, a L3 cache of 6 MB shared among all cores whose line size is 64 bytes. It has CentOS 7.6 installed while the Virtual Machines in our experiments have CentOS 7.6 minimal and both have one virtual core. The hypervisor WAs KVM. This information is summarized in Table 1 The target of our experiments was the 1.0.1f OpenSSL release, which includes the T-Table implementation of AES.

In the previous section, we present attacks that work for different lengths of AES keys (128 or 256 bits). These attacks assume the knowledge of unused cache lines to discard possible key values and that the ciphertexts are known by the attacker. We performed the experiments in two different scenarios (virtualized and non-virtualized environments) considering different techniques to retrieve information about the cache utilization (*Prime+Probe*, *Prime+Abort*, *Flush+Reload* and *Flush+Flush*) and different key lengths. Table 2 shows the mean number of encryptions that had to be monitored to obtain the full key in non-virtualized environments (the victim and the attacker processes run in the same OS). Similarly, Table 3 shows the results for virtualized environments (the victim runs in one VM and the attacker in a different one sharing the same host).

Table 2. Mean number of samples required to retrieve the whole key in non-virtualized environments.

	Flush+Reload	Flush+Flush	Prime+Probe	Prime+Abort
AES 128	3000	15,000	14,000	3500
AES 256	8000	35,000	38,000	8000

Table 3. Mean number of samples required to retrieve the whole key in virtualized environments (cross-VM).

	Flush+Reload	Flush+Flush	Prime+Probe
AES 128	10,000	40,000	45,000
AES 256	28,000	100,000	110,000

When applying the *Prime+Abort* technique, we wanted to monitor four different lines. We needed to define four transactional regions to distinguish between them, which also means we used more resources. Note that we do not include the results for the *Prime+Abort* technique in virtualized environments. The TSX instructions were virtualized, thus it should be possible to perform *Prime+Abort* in a cross-VM scenario. However, with our experimental setup consisting on an attacker preparing the cache, triggering an encryption and waiting for the result, we got too many aborts. Since all of them have similar causes, distinguishing between the aborts that were due to the encryption and the ones that were noise was difficult. For this reason, and since we considered that our point was already proved, we did not evaluate the data collected with the *Prime+Abort* technique in the cross-VM scenario.

For a fair comparison between the access and our non-access approach, we also retrieved the key for the access approach considering the key of 128 bits. Table 4 shows the results for both approaches and the percentage of improvement regarding the number of traces required to successfully retrieve the secret key. In all cases, the number of samples was reduced significantly.

Table 4. Mean number of samples required to retrieve the whole 128 bits key with the access approach (first column), our approach (second column) and the improvement our approach means. (V) means virtualized scenario.

	Access	Non-Access	Improvement
Flush+Reload	100,000	3000	97%
Flush+Flush	250,000	16,000	94%
Prime+Probe	190,000	14,000	93%
Prime+Abort	110,000	3500	97%
Flush+Reload(V)	400,000	10,000	98%
Flush+Flush(V)	1,120,000	40,000	96%
Prime+Probe(V)	1,250,000	45,000	96%

We noticed that, when monitoring more than one line at a time, each table provided a different amount of information. This fact can be observed in Figure 5, where we represent a color map for each byte of the key candidates array for that byte. Dark colors represent fewer discarded values (correct key values) while light colors represent most discarded values. With a uniform distribution of the input plaintexts, the minimum score was more remarkable as we increased the number of encryptions. The figure shows that there is a pattern for each byte associated with the same table. We retrieved a different number of useful samples for each table. This was due to the prefetcher and our uniform access pattern that seemed to trigger it. This meant that more encryptions were required to get the full key.

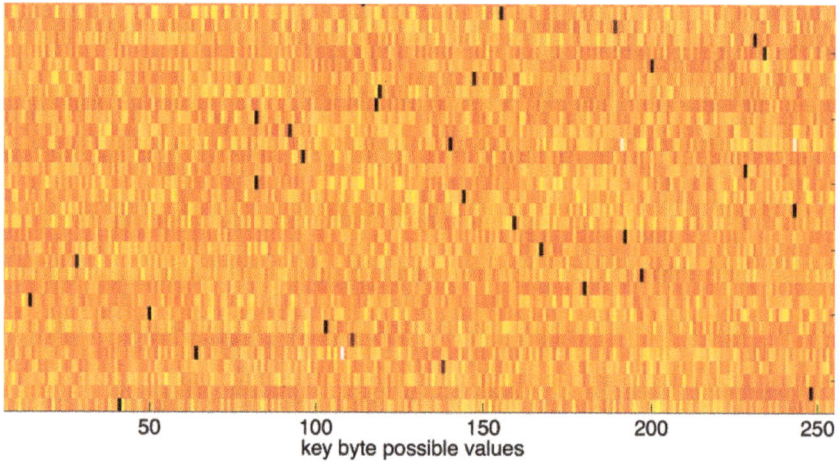

Figure 5. Color map of discarding scores distribution for the 32 bytes of the key retrieved with the *Flush+Reload* technique. Correct key values are the darkest ones.

We could completely recover the full AES 256 key using the EBD attack. Comparing the number of traces required to obtain the 128 bit key and the number of traces required to obtain the 256 bit key, AES-256 happened to be only three times harder to attack with our approach. The different probabilities of non-accessing a line explained our results: 2.69% for AES-256 and 7.5% for AES-128, whichin theory indicates 2.8 × more encryptions being needed to get the same information. In practice, we measured 5.29%, 3.87%, 4.55% and 4.31% of cache misses within our samples for AES-128 and each of the lines, respectively, and 1.52%, 0.93%, 1.31% and 1.02% for AES-256 for *Flush+Reload*. These results allow us to discuss the results in [49]. They state that, although larger key lengths translate into an exponential increase in the complexity of a brute force approach, this cannot be applied to side-channel attacks. In addition, using the two last rounds of the encryption algorithm (a version that uses a different table for computing the last round of the encryption), they arrive at the conclusion that AES-256 only provides an increase in complexity of 6–7 compared to cache-based attacks on AES-128. We, in contrast, demonstrate that this factor is equal to 3.

In the case of the EBD attack, we did not find similar results or experimental setups to compare with. All studied attacks can be described as eminently theoretical. In the literature, Aciiçmez et al. [50] stated that their attack can be applied to the second round of the encryption, but they did not provide experimental results. In 2009, Biryukov et al. [51] presented a distinguisher and related key attack on AES-256 with 2^{131} data complexity. Later, they slightly improved their results [52], achieving 2^{119} data complexity. In [53], they described an attack using two related keys to recover the 256-bit key, but their target cipher was the nine-round version of AES-256 and used chosen-plaintexts. Another attack [54] runs in a related-subkey scenario with four related keys targeting the 13-round AES-256; it has 2^{76} time and data complexity. Bogdanov et al. [55] presented a technique of cryptanalysis with bicliques, also leading to a non-practical complexity recovering the key. Kim [56] explained also theoretically how to apply differential fault analysis to reduce the required number of faults in AES-192 and AES-256.

As stated above, each retrieved sample that carries information about non-accesses to the T-Tables reduces the key search space. This means that, if the attack is interrupted for some reason before it finishes and the attacker decides to apply brute force to retrieve the key, the number of keys he has to check is much smaller. In Figure 6, we present the base 2 logarithm of the key search space vs. the total number of samples retrieved, i.e., the number of bits that remain unknown after each performed encryption. Note that, in the case of the encryption with the 256 bit key, we also begin with 128 unknown bits; as explained above, we get the full key in two round using the same data and in each round we recover 128 bits. Thus, the figure represents the mean number of unknown bits per round.

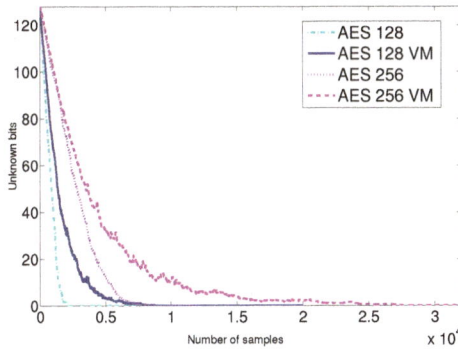

Figure 6. Key bits that remain unknown after each encryption for the *Flush+Reload* technique.

To put these data in context, we also measured the time the two processes (victim and attacker) need to interact until we can obtain the full key. This time is referred to the measurement collection step, since the information processing is performed offline. Table 5 shows these mean times.

Table 5. Mean times required for gathering the data of Figure 6.

	Attack Duration
AES-128	43 ms
AES-128 (VM)	558 ms
AES-256	135 ms
AES-256 (VM)	1732 ms

Assuming the number of collected samples varies linearly with the time, these results show that, if we want to detect such attacks before they succeed, it is necessary to act quickly. Detection times have to be in the order of milliseconds. Chiappetta et al. [19] let the victim run for 100 ms, which means our approach would succeed before they detect the attack. Even detection times of around 10 ms imply that around half of the key would be leaked. In CloudRadar [20], detection times are approximately 5 ms, then the hypervisor would need to migrate the machine. CacheShield [21] triggers an alarm after 4.48 ms, and would have to trigger later a countermeasure. It is realistic to assume that, even with any of these countermeasures enabled, victim and attacker can interact for at least 8–10 ms. This means part of the key would be leaked anyway, whereas there would be no leakage with the access approach. Note that we are assuming that the victim is encrypting random plaintext. If the attacker can make the victim encrypt chosen plaintext in such a way the number of non-accessed lines increases, our approach would succeed much more quickly and trigger almost no cache misses, rendering previous countermeasures ineffective.

We demonstrated that our non-access approach retrieves the AES key using fewer samples than the traditional access approach in Intel processors that mount an inclusive cache. However, AMD and ARM processors [10–12] that are vulnerable to cache attacks can also benefit from our attack. We believe it is possible to achieve a similar reduction in the number of samples required to succeed in these processors. It was also recently demonstrated that non-inclusive caches are also vulnerable to cache attacks [57]. As a result, our approach is feasible for non-inclusive caches. As long as the attacker is able to evict the victim's data from the cache, he will be able to use our approach to improve the efficiency of the attack against AES.

The T-Table implementation has been replaced by the S-Box implementation in software. In this case, the encryption is performed using a unique table holding 256 byte values. An attacker can also use our approach to target this implementation. If he is able to flush a line in the last round of the encryption, and given that the 16 accesses to the table per round represent a non-accessing probability of 1% (the attack is feasible), each non-accessed line would allow us to discard 64 values out of 256 possible of each byte of the key. Note that our presented attack against the T-table implementation discards 16 values of 256 possible for 4 bytes of the key. Thus, the estimated leakage of the theoretical attack to the last round of the new implementation is about 75 times higher than the presented attack to the T-table implementation.

5. Conclusions

The massive adoption of cloud computing has introduced numerous advantages, but also new challenging security and privacy risks. One of these risks is the information leakage derived from resource sharing among VMs and, specifically, from the shared cache memory. This risk is also present in non-virtualized environments where all the running processes share the cache.

This work introduces a novel approach to exploit the information gained from the cache by means of existing cache attacks: *Flush+Reload*, *Flush+Flush*, *Prime+Probe* and *Prime+Abort*. We focus on the cache lines that have not been used and target the AES T-Table implementation to prove our point. Even when the T-table version of the AES algorithm is known to be vulnerable to side-channel attacks, it is still available in newer OpenSSL versions (compiling with the no-asm flag). Using the non-access attack, we demonstrate a reduction that varies from 93% to 98% in the amount of encryptions required to obtain an AES-128 key. This improvement is crucial to develop a practical attack on real servers, where the AES key is usually a session key; i.e., it is periodically renewed using key-exchange algorithms. Besides, we demonstrate that, if there is a detection system trying to detect cache attacks running, the amount of leaked information will depend on the detection time, but our approach recovers information from the very beginning of the attack.

Moreover, we present a practical attack in both time and complexity against the 256-bit version of AES: the encryption-by-decryption (EBD) cache attack. We implemented the attack on a real experimental setup and successfully obtained the encryption key in a matter of milliseconds, requiring only 10,000 encryptions in the best scenario. EBD retrieves the key by breaking the attack into two 128-bit stages. In the first stage, non-accessed lines are used to obtain the last 128 bits of the key by discarding key values when the data has not been used. In the second stage, EBD takes the 128-bit subkey obtained in the first stage and uses it with the decryption function (performing part of an AES decryption) to reach an state equivalent to an encryption state from which the non-accessed lines can serve to obtain the remaining 128 bits of the key. Our results show that AES-256 is only three times harder to attack than AES-128.

Author Contributions: Conceptualization, S.B., P.M., J.-M.d.G. and J.M.M.; Data curation, S.B.; Funding acquisition, S.B. and J.M.M.; Investigation, S.B.; Methodology, S.B. and P.M.; Software, S.B.; Supervision, J.M.M.; Writing—original draft, S.B. and P.M.; and Writing—review and editing, J.-M.d.G.

Funding: This research was funded by Spanish Ministry of Economy and Competitiveness grant numbers TIN-2015-65277-R, AYA2015-65973-C3-3-R and RTC-2016-5434-8.

References

1. Osvik, D.A.; Shamir, A.; Tromer, E. Cache Attacks and Countermeasures: The Case of AES. In Proceedings of the Topics in Cryptology—CT-RSA, San Jose, CA, USA, 13–17 February 2006; Springer: Heidelberg/Berlin, Germany, 2006; Volume 3860, pp. 1–20.
2. Yarom, Y.; Falkner, K. FLUSH+RELOAD: A High Resolution, Low Noise, L3 Cache Side-Channel Attack. In Proceedings of the USENIX Security Symposium, San Diego, CA, USA, 20–22 August 2006; USENIX Association: Berkeley, CA, USA, 2014; pp. 719–732.

3. Gruss, D.; Maurice, C.; Wagner, K. Flush+Flush: A Stealthier Last-Level Cache Attack. *CoRR* **2015**._14. [CrossRef]

4. Disselkoen, C.; Kohlbrenner, D.; Porter, L.; Tullsen, D. Prime+Abort: A Timer-Free High-Precision L3 Cache Attack using Intel TSX. In Proceedings of the 26th USENIX Security Symposium (USENIX Security 17), Vancouver, BC, Canada, 16–18 August 2017; USENIX Association: Berkeley, CA, USA, 2017; pp. 51–67.

5. Bernstein, D.J. Cache-Timing Attacks on AES 2005. Available online: http://palms.ee.princeton.edu/system/files/Cache-timing+attacks+on+AES.pdf (accessed on 28 February 2019) .

6. Weiß, M.; Heinz, B.; Stumpf, F. A Cache Timing Attack on AES in Virtualization Environments. In Proceedings of the International Conference on Financial Cryptography and Data Security—FC, Kralendijk, Bonaire, 27 Februray–2 March 2012; Springer: Heidelberg/Berlin, Germany, 2012; Volume 7398, pp. 314–328.

7. Brumley, B.B.; Hakala, R.M. Cache-Timing Template Attacks. In Proceedings of the Advances in Cryptology—ASIACRYPT, Tokyo, Japan, 6–10 December 2009; Springer: Heidelberg/Berlin, Germany, 2009; Volume 5912, pp. 667–684.

8. Chen, C.; Wang, T.; Kou, Y.; Chen, X.; Li, X. Improvement of trace-driven I-Cache timing attack on the {RSA} algorithm. *J. Syst. Softw.* **2013**, *86*, 100–107. [CrossRef]

9. Inci, M.S.; Gulmezoglu, B.; Irazoqui, G.; Eisenbarth, T.; Sunar, B. Seriously, Get off My Cloud! Cross-VM RSA Key Recovery in a Public Cloud. Cryptology ePrint Archive, Report 2015/898. 2015. Available online: http://eprint.iacr.org/ (accessed on 28 February 2019).

10. Irazoqui, G.; Eisenbarth, T.; Sunar, B. Cross Processor Cache Attacks. In Proceedings of the 11th ACM on Asia Conference on Computer and Communications Security (ASIA CCS '16), Xi'an, China, 30 May–3 June 2016; ACM: New York, NY, USA, 2016; pp. 353–364.

11. Lipp, M.; Gruss, D.; Spreitzer, R.; Mangard, S. ARMageddon: Last-Level Cache Attacks on Mobile Devices. *arXiv* **2015**, arXiv:1511.04897.

12. Green, M.; Rodrigues-Lima, L.; Zankl, A.; Irazoqui, G.; Heyszl, J.; Eisenbarth, T. AutoLock: Why Cache Attacks on ARM Are Harder Than You Think. In Proceedings of the 26th USENIX Security Symposium (USENIX Security 17), Vancouver, BC, Canada, 16–18 August 2017; USENIX Association: Berkeley, CA, USA, 2017; pp. 1075–1091.

13. Ristenpart, T.; Tromer, E.; Shacham, H.; Savage, S. Hey, you, get off of my cloud: exploring information leakage in third-party compute clouds. In Proceedings of the ACM Conference on Computer and Communications Security (CCS), Chicago, IL, USA, 9–13 November 2009; ACM: New York, NY, USA, 2009; pp. 199–212.

14. Zhang, Y.; Juels, A.; Oprea, A.; Reiter, M.K. HomeAlone: Co-residency Detection in the Cloud via Side-Channel Analysis. In Proceedings of the IEEE Symposium on Security and Privacy, S&P, Berkeley, CA, USA, 22–25 May 2011; pp. 313–328.

15. Wang, Z.; Lee, R.B. New Cache Designs for Thwarting Software Cache-based Side Channel Attacks. In Proceedings of the 34th Annual International Symposium on Computer Architecture, ISCA '07, San Diego, CA, USA, 9–13 June 2007; ACM: New York, NY, USA, 2007; pp. 494–505.

16. Kim, T.; Peinado, M.; Mainar-Ruiz, G. STEALTHMEM: System-Level Protection Against Cache-Based Side Channel Attacks in the Cloud. In Proceedings of the 21st USENIX Security Symposium (USENIX Security 12), Bellevue, WA, USA, 8–10 August 2012; USENIX Association: Berkeley, CA, USA, 2012; pp. 189–204.

17. Liu, F.; Ge, Q.; Yarom, Y.; Mckeen, F.; Rozas, C.; Heiser, G.; Lee, R.B. CATalyst: Defeating last-level cache side channel attacks in cloud computing. In Proceedings of the 2016 IEEE International Symposium on High Performance Computer Architecture (HPCA), Barcelona, Spain, 12–16 March 2016; pp. 406–418.

18. Briongos, S.; Malagón, P.; Risco-Martín, J.L.; Moya, J.M. Modeling Side-channel Cache Attacks on AES. In Proceedings of the Summer Computer Simulation Conference, SCSC '16, Montreal, QC, Canada, 24–27 July 2016; Society for Computer Simulation International: San Diego, CA, USA, 2016; pp. 37:1–37:8.

19. Chiappetta, M.; Savas, E.; Yilmaz, C. Real time detection of cache-based side-channel attacks using hardware performance counters. *Appl. Soft Comput.* **2016**, *49*, 1162–1174. [CrossRef]

20. Zhang, T.; Zhang, Y.; Lee, R.B. CloudRadar: A Real-Time Side-Channel Attack Detection System in Clouds. In Proceedings of the 19th International Symposium on Research in Attacks, Intrusions, and Defenses, RAID 2016, Paris, France, 19–21 September 2016; Monrose, F., Dacier, M., Blanc, G., Garcia-Alfaro, J., Eds.; Springer International Publishing: Cham, Switzerland, 2016; pp. 118–140.

21. Briongos, S.; Irazoqui, G.; Malagón, P.; Eisenbarth, T. CacheShield: Detecting Cache Attacks Through Self-Observation. In Proceedings of the Eighth ACM Conference on Data and Application Security and Privacy, CODASPY '18, Tempe, AZ, USA, 19–21 March 2018; ACM: New York, NY, USA, 2018; pp. 224–235.

22. Maurice, C.; Le Scouarnec, N.; Neumann, C.; Heen, O.; Francillon, A. Reverse Engineering Intel Last-Level Cache Complex Addressing Using Performance Counters. In Proceedings of the Research in Attacks, Intrusions, and Defenses, Kyoto, Japan, 2–4 November 2015; Bos, H., Monrose, F., Blanc, G., Eds.; Springer International Publishing: Cham, Switzerland, 2015; pp. 48–65.

23. Irazoqui, G.; Eisenbarth, T.; Sunar, B. Systematic Reverse Engineering of Cache Slice Selection in Intel Processors. In Proceedings of the 2015 Euromicro Conference on Digital System Design, DSD 2015, Madeira, Portugal, 26–28 August 2015; pp. 629–636.

24. Arcangeli, A.; Eidus, I.; Wright, C. Increasing memory density by using KSM. In Proceedings of the Linux Symposium, Montreal, QC, Canada, 13–17 July 2009; pp. 19–28.

25. Kernel Samepage Merging. 2015. Available online: http://kernelnewbies.org/Linux_2_6_32#head-d3f32e41df508090810388a57efce73f52660ccb/ (accessed on 28 February 2019).

26. Hu, W. Lattice scheduling and covert channels. In Proceedings of the IEEE Symposium on Research in Security and Privacy, Oakland, CA, USA, 4–6 May 1992; IEEE Computer Society: Washington, DC, USA, 1992; pp. 52–61.

27. Kocher, P.C. Timing Attacks on Implementations of Diffie-Hellman, RSA, DSS, and Other Systems. In Proceedings of the Advances in Cryptology—CRYPTO, Santa Barbara, CA, USA, 18–22 August 1996; Springer: Heidelberg/Berlin, Germany, 1996; Volume 1109, pp. 104–113.

28. Kelsey, J.; Schneier, B.; Wagner, D.; Hall, C. Side Channel Cryptanalysis of Product Ciphers. *J. Comput. Secur.* **2000**, *8*, 141–158. [CrossRef]

29. Page, D. Theoretical Use of Cache Memory as a Cryptanalytic Side-Channel. Cryptology ePrint Archive, Report 2002/169. 2002. Available online: http://eprint.iacr.org/ (accessed on 28 February 2019).

30. Tsunoo, Y.; Saito, T.; Suzaki, T.; Shigeri, M.; Miyauchi, H. Cryptanalysis of DES Implemented on Computers with Cache. In Proceedings of the Cryptographic Hardware and Embedded Systems—CHES, Cologne, Germany, 8–10 September 2003; Springer: Heidelberg/Berlin, Germany, 2003; Volume 2279, pp. 62–76.

31. Irazoqui, G.; Inci, M.S.; Eisenbarth, T.; Sunar, B. Fine grain Cross-VM Attacks on Xen and VMware Are Possible! In Proceedings of the 2014 IEEE Fourth International Conference on Big Data and Cloud Computing, (BDCLOUD '14), Sydney, Australia, 3–5 December 2014; IEEE Computer Society: Washington, DC, USA, 2014; pp. 737–744.

32. Percival, C. Cache missing for fun and profit. In Proceedings of the BSDCan 2005, Ottawa, ON, Canada, 13–14 May 2005.

33. Gullasch, D.; Bangerter, E.; Krenn, S. Cache Games—Bringing Access-Based Cache Attacks on AES to Practice. In Proceedings of the IEEE Symposium on Security and Privacy, S&P, Berkeley, CA, USA, 22–25 May 2011; pp. 490–505.

34. Gruss, D.; Spreitzer, R.; Mangard, S. Cache Template Attacks: Automating Attacks on Inclusive Last-Level Caches. In Proceedings of the USENIX Security Symposium, Washington, DC, USA, 12–14 August 2015; USENIX Association: Berkeley, CA, USA, 2015; pp. 897–912.

35. Irazoqui, G.; Inci, M.S.; Eisenbarth, T.; Sunar, B. Lucky 13 Strikes Back. In Proceedings of the 10th ACM Symposium on Information, Computer and Communications Security, ASIA CCS '15, Singapore, 14–17 April 2015; ACM: New York, NY, USA, 2015; pp. 85–96.

36. Irazoqui, G.; Inci, M.S.; Eisenbarth, T.; Sunar, B. Wait a Minute! A fast, Cross-VM Attack on AES. In Proceedings of the Research in Attacks, Intrusions and Defenses Symposium, RAID, Gothenburg, Sweden, 17–19 September 2014; Springer: Heidelberg/Berlin, Germany, 2014; Volume 8688, pp. 299–319.

37. Yarom, Y.; Benger, N. Recovering OpenSSL ECDSA Nonces Using the FLUSH+RELOAD Cache Side-Channel Attack; Cryptology ePrint Archive, Report 2014/140. 2014. Available online: http://eprint.iacr.org/ (accessed on 28 February 2019).

38. Irazoqui, G.; Inci, M.S.; Eisenbarth, T.; Sunar, B. Know Thy Neighbor: Crypto Library Detection in Cloud. In Proceedings of the Privacy Enhancing Technologies 2015, Philadelphia, PA, USA, 30 June–2 July 2015; pp. 25–40.

39. Zhang, Y.; Juels, A.; Reiter, M.K.; Ristenpart, T. Cross-Tenant Side-Channel Attacks in PaaS Clouds. In Proceedings of the 2014 ACM SIGSAC Conference on Computer and Communications Security, CCS '14, Scottsdale, AZ, USA, 3–7 November 2014; ACM: New York, NY, USA, 2014; pp. 990–1003.

40. Kulah, Y.; Dincer, B.; Yilmaz, C.; Savas, E. SpyDetector: An approach for detecting side-channel attacks at runtime. *Int. J. Inf. Secur.* **2018**. [CrossRef]

41. Liu, F.; Yarom, Y.; Ge, Q.; Heiser, G.; Lee, R.B. Last-Level Cache Side-Channel Attacks are Practical. In Proceedings of the IEEE Symposium on Security and Privacy (S&P), San Jose, CA, USA, 17–21 May 2015; pp. 605–622.

42. Aciiçmez, O.; Schindler, W. A Vulnerability in RSA Implementations Due to Instruction Cache Analysis and its Demonstration on OpenSSL. In Proceedings of the Topics in Cryptology—CT-RSA 2008, San Francisco, CA, USA, 8–11 April 2008; Springer: Heidelberg/Berlin, Germany, 2008; pp. 256–273.

43. Yarom, Y.; Ge, Q.; Liu, F.; Lee, R.B.; Heiser, G. Mapping the Intel Last-Level Cache. Cryptology ePrint Archive, Report 2015/905. 2015. Available online: http://eprint.iacr.org/ (accessed on 28 February 2019).

44. Apecechea, G.I.; Eisenbarth, T.; Sunar, B. S$A: A Shared Cache Attack That Works across Cores and Defies VM Sandboxing—And Its Application to AES. In Proceedings of the 2015 IEEE Symposium on Security and Privacy, SP 2015, San Jose, CA, USA, 17–21 May 2015; pp. 591–604.

45. İnci, M.S.; Gulmezoglu, B.; Irazoqui, G.; Eisenbarth, T.; Sunar, B. Cache Attacks Enable Bulk Key Recovery on the Cloud. In Proceedings of the Cryptographic Hardware and Embedded Systems—CHES 2016: 18th International Conference, Santa Barbara, CA, USA, 17–19 August 2016.

46. Daemen, J.; Rijmen, V. *The Design of Rijndael: AES—The Advanced Encryption Standard*; Information Security and Cryptography; Springer: Heidelberg/Berlin, Germay, 2002.

47. Gülmezoglu, B.; Inci, M.S.; Irazoqui, G.; Eisenbarth, T.; Sunar, B. A Faster and More Realistic Flush+Reload Attack on AES. In Proceedings of the International Workshop on Constructive Side-Channel Analysis and Secure Design—COSADE, Berlin, Germany, 13–14 April 2015; Springer: Heidelberg/Berlin, Germay, 2015; Volume 9064, pp. 111–126.

48. Intel. Intel® 64 and IA-32 Architectures Optimization Reference Manual (Section 2.1.1.2). 2017. Available online: https://software.intel.com/sites/default/files/managed/9e/bc/64-ia-32-architectures-optimization-manual.pdf (accessed on 28 February 2019).

49. Neve, M.; Tiri, K. On the Complexity of Side-Channel Attacks on AES-256—Methodology and Quantitative Results on Cache Attacks. Cryptology ePrint Archive, Report 2007/318. 2007. Available online: http://eprint.iacr.org/ (accessed on 28 February 2019).

50. Aciiçmez, O.; Schindler, W.; Koç, Ç.K. Cache Based Remote Timing Attack on the AES. In Proceedings of the Topics in Cryptology—CT-RSA, San Francisco, CA, USA, 5–9 February 2007; Springer: Heidelberg/Berlin, Germany, 2007; Volume 4377, pp. 271–286.

51. Biryukov, A.; Khovratovich, D.; Nikolić, I. Distinguisher and Related-Key Attack on the Full AES-256. In Proceedings of the Annual International Cryptology Conference on Advances in Cryptology, Santa Barbara, CA, USA, 16–20 August 2009; Springer: Heidelberg/Berlin, Germany, 2009; pp. 231–249.

52. Biryukov, A.; Khovratovich, D. Related-Key Cryptanalysis of the Full AES-192 and AES-256. Cryptology ePrint Archive, Report 2009/317. 2009. Available online: http://eprint.iacr.org/ (accessed on 28 February 2019).

53. Biryukov, A.; Dunkelman, O.; Keller, N.; Khovratovich, D.; Shamir, A. Key Recovery Attacks of Practical Complexity on AES-256 Variants with up to 10 Rounds. In Proceedings of the Advances in Cryptology—EUROCRYPT, French Riviera, France, 30 May–3 June 2010; Springer: Heidelberg/Berlin, Germany, 2010; Volume 6110, pp. 299–319.

54. Biryukov, A.; Khovratovich, D. Feasible Attack on the 13-round AES-256. Cryptology ePrint Archive, Report 2010/257. 2010. Available online: http://eprint.iacr.org/ (accessed on 28 February 2019).

55. Bogdanov, A.; Khovratovich, D.; Rechberger, C. Biclique Cryptanalysis of the Full AES. In Proceedings of the Advances in Cryptology—ASIACRYPT, Seoul, Korea, 4–8 December 2011; Springer: Heidelberg/Berlin, Germany, 2011; Volume 7073, pp. 344–371.

56. Kim, C.H. Differential fault analysis of AES: Toward reducing number of faults. *Inf. Sci.* **2012**, *199*, 43–57. [CrossRef]
57. Yan, M.; Sprabery, R.; Gopireddy, B.; Fletcher, C.; Campbell, R.; Torrellas, J. Attack directories, not caches: Side channel attacks in a non-inclusive world. In Proceedings of the 2019 IEEE Symposium on Security and Privacy (SP) (2019), San Fransisco, CA, USA, 20–22 May 2019.

*applied
sciences*

MDPI

Article

Fast and Secure Implementation of Modular Exponentiation for Mitigating Fine-Grained Cache Attacks

Youngjoo Shin

School of Computer and Information Engineering, Kwangwoon University, 20 Kwangwoon-ro, Nowon-gu, Seoul 01897, Korea; yjshin@kw.ac.kr; Tel.: +82-2-940-5130

Received: 5 July 2018; Accepted: 3 August 2018; Published: 5 August 2018

Abstract: Constant-time technique is of crucial importance to prevent secrets of cryptographic algorithms from leakage by cache attacks. In this paper, we propose Permute-Scatter-Gather, a novel constant-time method for the modular exponentiation that is used in the RSA cryptosystem. On the basis of the scatter-gather design, our method utilizes pseudo-random permutation to obfuscate memory access patterns. Based on this strategy, the resistance against fine-grained cache attacks is ensured, i.e., providing the higher level of security than the existing scatter-gather implementations. Evaluation shows that our method outperforms the OpenSSL library at most 11% in the mainstream Intel processors.

Keywords: cache attack; cache side-channel attack; constant-time cryptographic algorithm; rsa cryptosystem; scatter-gather implementation; modular exponentiation

---

1. Introduction

Cache attacks, such as Prime+Probe [1–4] and Flush+Reload [5–10], exploit the usage of CPU cache as a side channel to infer secret information of victim applications. Due to its high resolution, the cache attack is very effective in attacking cryptographic algorithms [11–15]. By monitoring secret-dependent patterns in memory access or control flow, an adversary can successfully extract private keys in an implementation of the cryptographic algorithms. Thus, it is necessary to consider constant-time programming when implementing cryptographic software secure against cache attacks. The constant-time programming is an implementation technique that ensures the cryptographic algorithm has constant patterns during the execution irrespective of an input (i.e., secret) in its implementation.

Scatter-gather [16] is a constant-time programming technique for the RSA algorithm [17], which is used in OpenSSL library [18]. The RSA encryption/decryption (or sign/verify) are basically performed as modular exponentiation, in which the exponent is a private key (or a singing key). For computational efficiency, several multipliers are pre-computed, stored as a table in the memory and accessed later during the exponentiation. In a naive lookup-based implementation, multipliers are located in separate memory lines, so accessing them would cause observable unique access patterns, which is susceptible to cache attacks. Scatter-gather technique revises the arrangement of multipliers on the table so that any multipliers are accessed with the constant pattern.

The current implementation of the scatter-gather technique has the underlying assumption that cache adversaries only observe the access pattern at the granularity of cache line (i.e., 64 bytes) [19,20]. However, such assumption was broken as more fine-grained cache attack has been recently discovered. This cache attack, dubbed *Cache-bleed* [21], exploits the cache-bank conflict between hyper-threads to observe the secret-dependent access pattern at the bank level during the gathering phase.

In this paper, we propose Permute-Scatter-Gather, a novel constant-time method for the RSA modular exponentiation, which is resistant against fine-grained cache attacks. Based on the scatter-gather design, our technique employs a pseudo-random permutation for locating multipliers in a scattered memory layout. Such permutation actually obfuscates the memory access pattern, thus prevents any adversaries even mounting fine-grained cache attacks from inferring the secret from the observations. Furthermore, our novel technique for *constant-time permutation* allows the permutation itself to have the constant-time property, making more secure against cache attacks.

Our evaluation shows that the proposed method outperforms the existing countermeasure, implemented in the recent version of OpenSSL, at most 11% in the mainstream processors. It is also shown that the Permute-Scatter-Gather can be easily adopted with the OpenSSL without significant effort, increasing the practicality of the proposed method.

The rest of this paper is organized as follows. Background is presented in Section 2. Details on the Permute-Scatter-Gather, and their evaluations are given in Section 3, and Section 4, respectively. Finally, we conclude the paper in Section 5.

2. Background

2.1. Scatter-Gather Implementation

The main operation of RSA decryption (or sign) is the modular exponentiation; calculate b^e mod n for a secret exponent e. OpenSSL library performs the modular exponentiation by a fixed-window exponentiation algorithm [22] (See Algorithm 1). In a pre-computation phase, the algorithm computes a set of multipliers $m_i = m^j b$ mod n for $0 \leq j < 2^w$, where w is a window size. In an exponentiation phase, it scans each fraction of e of size w from $e_{\lceil k/w \rceil}$ to e_0. For each digit e_i, it multiplies r, the intermediate result from squaring, by the pre-computed multiplier m_{e_i}. In the OpenSSL library, the window size is set to $w = 5$, so there are 32 multipliers in total.

Algorithm 1 Fixed-window exponentiation

Require: k-bit exponent $e = \sum_{i=0}^{\lceil k/w \rceil} e_i \cdot 2^{wi}$, window size w, base b, modulus n
Ensure: b^e mod n
  1: **procedure** EXPONENTIATION(w, b, n, e)
  2:      // Pre-computation phase
  3:      $m_0 \leftarrow 1$
  4:      **for** $i \leftarrow 1$ **to** $2^w - 1$ **do**
  5:          $m_i \leftarrow m_{i-1} \cdot b$ mod n
  6:      **end for**
  7:
  8:      // Exponentiation phase
  9:      $r \leftarrow 1$
10:      **for** $i \leftarrow \lceil k/w \rceil - 1$ **to** 0 **do**
11:          **for** $j \leftarrow 1$ **to** w **do**
12:              $r \leftarrow r^2$ mod n
13:          **end for**
14:          $r \leftarrow r \cdot m_{e_i}$ mod n
15:      **end for**
16:      **return** r
17: **end procedure**

Scatter-gather implementation is a constant-time programming technique to avoid secret-dependent access at the cache line granularity [23]. Instead of storing multipliers consecutively in memory, it scatters each multiplier across multiple cache lines (Figure 1). When using the multiplier (i.e., in gathering phase), the fragments of the required multiplier are gathered to a buffer for the multiplication.

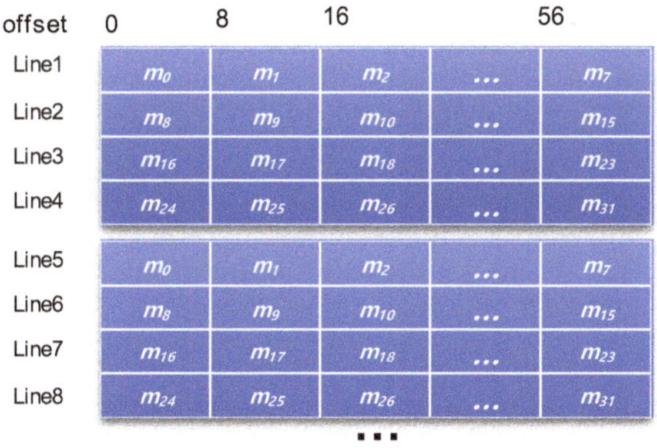

Figure 1. Memory layout of the multiplier table in OpenSSL.

2.2. Fine-Grained Cache Attack and Its Countermeasure

2.2.1. Fine-Grained Cache Attack

In Intel processors, a cache line is divided into multiple cache banks, each of which has part of the line specified by the line offset. In such cache design, concurrent requests to the same line can be served in parallel if the requested offsets are on the different banks. However, requests to the same bank would cause a cache line conflict, resulting in observable execution delay [24,25]. Such conflict at a cache line introduces fine-grained cache attacks such as Cache-bleed [21]. This kind of attacks exploits a bank level timing channel introduced by the cache line conflict. The granularity of the channel allows distinguishing between memory accesses within the same cache line.

With this attack, an adversary can infer which multipliers are accessed during the gathering phase in the exponentiation. It was shown that the scatter-gather implementation of OpenSSL library of the version 1.0.2f is vulnerable to the fine-grained cache attack, allowing the full recovery of RSA private keys [21].

2.2.2. Constant-Time Gather Procedure

The root cause of the OpenSSL's vulnerability to fine-grained cache attacks comes from that with the bank-level granularity, it has secret-dependent memory access in gathering phase. To mitigate the attack, the vulnerable version of the OpenSSL library (i.e., the version 1.0.2f) has been patched in the later version 1.0.2g so that all secret-dependent accesses are eliminated. More specifically, in the modified gathering process, all the multipliers laid on a single memory line are loaded into four 128-bit SSE (Streaming SIMD Extensions) registers (e.g., xmm0-xmm3). The relevant multiplier is then selected among them by masking the register values accordingly. The masks are necessarily calculated on-the-fly based on the index of the multiplier to be used.

The OpenSSL's countermeasure requires modifications of two gathering functions, bn_gatter5() and bn_mul_mont_gather5(), in the source file bn/x86_64-mont5.s.

This results in 10–20% performance drops of the modular exponentiation in RSA algorithms.

3. Permute-Scatter-Gather Implementation

In this section, we give details on the Permute-Scatter-Gather (or Permute-SG in short), the proposed method for secure modular exponentiation against fine-grained cache attacks, which is also faster than constant-time gather procedure.

3.1. Threat Model

Cache attacks often target secret keys of a victim process performing an encryption algorithm. In this paper, we assume that an adversary is a process which is co-resident on the same machine as the victim process. Due to the memory protection provided by modern operating systems, an adversary process is prohibited to view the content of the victim's memory. Despite of the process isolation, however, a logical processor is shared among processes, by which the adversary can exploit the cache-bank conflict. By mounting the fine-grained cache attack, the adversary tries to learn about the victim's secret key. We also assume that the adversary is able to execute arbitrary programs on a processor core shared with the victim process. However, as we mentioned above, the adversary does not have access to the victim's memory space.

In our threat model, we do not require that the target executable binary (e.g., OpenSSL library) running in the victim be kept secret. That is, the adversary has sufficient information on a logical structure of the binary such as the control flow and the exploitable locations. However, the adversary has no information about the runtime states (e.g., secret keys or permutation tables) of the executable, which are located on the data section of the binary in the victim's process.

3.2. Overview and Design Goals

The idea of the Permute-SG is basically to unlink the index of a multiplier from its memory location, thereby making it infeasible to figure out the multiplier used during the exponentiation. For this, the proposed method obfuscates the memory locations of the multipliers through a pseudo-random permutation. Specifically, given an index idx and a pseudo-random permutation P, the location of the multiplier is determined by the permuted index idx' = P(idx). In this way, all the 32 multipliers are rearranged in the table according to their permuted indices. By mounting cache attacks, an adversary might get the trace of P(idx). However, he/she cannot infer which multipliers are actually used from the obtained trace.

We construct the Permute-SG technique with consideration of achieving the following design goals:

- *Resistance against fine-grained cache attacks.* No information about the actually accessed multiplier should be revealed to adversaries who can observe memory accesses with bank-level granularity.
- *Computational efficiency.* Performance degradation in modular exponentiation due to applying this method should be minimized.
- *Adaptability.* It should be easily integrated into the existing implementation (e.g., OpenSSL library) without significant modification of source codes.

3.3. Implementation

The Permute-SG is augmented with ease to the OpenSSL's scatter-gather implementation (i.e., the version of 1.0.2f). The procedure of the Permute-SG for the modular exponentiation is performed through the following steps:

1. Permute step: In this step, a permutation P is randomly generated from $\mathcal{P}$, the set of all permutations. The generation process is conducted along with the precomputation phase of modular exponentiation algorithm (Algorithm 1).
2. Scatter step: This step is the same as the scatter procedure in the OpenSSL, except that the scattering location of a multiplier with an index idx is determined by P(idx).

3. Gather step: This step is the same as the gather procedure in the OpenSSL, except that the gathering location of a multiplier with an index idx is determined by P(idx).

3.3.1. Challenging Issue

As described above, we can easily integrate the Permute-SG technique into the OpenSSL library, thus adaptability, one of our design goals, is trivially achieved. However, it is not trivial to achieve the other two design goals together when implementing the technique. That is, evaluating P with an index idx is a time consuming operation and it occurs at every scatter and gather step. This may lead to the significant performance degradation. The optimal solution is to implement the evaluation procedure using a permutation table. By looking up the table with idx, the value of P(idx) can be retrieved just within a few CPU cycles. For security perspective, however, the lookup operation with the permutation table is subject to the fine-grained cache attack. This is because the memory access to the table reveals the index of the used multiplier during exponentiation. Therefore, implementing the permutation with regard to efficiency and security is a challenging problem.

3.3.2. Constant-Time Permutation

We overcome the challenging problem by implementing *constant-time permutation*. It is a lookup-based technique that always has constant memory access pattern irrespective of the accessed index, thus revealing no information to adversaries. For the computational efficiency, the constant-time permutation is implemented in a x86 assembly. Since a memory access is a costly operation, the number of access needs to be minimized for the constant-time lookup procedure. We achieve this by utilizing only a single SSE load instruction. By doing so, the memory access time for the lookup can be confined to just a single CPU cycle in the case of the table being loaded to a L1 cache [26].

To load a permutation table into a SSE register by a single load instruction, we have to fit the size of the table within the width of the register. In most Intel x86 processors, SSE registers are 128 bits in length (Recent Intel processors support Advanced Vector Extension (AVX), in which the size of registers are more than 128 bits in length. For our technique to be widely deployed, we only consider SSE instructions in this paper). Please note that there are 32 multipliers in total, and thus the size of each index should be at least 5 bits in length. This indicates that a room of 160 bits is needed in the table to store all the indicies, which is larger than the size of the SSE register. We solve this problem in a way that the four leftmost bits of the index are stored in the table instead of all the bits being stored. This makes the four bits of the index to be permuted while the remaining rightmost bit is left unchanged during the permutation process.

Figure 2 illustrates the process of constant-time permutation. We have PermTab, an array with a length of 128 bits, which is divided into two 64-bit permutation tables, $PermTab_H$ and $PermTab_L$. The address of PermTab is 16 bytes aligned so that a single load instruction can load both tables into a SSE register. Two pseudo-random permutations P_0 and P_1, which are generated independently in the Permute step, are set up to those tables respectively. Each table contains a permuted list of partial indices of 4 bits in its slots $s_0, s_1, ..., s_{15}$ according to the permutation.

In the permutation process, the value of the four leftmost bits of idx, denoted by X in Figure 2, is used to lookup the values of the corresponding slots in the tables simultaneously. For instance, the case of X = 2 would make concurrent lookups to $PermTab_H$ and $PermTab_L$ with the same slot s_2, resulting $P_0(X)$ and $P_1(X)$. The remaining rightmost bit of idx, denoted by Y, is used to select the one among them. As a result, the permuted index idx' is constructed from $P_Y(X)$ and Y, where $P_Y \in \{P_0, P_1\}$ as shown in Figure 2. The memory location of the multiplier is then determined by the permuted index idx'.

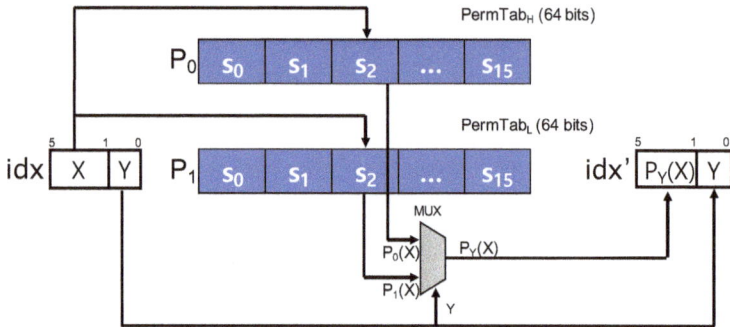

Figure 2. The process of constant-time permutation

Listing 1 presents the implementation of the constant-time permutation. The source code is written in perlasm, a x86 assembly language in the form of a perl script. In lines 1–2, the 16 bytes array of PermTab, which comprises $PermTab_H$ and $PermTab_L$, is loaded into a xmm1 register. In lines 3–9, the slots from $PermTab_H$ and $PermTab_L$, corresponding to the four leftmost bits of the index idx (denoted by \$idx), are selected in the xmm1 register, and values in those slots are loaded to r10 and r11 registers, respectively. In lines 10–17, one of the values is chosen from r10 and r11 according to the rightmost bit of \$idx, and saved to rax register. Finally, in lines 18–20, the permuted index idx' is produced from the value in rax and the rightmost bit of idx, and then loaded into \$idx as an output.

Listing 1: The assembly of constant-time permutation.

```
1:  lea     .LPermTab(%rip),%rax
2:  movdqa  0(%rax), %xmm1
3:  mov     $idx, %rax
4:  shr     \$1, %rax
5:  shl     \$2, %rax
6:  mov     %rax, %xmm0
7:  psrlq   %xmm0, %xmm1
8:  pextrq  \$1, %xmm1, %r10
9:  mov     %xmm1, %r11
10: and     \$1, $idx
11: not     $idx
12: add     \$1, $idx
13: mov     %r11, %rax
14: xor     %r10, %r11
15: and     $idx, %r11
16: xor     %r11, %rax
17: and     \$15, %rax
18: and     \$1, $idx
19: shl     \$1, %rax
20: add     %rax, $idx
```

4. Evaluation

4.1. Resistance Against Fine-Grained Cache Attacks

Suppose that an application $\mathcal{V}$ executes a modular exponentiation which is implemented with the Permute-SG technique. $\mathcal{V}$ might be a RSA application that performs a decryption with a RSA private key. By leveraging fine-grained cache attacks, an adversary $\mathcal{A}$ attempts to know the information of the multiplier (i.e., the index idx) which is used when $\mathcal{V}$ conducts the gathering phase (i.e., Gather step in Section 3.3). $\mathcal{A}$ may observe the memory offset accessed by $\mathcal{V}$ at fine-grained granularity. The offset, however, only reveals the information of P(idx). Unless $\mathcal{A}$ knows the permutation P, he/she cannot infer idx from P(idx).

$\mathcal{A}$ may attempt to learn idx by observing the memory access to the array PermTab. As described in Section 3.3, the access to the permutation table occurs in a single load instruction (Line 2 in Listing 1) and is independent on the value of idx. Therefore, it is infeasible to know the index of the accessed multiplier by observing access to the permutation table.

4.2. Adaptability

The Permute-SG is designed to be easily augmented to the existing scatter-gather implementation of the OpenSSL library. As described in Section 3.3, the modification is only required in the library at the part of the precomputation of modular exponentiation as well as the part of locating the multiplier in Scatter and Gather steps.

4.3. Computational Efficiency

We conducted some benchmarks to evaluate the computational efficiency of the proposed method. For this, an OpenSSL library of the version 1.0.2f is modified by replacing its scatter-gather part with our Permute-SG implementation. We selected this version since it is vulnerable to fine-grained cache attacks [21]. The benchmarks were performed on a server equipped with a Xeon E5-2620v4 processor (Broadwell) and a PC with a Core i7-7820HQ processor (Kaby Lake). We used a benchmarking tool included in the OpenSSL framework, and measured the speed of the RSA signing and verifying operations for each implementation.

Table 1 and Figure 3 show the benchmarking results. The terms 'SG' and 'SG-Const' refer to the unmodified OpenSSL libraries of version 1.0.2f and 1.0.2g, respectively. Both have the scatter-gather implementation, of which the SG is vulnerable to fine-grained cache attacks while the SG-Const has a countermeasure with constant-time gather procedure (See Section 2.2.2). In Figure 3, the benchmarking results are illustrated in a relative manner to give an intuitive comparison. The SG shows the fastest performance result, which comes at the cost of lacking the countermeasure against the fine-grained cache attacks. Among the implementations with the countermeasure, the Permute-SG is the fastest in all the benchmarking cases. In Broadwell processor, the Permute-SG shows almost the same performance as the SG, and is 11% faster than the SG-Const for signing operation in RSA 4096-bits. Because of the microarchitectural difference, the Permute-SG shows a little performance degradation in Kaby Lake processor, in which it still outperforms the SG-Const. It is worth noting that in RSA 1024-bits, all the implementations show the same performance, because the scatter-gather is only applied to more than RSA 2048-bits in OpenSSL.

Table 1. The result of performance evaluation.

(a) Benchmark on Xeon E5-2620v4 (Broadwell)

RSA Bits	SG		SG-Const		Permute-SG	
	sign/s	verify/s	sign/s	verify/s	sign/s	verify/s
1024	6698.2	96,683.1	6702.4	96,869.6	6555.4	96,262.6
2048	903.8	28,983.1	868.3	27,985.6	902.7	28,934.8
4096	126.4	7835.7	113.3	7400	125.7	7831.7

(b) Benchmark on Core i7-7820HQ (Kaby Lake)

RSA Bits	SG		SG-Const		Permute-SG	
	sign/s	verify/s	sign/s	verify/s	sign/s	verify/s
1024	7442.1	108,685.8	7101.9	97,294.2	7544.7	103,288.5
2048	999.6	31,684.9	889.0	28,972.2	911.6	29,650.9
4096	140.9	8748.2	116.3	7877.3	128.9	7977.5

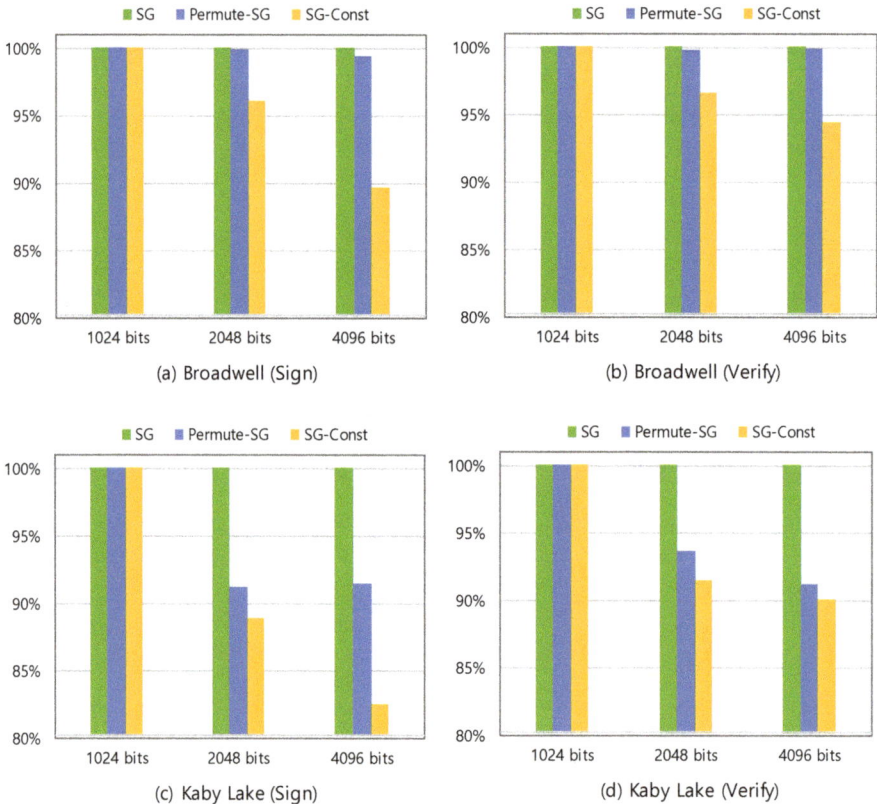

(a) Broadwell (Sign)

(b) Broadwell (Verify)

(c) Kaby Lake (Sign)

(d) Kaby Lake (Verify)

Figure 3. Comparison in the benchmarking results of scatter-gather implementations

5. Conclusions

In this paper, we proposed Permute-Scatter-Gather, a novel constant-time method for the modular exponentiation in the RSA cryptosystem. Based on the scatter-gather design, we utilized pseudo-random permutation in the construction to obfuscate memory access patterns so as to mitigate fine-grained cache attacks. Throughout rigorous evaluations, we showed that our method provides the required security, computational efficiency as well as adaptability, making it practicable in real world applications.

Author Contributions: Y.S. wrote this article. He performed analysis on the proposed method in terms of both security and performance as well.

Funding: This work was supported by the National Research Foundation of Korea (NRF) grant funded by the Korea government (MSIP) (No.2017R1C1B5015045) and by the MISP (Ministry of Science, ICT & Future Planning), Korea, under the National Program for Excellence in SW supervised by the IITP (Institute for Information & communications Technology Promotion) (2017-0-00096).

Conflicts of Interest: The authors declare no conflicts of interest.

References

1. Liu, F.; Yarom, Y.; Ge, Q.; Heiser, G.; Lee, R.B. Last-level cache side-channel attacks are practical. In Proceedings of the 2015 IEEE Symposium on Security and Privacy, San Jose, CA, USA, 18–20 May 2015; pp. 605–622.
2. Irazoqui, G.; Eisenbarth, T.; Sunar, B. S$A: A shared cache attack that works across cores and defies VM sandboxing—And its application to AES. In Proceedings of the 2015 IEEE Symposium on Security and Privacy, San Jose, CA, USA, 17–21 May 2015; pp. 591–604.
3. Gulmezoglu, B.; Irazoqui, G.; Eisenbarth, T.; Sunar, B. Cache Attacks Enable Bulk Key Recovery on the Cloud. In Proceedings of the International Conference on Cryptographic Hardware and Embedded Systems (CHES 2016), Santa Barbara, CA, USA, 17–19 August 2016; pp. 368–388.
4. Disselkoen, C.; Kohlbrenner, D.; Porter, L.; Tullsen, D. PRIME+ABORT: A Timer-Free High-Precision L3 Cache Attack using Intel TSX. In Proceedings of the 26th USENIX Security Symposium, Vancouver, BC, Canada, 16–18 August 2017; pp. 51–67.
5. Yarom, Y.; Falkner, K. Flush + Reload: A High Resolution , Low Noise , L3 Cache Side-Channel Attack. In Proceedings of the 23rd USENIX Security Symposium, San Diego, CA, USA, 20–22 August 2014; pp. 719–732.
6. Yarom, Y.; Benger, N. *Recovering OpenSSL ECDSA Nonces Using the Flush + Reload Cache Side-Channel Attack*; IACR Cryptol. ePrint Archive, Report 2014/140; International Association for Cryptologic Research: Santa Barbara, CA, USA, 2014.
7. Berk, G.; Inci, M.S.; Irazoqui, G.; Eisenbarth, T.; Sunar, B. A Faster and More Realistic Flush + Reload Attack on AES. In Proceedings of the International Workshop on Constructive Side-Channel Analysis and Secure Design (COSADE 2015), Graz, Austria, 14–15 April 2015; pp. 111–126.
8. Gruss, D.; Spreitzer, R.; Mangard, S. Cache Template Attacks: Automating Attacks on Inclusive Last-Level Caches. In Proceedings of the 24th USENIX Security Symposium, Washington, DC, USA, 12–14 August 2015; pp. 897–912.
9. Gruss, D.; Maurice, C.; Wagner, K.; Mangard, S. Flush+Flush: A Fast and Stealthy Cache Attack. In Proceedings of the 13th International Conference on Detection of Intrusions and Malware, and Vulnerability Assessment (DIMVA 2006), Berlin, Germany, 13–14 July 2016; pp. 279–299.
10. Zhang, Y.; Juels, A.; Reiter, M.K.; Ristenpart, T. Cross-Tenant Side-Channel Attacks in PaaS Clouds. In Proceedings of the 2014 SIGSAC ACM Conference on Computer and Communications Security (CCS 2014), Scottsdale, AR, USA, 3–7 November 2014; pp. 990–1003.
11. Ge, Q.; Yarom, Y.; Cock, D.; Heiser, G. A Survey of Microarchitectural Timing Attacks and Countermeasures on Contemporary Hardware. *J. Cryptogr. Eng.* **2018**, *8*, 1–27. [CrossRef]
12. Garman, C.; Green, M.; Kaptchuk, G.; Miers, I.; Rushanan, M. Dancing on the Lip of the Volcano: Chosen Ciphertext Attacks on Apple iMessage. In Proceedings of the 25th USENIX Security Symposium is sponsored by USENIX, Austin, TX, USA, 10–12 August 2016.
13. García, C.P.; Brumley, B.B. Constant-Time Callees with Variable-Time Callers. In Proceedings of the 26th USENIX Security Symposium, Vancouver, BC, Canada, 16–18 August 2017; pp. 83–98.

14. Genkin, D.; Valenta, L.; Yarom, Y. May the Fourth Be With You: A Microarchitectural Side Channel Attack on Several Real-World Applications of Curve25519. In Proceedings of the 2017 ACM SIGSAC Conference on Computer and Communications Security (CCS 2017), Dallas, TX, USA, 30 October–3 November 2017; pp. 845–858.

15. Kaufmann, T.; Pelletier, H.; Vaudenay, S.; Villegas, K. When constant-time source yields variable-time binary: Exploiting curve25519-donna built with MSVC 2015. In Proceedings of the 15th International Conference on Cryptology and Network Security (CANS 2016), Milan, Italy, 14–16 November 2016; pp. 573–582.

16. Gopal, V.; Guilford, J.; Ozturk, E.; Feghali, W.; Wolrich, G.; Dixon, M. Fast and Constant-Time Implementation of Modular Exponentiation. In Proceedings of the Embedded Systems and Communications Security, Niagara Falls, NY, USA, 27 September 2009.

17. Ronald, L.; Rivest, A.S.; Adleman, L. A method for obtaining digital signatures and public-key cryptosystems. *Commun. ACM* **1978**, *21*, 120–126.

18. OpenSSL. *OpenSSL, Cryptography and SSL/TLS Toolkit.* Available online: https://www.openssl.org/ (accessed on 5 August 2018).

19. Brickell, E. Technologies to improve platform security. In Proceedings of the Workshop on Cryptographic Hardware and Embedded Systems, Nara, Japan, 28 September–1 October 2011.

20. Brickell, E. The impact of cryptography on platform security. In Proceedings of the CT-RSA 2012, San Francisco, CA, USA, 27 February–2 March 2012.

21. Yarom, Y.; Genkin, D.; Heninger, N. CacheBleed: A Timing Attack on OpenSSL Constant Time RSA. *J. Cryptogr. Eng.* **2017**, *7*, 99–112. [CrossRef]

22. Bos, J.; Coster, M. Addition chain heuristics. In Proceedings of the Conference on the Theory and Application of Cryptology CRYPTO, Santa Barbara, CA, USA, 20–24 August 1989.

23. Ernie Brickell, G.G.; Seifert, J.P. Mitigating cache/timing based side-channels in AES and RSA software implementations. In Proceedings of the RSA Conference 2006 Session DEV-203, San Jose, CA, USA, 13–17 February 2006.

24. Fog, A. *The Microarchitecture of Intel, AMD and via CPUs: An Optimization Guide for Assembly Programmers and Compiler Makers*; Technical University of Denmark: Lyngby, Denmark, 2016.

25. Intel. *Intel 64 and IA-32 Architectures Optimization Reference Manual (April 2018)*; Intel: Santa Clara, CA, USA, 2018.

26. Intel. *Intel 64 and IA-32 Architectures Software Developer Manuals (March 2018)*; Intel: Santa Clara, CA, USA, 2018.

applied
sciences

MDPI

Article

Chaos-Based Physical Unclonable Functions

Krzysztof Gołofit * and Piotr Z. Wieczorek *

Institute of Electronic Systems, Faculty of Electronics and Information Technology,
Warsaw University of Technology, Nowowiejska 15/19, 00-665 Warsaw, Poland
* Correspondence: K.Golofit@elka.pw.edu.pl (K.G.); P.Z.Wieczorek@elka.pw.edu.pl (P.Z.W.);
  Tel.: +48-22-234-7634 (K.G.); +48-22-234-7336 (P.Z.W.)

Received: 30 January 2019; Accepted: 5 March 2019; Published: 9 March 2019

Abstract: The concept presented in this paper fits into the current trend of highly secured hardware authentication designs utilizing Physically Unclonable Functions (PUFs) or Physical Obfuscated Keys (POKs). We propose an idea that the PUF cryptographic keys can be derived from a chaotic circuit. We point out that the chaos theory should be explored for the sake of PUFs as a natural mechanism of amplifying random process variations of digital circuits. We prove the idea based on a novel design of a chaotic circuit, which utilizes time in a feedback loop as an analog continuous variable in a purely digital system. Our design is small and simple, and therefore feasible to implement in inexpensive reprogrammable devices (not equipped with digital clock manager, programmable delay line, phase locked loop, RAM/ROM memory, etc.). Preliminary tests proved that the chaotic circuit PUFs work in both advanced Field-Programmable Gate Arrays (FPGAs) as well as simple Complex Programmable Logic Devices (CPLDs). We showed that different PUF challenges (slightly different implementations based on variations in elements placement and/or routing) have provided significantly different keys generated within one CPLD/FPGA device. On the other hand, the same PUF challenges used in a different CPLD/FPGA instance (programmed with precisely the same bit-stream resulting in exactly the same placement and routing) have enhanced differences between devices resulting in different cryptographic keys.

Keywords: physically unclonable function; chaos theory; chaotic circuit; FPGA; CPLD; challenge-response authentication; hardware security; side-channel attacks; cryptographic keys

---

1. Introduction

Modern cryptography is facing progressively more attacks directed not on cryptographic algorithms, but on their implementations—even the secured ones [1]. Among many kinds of side-channel attacks (SCAs), there are various ways of retrieving information from memories, where the cryptographic keys are kept [2,3], and therefore there is a struggle for securing the memories against SCAs [4]. At the same time, there is an increasing demand for secure cryptography with an application in small and inexpensive circuits (like Internet of Things devices, wearables, implantable medical devices, etc. [5,6]). These are the reasons why PUFs are drawing more and more attention in modern secure electronics—they can make it possible to create "a vault" for cryptographic keys, without building an actual vault [7,8].

The advantages of particular PUFs (as well the chaotic PUF presented in this paper) in the fields of cryptography and security include the following:

- PUF keys are usually not present in the system (cryptographic keys are not kept in any volatile memory, non-volatile memory, latches nor registers);
- a key is temporarily activated (re-generated) when it is required in the system;
- a key can be activated only by its owner (by the use of owner's initiation vector—the *PUF challenge*);

- keys are unique and different for every instance of a similar device (programmed in the same way, with the use of the same code, and operating on the same data);
- keys cannot be copied, cloned nor extracted from the device, as well as they are tamper-proof (any attempt of tampering should destroy the keys).

Such features can be utilized in various attractive ways:

- unambiguous and incontestable identification of a unit;
- authentication, digital signature, encryption/decryption;
- owner/manufacturer authentication (e.g., for the use of certified updates, preventing hacking);
- immunity to spoofing, cloning, reverse engineering, and man-in-the-middle attacks.

The advantages and applications of PUFs initiated a search for methods of harvesting of PUF keys from physical processes. Among proposed solutions, we can find: ring oscillators [7,9], transient effect ring oscillators [10], dynamic ring oscillators [11], ordering-based ring oscillator [12], convergence time of bistable rings [8], sneak paths in the resistive X-point array [13], power consumption differences of Advanced Encryption Standard Sbox inversion functions [14], occurrence of metastability [15], static memory [16,17], dynamic memory [18,19], switching behavior of emerging magneto-resistive memory devices [20], switching of resistive random access memory [21], reduction–oxidation resistive switching memories [22], decay-based Dynamic Random Access Memory [23], locally enhanced defectivity [24], combination of multiplexers and arbiters [25], wireless sensors [26], Complementary Metal-Oxide Semiconductor image sensors [27], nonlinearities of data converters [28], mismatch of capacitor ratios [29], primitive shifting permutation network (barrel shifter) [30], cellular neural networks [31], customized dynamic two-stage comparator [32], and many others.

Among the ideas, there are various ways of using ring oscillators for the sake of PUFs (containing an odd number of inverters); however, a ring consisting of an even number of inverters can stabilize in only one of two states when powered up or, more generally, when it is initiated from an unstable state. Such an architecture is called bistable ring PUF (BR-PUF) [33–35]; however, in this particular application, the inverters were replaced with more suitable cells that provide an easy cell reset (by the use of NOR gates or a dedicated architecture) as well as the ability to chose one of two gates (for the sake of PUF challenges). Nevertheless, there is a complex behavior involved (complex feedback situation causes oscillations that may take a long time until the whole ring converges to a stable state), which depends on the process variation mismatch of, e.g., the threshold voltage and carrier mobility of transistors, and noise. In other words, the idea amplifies an instance process variations and converts it to a PUF key.

The definition of chaos applied to deterministic dynamical systems involves sensitive dependence on initial conditions ([36], p. 736). If these initial conditions are purely (or mainly) hardware based, it is grounds for a PUF. Chaotic circuits offer high quality randomness, but they require either full custom or discrete implementations of analogue circuits (e.g., [37–39]). Digital chaotic implementations (incorporating reprogrammable devices) suffer from limited computational precision resulting in recurring sequences and the pseudo-random output [40–42]. However, deterministic circuits can be very simple in design and the chaotic process can produce time series, which seem to be unpredictable to the observer, due to the sophisticated dynamic behavior in the limited observation time [43]. It turns out that chaotic systems described with simple linear one dimensional formulas can produce very complex circuit behavior [44]. In such systems, the "unpredictability" results from the sensitivity to an initial condition, which affects the circuit's state in time. In this paper, we propose the solution, in which the PUF keys are harvested from a chaotic circuit. The proposed circuit recursively amplifies instance differences of their electronic devices over a time.

2. Chaos-Based PUFs

In order to tackle the described issues and join the advantages of analog chaotic signals with digital simplicity, it would be very valuable to identify an analog continuous variable that could be utilized in purely digital circuits. Therefore, we propose a concept of system with a continuous time variable (δ) that manifests chaotic behavior. For this purpose, we base our PUF on recently introduced concept [45] with switchable chain ring oscillators (SCROs)—the idea incorporates a pair of SCROs (SCRO1 and SCRO2) as shown in Figure 1.

An SCRO consists of two switchable delay lines formed of inverters (DLa and DLb) and it operates at one of two frequencies (f_1, f_2) that are never equal:

$$\begin{cases} f_1 = \frac{1}{2\tau_a}, & f_2 = \frac{1}{2(\tau_a+\tau_b)}, \\ f_1 = \frac{1}{2(\tau_a+\tau_b)}, & f_2 = \frac{1}{2\tau_a}, \end{cases} \tag{1}$$

where τ_a and τ_b delays correspond to the DLa and DLb propagation times. Rising slopes at SCRO outputs are detected by a phase detector (PD), which acts as an arbiter providing logical $s[m] = 1$ when the rising slope of SCRO1 appears before SCRO2 ($\delta[m] > 0$ for the m-th comparison). The logical $s[m] = 0$ occurs in the other case ($\delta[m] < 0$). The simplest PD implementation consists of one D-flip-flop.

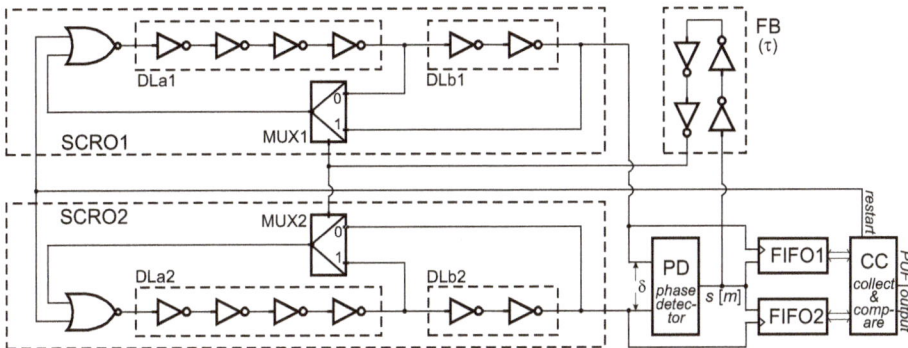

Figure 1. The block diagram of the proposed chaos-based PUF circuit.

The feedback signal from the PD instantly aims to adjust SCRO phase (δ) but never succeeds. Moreover, the feedback always comes a little late due to delays (τ) of the feedback loop (FB). The higher feedback delay results in the greater range of possible SCROs phases. If τ was negligible, the circuit would operate in a periodic mode:

$$\delta[m+1] = \begin{cases} \delta[m] - 2\tau_b, & \delta[m] > 0, \\ \delta[m] + 2\tau_b, & \delta[m] < 0, \end{cases} \tag{2}$$

but, if the τ delay exceeds $\tau_a - \tau_b$ (i.e., $\tau_a + \tau_b > \tau > \tau_a - \tau_b$), the phase correction signal (the slope of a logical value change) for the higher frequency SCRO will not arrive on time (both SCROs will work on the same frequency for a moment). Moreover, if the τ delay exceeds $\tau_a + \tau_b$, the phase correction appears too late also for the other SCRO (the one operating at lower frequency) and there will be no phase correction on time. An example series of δ corrections for consecutive m steps ($\delta[m]$) obtained in the Xilinx CoolRunner II (XC2C256) (San Jose, CA, USA) is shown in Figure 2.

Three realizations start from the same state (phase) and a wrong phase correction value ($m = 1$). They properly correct their phases in the next four steps ($m : 2, 3, 4, 5$). Steps 6 and 7 (8 and 9; 10 and 11, etc.) show the case where the phase correction came too late for one of the SCROs.

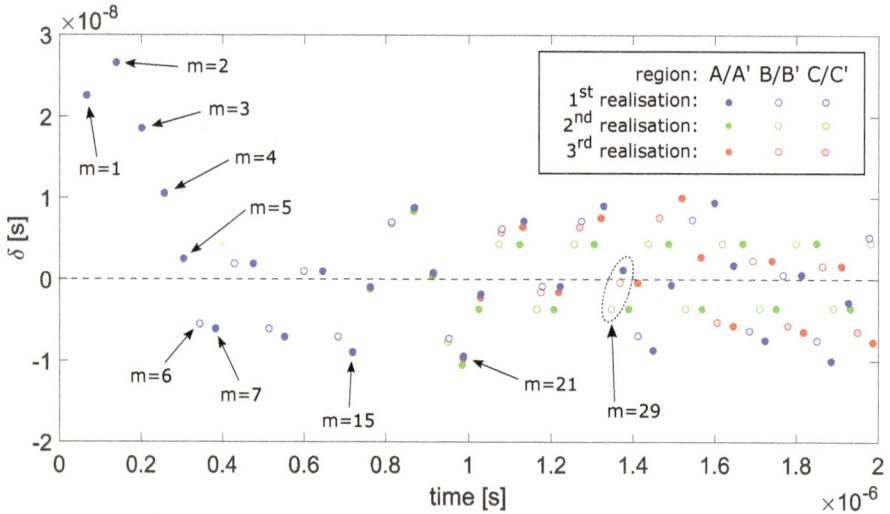

Figure 2. Deterministic chaotic $\delta[m]$ time series (in CPLD XC2C256).

Usually, a chaotic 1D map is the easiest way to visualize chaotic behavior [46]; therefore, Figure 3 shows a graph $\{\delta[m+1], \delta[m]\}$ of consecutive m steps.

One can see that there are three types of regions (A, B, C) associated with the circuit's behavior depending on δ and complementary regions (A', B', C') since the phase adjustments are identical for the both positive and negative δ values. The standard regions of operation (A and A') can be understood as proper δ adjustments (as if there was no influence of τ delay). Regions B and B' correspond to the case when the adjustment in two subsequent steps does not apply (mainly due to the τ delay). The classification of the steps to the regions was also marked in Figure 2.

There is an important phenomenon that accelerates divergence of a system's chaotic trajectory—mainly inconsistency of a logical level at multiplexer's inputs (MUX1 or MUX2).

If the switching of an SCRO occurs at one of the slopes, it may cause a voltage discontinuity at the MUX output. This glitch, after it propagates through a number of inverters, causes a slight shift of the rising edge in one of SCROs, in which the inconsistency occurred. These states' inconsistencies can be observed as extensions of linear B and B' regions and were marked as C and C' in Figure 3 (as well as in Figure 2). Another phenomenon occurs when the rising edges of SCROs are close enough to each other ($\delta \approx 0$). In this case, the PD classification result (of SCRO slope priority) may come a little late due to occurrences of metastability [15]. Such events are quite rare (but possible); nevertheless, they also influence the delay of the feedback correction signal (see [45]). In very rare cases of $\delta \approx 0$, the PD classification result (s) can be even wrong (a random value, due to metastability occurrence).

Such a chaotic system is very sensitive to tolerances of parameters of all electronic devices used in the design. Its sensitivity originates from all the physical parameters affected by process variations (transistors and paths geometry, material heterogeneity and as a result the electrical parameters). For this reason, any change in the circuit's structure (e.g., the use of a different inverter or different path connecting the same elements) influences its behavior—it also opens the door to creation of various challenge vectors (in the PUF's *challenge–response* system). It is also the reason why every instance of an identical circuit (having exactly the same placement and routing) behaves in a different way. One can see in the example in Figure 2 that, after several steps ($m : 15$–20), the circuit's trajectory starts different paths and, beginning from $m = 29$, the PD classifications also differ. This step ($m = 29$) is a specific turning point for this particular implementation (particular structure). If the change in the circuit's structure (for the sake of PUF's challenge) is not an option, the paths or inverters can be easily multiplexed—for example, in the way it is implemented in BR-PUFs [33].

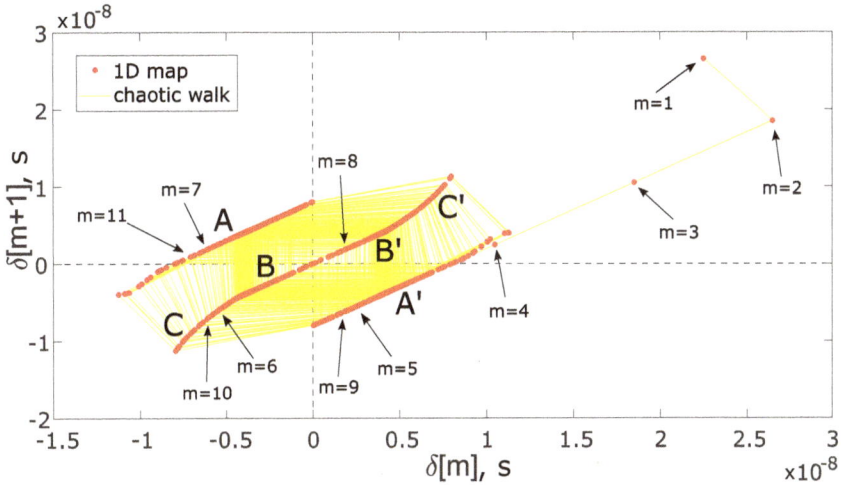

Figure 3. Example chaotic map of $\delta[m]$ with a few initial steps (m).

Since the SCROs can be considered as free running ring oscillators, there is obviously a phase walk present. At some point, it becomes to influence the circuit's behavior resulting in circuit's random state trajectory (as utilized in [45] for the use of true random number generator). If there was no phase walk (and other stochastic physical processes), the circuit would be a perfect deterministic chaotic system (with only the "initial conditions" determining its further behavior). However, after a period of time (a number of m steps, corresponding to m PD classifications), the circuit turns into a random number generator (RNG). Nevertheless, before it happens, and, after instance's trajectory begins to differ, there is a time window for a PUF extraction as moments of randomness initialization—depicted in Figure 4.

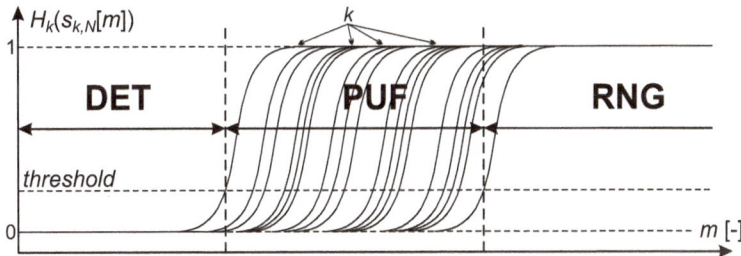

Figure 4. Model of a time-based window for PUF extraction.

Multiple runs ($n \in N$) of the same instances ($k \in K$) of the circuit in Figure 1 result in the same behavior till the moment when the stochastic physical processes start to randomize subsequent chaotic trajectories. It can be easily measured and observed by entropy values H_k obtained from various realizations of $s_N[m]$ (e.g., $|N| = 50$) for adjacent steps ($m = 0, 1, 2, ...$) within one of instances. For example, if the output produced by one instance (k) at a specific step (m) for each of the (N) realizations is the same (either 1 or 0), then $H_k = 0$. On the other hand, if the half of the realizations at this step have different output values than the other half, then $H_k = 1$. When $H_{th} = 0.5$ is reached, only a single bit out of nine has a different value (specifically $H_k = 0.50326$). Following the definition of Shannon's entropy [47], the order of ones and zeros does not matter. This way, three different ranges of operation can be distinguished depending on the values of H. The first manifests the deterministic behavior for all instances (DET in Figure 4). The second range manifests different moments (for different instances) for the circuit to operate in the non-deterministic way (PUF key extraction). In the third range all of the instances manifest random behavior (RNG—random number generator).

3. Behavioral Modeling

The chaotic circuit described with a 1D map must have a region of operation, in which the inclination coefficient of subsequent values of the state variable (δ) is $1 < k < 2$, where $\delta[m+1] = k\delta[m] + q$ and k, q are the constants which depend on the circuit parameters. The circuit shown in Figure 1 reveals a chaotic behavior due to the presence of C and C' regions with $k > 1$, as shown in Figure 3. These regions of operation result from the delay in the feedback loop (FB), which causes the logical state inconsistency in C and C' ranges of operation. In order to extensively verify the chaotic behavior of the system proposed in Figure 1, we have implemented its behavioral model in a Matlab Simulink environment (Matlab R2017a, The MathWorks, Inc., Natick, MA, USA). In the model, we have assumed linear and lumped output inverter resistance R_o and input capacitance C_o (of adjacent inverter attached in the ring), as shown in Figure 5.

These two lumped components are responsible for the finite rising/falling edges of signals. Subsequently, we have assumed the sigmoid transfer function (hyperbolic tangent) of direct current inverter characteristics, constant signal delay t_{pd}, and a Gaussian noise process $N(t)$ affecting the threshold level of the inverter (its transfer function). The $N(t)$ process of our model represents the phase walk of SCROs in their physical implementation, and allowed us to obtain non-deterministic circuit behavior, available in the physical implementations either in FPGA or CPLD. All the parameters R_o, C_o, t_{pd}, $N(t)$ (standard deviation) and inclination of the transfer function have been adjusted to obtain identical operation of SCROs in the Simulink model and physical implementation. The detailed method of parameter extraction is explained in [48]. This way, we have obtained identical phase-walk (and jitter), f_1 and f_2 frequencies of the model and the circuit physically implemented in a programmable device. With the use of the Simulink, we simulated the chaotic behavior of the circuit proposed in Figure 1. For this reason, we performed multiple transient analysis, in which the DLa1, DLb1, DLa2, DLb2, FB parameters (i.e., t_{pd} and R_o, C_o) were subjected to 1% dispersion (Monte Carlo analysis). During the analysis, the series of $\delta[m]$ values were registered for subsequent K sets of randomly generated DLa1, DLb1, DLa2, DLb2, FB parameters. Each transient analysis for a particular parameter set was repeated N times (realizations) in order to observe the non-deterministic circuit behavior of a particular parameter set. The experiment scheme can be described according to formula (3):

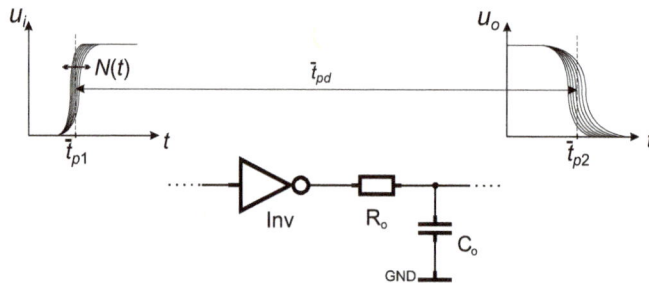

Figure 5. Model of an inverter.

$$
\delta_0[N, M] = \begin{cases}
\delta_{k=0,n=0}[0], & \delta_{k=0,n=0}[1], & \delta_{k=0,n=0}[2], & \cdots & \delta_{k=0,n=0}[M] \\
\delta_{k=0,n=1}[0], & \delta_{k=0,n=1}[1], & \delta_{k=0,n=1}[2], & \cdots & \delta_{k=0,n=1}[M] \\
\delta_{k=0,n=2}[0], & \delta_{k=0,n=2}[1], & \delta_{k=0,n=2}[2], & \cdots & \delta_{k=0,n=2}[M] \\
\vdots & \vdots & \vdots & \ddots & \vdots \\
\delta_{k=0,n=N}[0], & \delta_{k=0,n=N}[1], & \delta_{k=0,n=N}[2], & \cdots & \delta_{k=0,n=N}[M],
\end{cases}
$$

$$
\delta_1[N, M] = \begin{cases}
\delta_{k=1,n=0}[0], & \delta_{k=1,n=0}[1], & \delta_{k=1,n=0}[2], & \cdots & \delta_{k=1,n=0}[M] \\
\delta_{k=1,n=1}[0], & \delta_{k=1,n=1}[1], & \delta_{k=1,n=1}[2], & \cdots & \delta_{k=1,n=1}[M] \\
\delta_{k=1,n=2}[0], & \delta_{k=1,n=2}[1], & \delta_{k=1,n=2}[2], & \cdots & \delta_{k=1,n=2}[M] \\
\vdots & \vdots & \vdots & \ddots & \vdots \\
\delta_{k=1,n=N}[0], & \delta_{k=1,n=N}[1], & \delta_{k=1,n=N}[2], & \cdots & \delta_{k=1,n=N}[M],
\end{cases}
\tag{3}
$$

$$
\vdots
$$

$$
\delta_K[N, M] = \begin{cases}
\delta_{k=K,n=0}[0], & \delta_{k=K,n=0}[1], & \delta_{k=K,n=0}[2], & \cdots & \delta_{k=K,n=0}[M] \\
\delta_{k=K,n=1}[0], & \delta_{k=K,n=1}[1], & \delta_{k=K,n=1}[2], & \cdots & \delta_{k=K,n=1}[M] \\
\delta_{k=K,n=2}[0], & \delta_{k=K,n=2}[1], & \delta_{k=K,n=2}[2], & \cdots & \delta_{k=K,n=2}[M] \\
\vdots & \vdots & \vdots & \ddots & \vdots \\
\delta_{k=K,n=N}[0], & \delta_{k=K,n=N}[1], & \delta_{k=K,n=N}[2], & \cdots & \delta_{k=K,n=N}[M].
\end{cases}
$$

This way, we obtained $N \times K$ bit-strings of δ variable series sets and analyzed the standard deviations of the $\delta_{K,N}[0]$, $\delta_{K,N}[1]$, ..., $\delta_{K,N}[M]$ sets. After that, we were able to analyze the fluctuations of δ standard deviation in time $(\sigma_\delta(t))$, depending on the inter-class tolerance of circuit parameters. Figure 6 shows the δ standard deviation as a function of time (and $m \in M$) for particular instances $(k \in K)$.

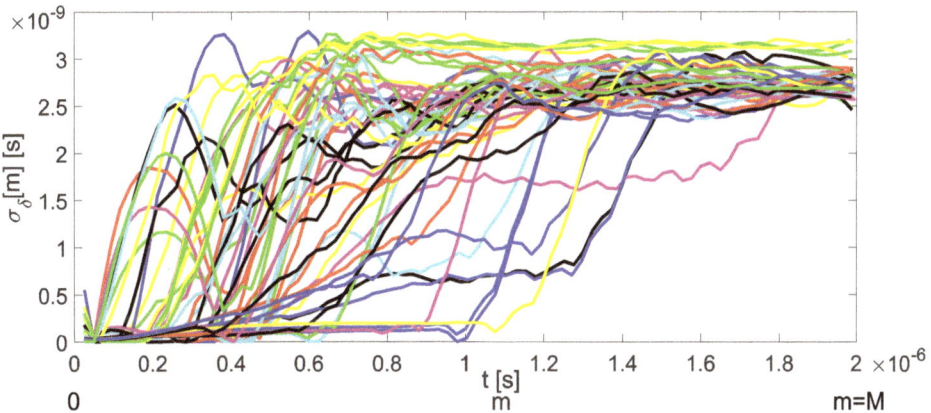

Figure 6. Standard deviation of δ as a function of time (m) for 50 different instances of circuit model $(K = 50)$.

One can see that the standard deviation of δ rapidly rises at a certain moment of time (m); however, the moment of rapid $(\sigma_\delta(t))$ increase depends on the particular system's parameter set (tolerance). The simulation results in Figure 6 clearly show that the time or the number of PD classifications $(0 < m < M)$ is a variable that distinguishes instances of the circuit, whereas this particular moment (m) divides the chaotic operation of the circuit in Figure 1 to either deterministic $(\sigma_\delta = 0)$ or non-deterministic $(\sigma_\delta > 0)$ operation.

Results in Figure 6 show the random behavior of the circuit's state variable δ.

The proposed chaotic circuit is asynchronous; therefore, δ is a continuous variable. In order to easily determine the moment of the abrupt change of the circuit's mode of operation (i.e., from deterministic to non-deterministic), it would be much easier to assess its mode according to the bit-string produced at the output of the PD block (see Figure 1). It is obvious that, when δ starts to act as a random variable, the PD circuit produces a random bit-string. An example of such a PD operation obtained in Simulink for a single instance is shown in Figure 7.

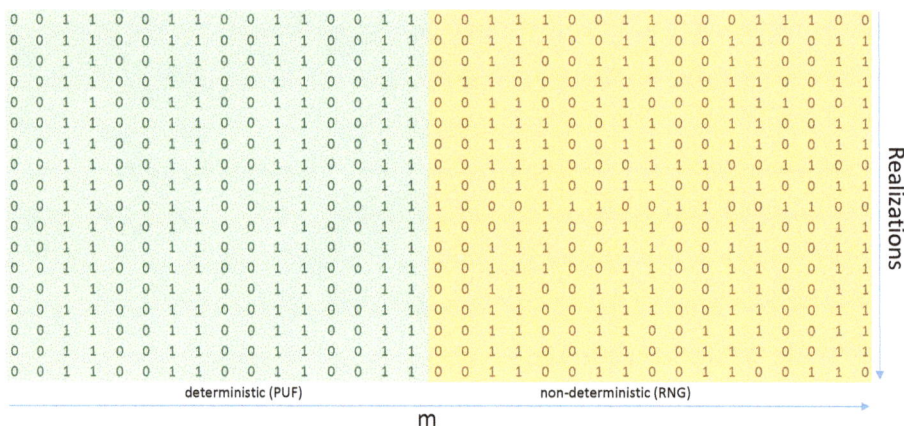

Figure 7. Bit-strings obtained for multiple realization in a single instance.

One can see that the PD bit-string generated in multiple realizations (within a single instance) can be divided into two regions, the first, in which each m-th element is independent from the realization, and the second, where the m-th bit-string element depends on the realization number (circuit run). In the first case, the Shanon entropy (calculated vertically) is approximately 0, whereas the second case yields Shannon entropy close to 1. Example results in Figure 7 show that the iteration number (m) that distinguishes the deterministic from non-deterministic operation could act as a PUF response.

A complete simulation according to (3) has been performed, whereas the Shannon entropy of bit-strings (see Figure 7) was used to determine the critical m iteration number (PUF response). The results of Monte Carlo simulation (with a use of a Savitzky–Golay filter due to sudden changes in values) are shown in Figure 8. The vertical lines indicate the moments dividing the deterministic from non-deterministic chaotic operations.

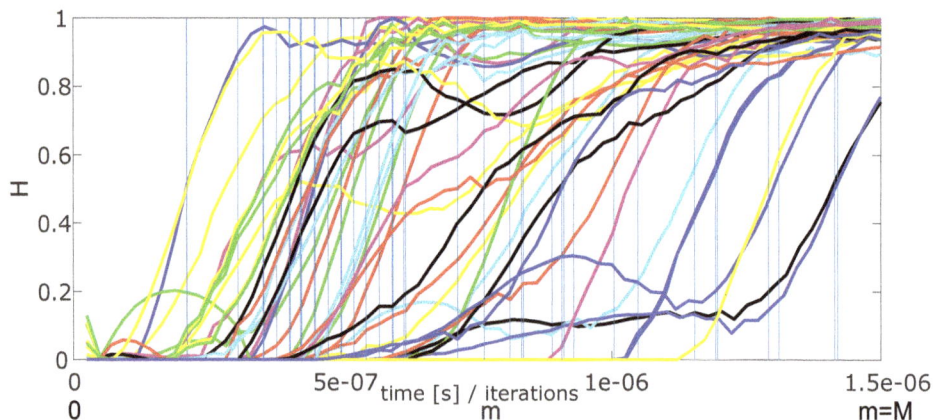

Figure 8. Shannon entropy (H) as a function of m for 50 instances.

The results in Figure 8 clearly show that, for an arbitrarily chosen H threshold (e.g., $H_{th} = 0.5$), the m value dividing types of operation acts as a PUF response (e.g., $\{m \mid H(m) = 0.5\}$). During simulations in Simulink, we have obtained critical m values ranging from 16 to 53, which yields 5.2 bits of a single PUF response.

We have also checked the temperature impact on the modeled device. For this purpose, the propagation delay of all modeled inverters and their jitter were subjected to temperature influence. Both the propagation delay and noise of inverters result from the physical phenomena typical for CMOS technology; however, in the case of jitter modeling of ring oscillators formed of inverters, it is hard to distinguish the influence of thermal Gaussian noise from the shot noise. For this reason, we have measured the inverter noise parameters by the measurement of jitter of physically implemented ring oscillators (in FPGA) in various temperatures (stabilized with a Peltier module). This way, we obtained the standard deviation of noise process for various temperatures in different devices (FPGA). The method was explained in details in [48]; moreover, it was also utilized for the measurements of thermal drift of average propagation delay, not resulting from the noise processes. Both the noise standard deviation and propagation delay thermal coefficients were used in the behavioral modeling (in Simulink), in order to evaluate the temperature impact on the critical m-value. We performed the Monte Carlo analysis of the circuit shown in Figure 1 with the experiment scheme (3) in 260 K. Furthermore, we repeated the experiment with the same set of randomly distributed tolerances of instances at 300 K. This way, we obtained two sets of critical m values, i.e., M_{270K} and M_{300K}. The T-test (*ttest2* in Matlab environment) applied to m variables from M_{270K} and M_{300K} corresponding to the same instances revealed that they fall into distributions with slightly different mean values. It turned out that a 40 K change in temperature results in -4.12 change in mean value of obtained m for the same instances. Therefore, the simulation yields $-0.103\frac{1}{K}$ critical m sensitivity (and therefore PUF sensitivity) to temperature.

4. Testing and Results

The preliminary verification was based on the implementations in five Xilinx Cool-Runner II CPLDs (three XC2C256 devices and two XC2C64) as well as nine devices of Xilinx Artix-7 XC7A100T FPGAs (CSG324ABX1625/1629). The $s_N[m]$ bit-strings were acquired with the use of the Texas Instruments DK-TM4C123G board (Dallas, TX, USA) connected to the PUF output as well as PD output signals were observed and measured with both oscilloscope and Agilent 53230A timer (currently Keysight, Santa Rosa, CA, USA). We have evaluated series of circuit's architectures by changing the numbers of inverters (implemented as look-up tables—LUTs) in each type of delay lines (DLa, DLb, FB). Consequently, we have selected for the following research the architecture that consists of: 13 LUTs in the DLa (DLa1, DLa2), six LUTs in the DLb (DLb1, DLb2) and nine LUTs in the FB (as τ). It is worth mentioning that the circuit's architecture, which can be seen just as numbers of standard delays in each of the chains (DLa, DLb, FB), is perfectly scalable and can be used in various devices made in various technology processes. For that reason, we have chosen for tests two quite faraway models of devices: 0.18 μm CPLDs and 28 nm FPGAs.

Every circuit was initiated with eight different PUF challenges (eight different SCRO elements placement or routing) and the implementations were exactly copied to each of instances ($k \in K$). The number of steps, after which the entropy (H_k) abruptly increases, was estimated as a distance between the system initialization and the moment when $H_k > 0.5$ (see Figure 4). The distance, measured as a number of consecutive δ comparisons was used as a PUF response. Analysis of five CPLDs and nine FPGAs indicated a linear correlation between challenges and responses within one device (intra-class correlation). On the other hand, the same implementations in different devices showed a uniqueness of each device (lack of inter-class correlation).

Figure 9 demonstrates a few examples of PD output bit-strings ($s_N[m]$) in a form of binary bitmaps—50 runs (n) of 200 steps (m) for eight example FPGA instances (K1–K8) and three challenges (C1, C2, C3). Derived keys are basically the numbers of steps (m) after which the deterministic chaos ends and the non-deterministic chaos begins. In the tests, the PUF time window (depicted in Figure 4) began at the value $m = 9$ and ended with $m = 197$ (therefore the observation window was shortened to $m = 200$ steps). Such a range (in binary representation) results in 7,6 bits of a key. The dispersion of the moment when H leads effectively to $\sim$6–7 bits of entropy with just a single challenge. The difference in the PUF time window between simulations and the physical implementations likely involves two factors. First, the model did not result from the actual technological spread that occurred at manufacture of devices. Second, the research involved only several devices—we cannot conclude about PUF statistical occurrences nor keys randomness based on such a small number of instances. For that reason, future research should follow.

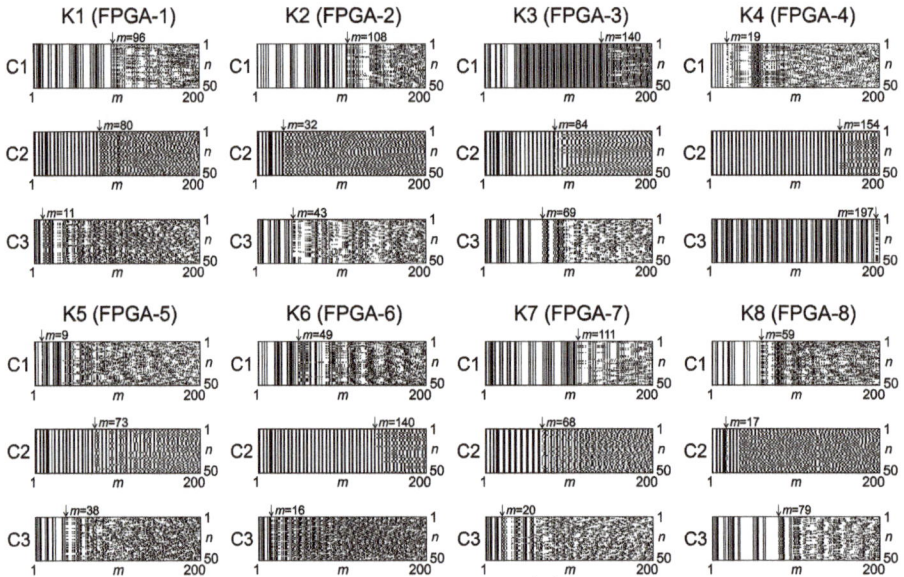

Figure 9. Phase detector output bit-strings ($s_N[m]$) and extracted PUF keys.

Taking into consideration practical aspects of retrieving PUF keys from a single instance, it is apparent that multiple bit-strings are required, but both the number of the streams as well as the required length of the stream vary. Since the 0.5 Shannon's entropy level is not very demanding in terms of variance or Hamming distance (the half value of H is reached when only one of nine bits differs), basically the first different bit-string can indicate the beginning of non-deterministic chaotic operation—as a matter of fact, the first different $s[m]$ value between the bit-strings. After the first difference occurs, the remaining part of the bit-string is redundant (nevertheless, it should be generated because of the vulnerability to timing SCAs). Consequently, the minimum number of bit-strings is two, whereas the maximum results from acceptable entropy uncertainty, and as a consequence, the acceptable critical m fluctuations. For this reason, we suggest to either generate a fixed number of bit-strings (e.g., $N = 10$) or to stop when the first bit-string differs from the others. In each of the procedures, there is always a possibility that a rare event for lower m may cause disturbance resulting in an incorrect m value. The error can be avoided by increasing the number of bit-strings, but it may be more efficient to ignore the result (detect the error with a simple cyclic redundancy check code) and repeat the key generation procedure. The number of bit-strings required to generate the m values for all of the examples presented in Figure 9 varies from 2 to 6—on average, 3.04 bit-strings per one generation.

In order to evaluate limitations of the proposed solution, we estimated critical m-values for eight Artix (Xilinx) and eight Cyclone 5 (Altera/Intel (Santa Clara, CA, USA)) devices, where each consisted of three independent PUFs. This way, we were able to estimate a histogram of critical m-values (keys), shown in Figure 10. The measured inter-class randomness of PUF estimated with standard deviation was σ = 52.6. Despite the limited probe size (16 devices × 3 PUFs), one can see in Figure 10 that the *m* distribution is asymmetric and the most likely *m* values are located in the 10–50 range (obtained also in simulation). The asymmetric *m* distribution results strictly from the influence of noise on the trajectory of chaotic system. In such a system, the location of subsequent points (see Figure 3) is strongly affected by the slight phase fluctuations (resulting from noise processes) during subsequent system states. In other words, the longer circuit operates, the higher the number of transitions near *C* and *C′* is. Therefore, it is unlikely to maintain the same deterministic behavior for high *m*-values in subsequent realizations by the system.

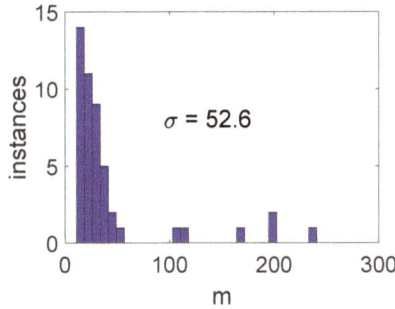

Figure 10. Inter-class PUF (*m*) distribution. Randomness of proposed PUF was measured on 16 devices (eight Xilinx Artix and eight Intel Cyclone V) with three PUFs each (48 instances).

In the proposed PUF solution, the chaotic operation results from *C* and *C′* ranges of operation, which ensure $1 < k < 2$ inclination in the chaotic map. The tolerances present in the system affect both the inclination of *C* and *C′* ranges (sections of chaotic map) and the length of these sections. Therefore, tolerances affect the probability of entering *C* and *C′* ranges of operation, and, in turn, the probability of entering different (random) trajectory paths.

The other parameter vital to PUFs is its reliability, which identifies the intra-class randomness. The 48 PUF instances used for randomness extraction in Figure 10 were subjected to challenge–response operation multiple times in the same operating conditions. This way, the *m* error distribution was obtained, which is shown in Figure 11. The mean value of obtained distribution μ ≈ 0, whereas the standard deviation σ = 6.94. It is worth mentioning that *m* error mainly results from the non-monotonic character of $H(m)$ dependence (see Figure 8), which affects the estimate of critical *m*-value based on the threshold level. Nevertheless, over 80% of intra-class *m* error corresponds to 2-bit Hamming distance, whereas the maximum Hamming distance corresponds to 3-bits.

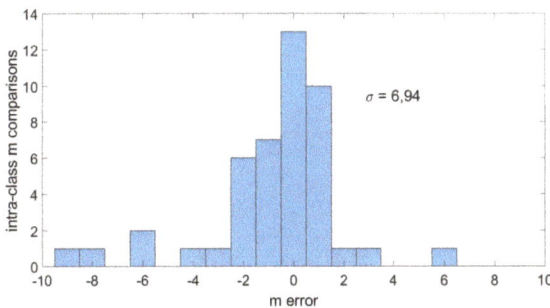

Figure 11. Intra-class PUF (*m*) distribution. Reliability of proposed PUF measured on 16 devices (eight Xilinx Artix and eight Intel Cyclone V) with three PUFs each (48 instances).

We have also investigated the influence of temperature on the critical m-value. For this purpose, 100 realizations (runs), of $m = 250$ steps long each (within each PUF instance) were used to estimate $H(m)$ dependence for two operating temperatures (i.e., 260 K and 300 K). Each PUF response requires multiple runs (realizations n) of chaotic system iterations (m). Therefore, each PUF response requires $m \times r$ iterations, where each m'th iteration length results from the corresponding SCRO frequency (1). It is obvious that each PUF response requires a $\frac{2mr}{f_1+f_2}$ interval; therefore, when 400 MHz average frequency of FPGAs is assumed, a single PUF response requires at least 25 µs without post-processing (e.g., entropy calculation).

Each 25 µs experiment (single PUF response extraction) was repeated multiple times in hardware (FPGA), in order to verify the PUF stability and reliability. The example $H(m)$ results for two FPGA instances at 260 K and 300 K are shown in Figure 12. One can see that, despite the temperature change, the critical m-value (PUF response) is rather invariant over temperature (Figure 12a–d). It turns out that the temperature mainly affects the amplitude of local H fluctuations; however, the global moment of rapid entropy increase remains constant in a particular device. The results in Figure 12 show the temperature influence on the $H(m)$ dependence, whereas, to evaluate the temperature impact on multiple instances, we needed to extract critical m-values (PUF responses) from multiple PUFs implemented in eight Artix and eight Cyclone V devices in 300 K and 260 K, respectively.

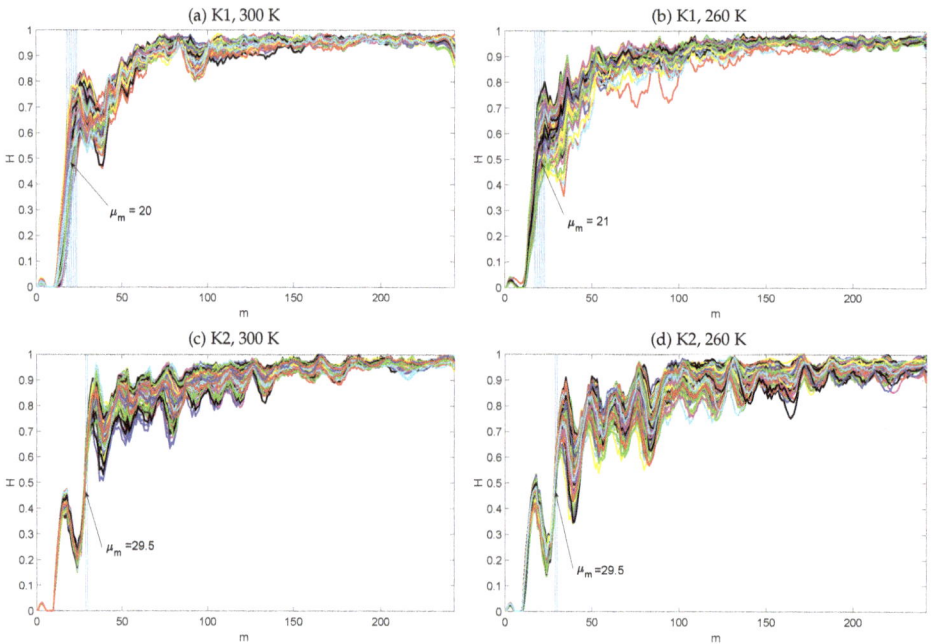

Figure 12. Example entropy vs. m relationships (100 circuit runs) of two circuits K1 (**a,b**) and K2 (**c,d**) at 300 K (**a,c**) and at 260 K (**b,d**).

The PUF responses ($\{m \mid H(m) = 0.5\}$) corresponding to particular instances were used to build a regression plot in Figure 13. One can see that PUF responses can be easily approximated with a linear function. The inclination coefficient (0.87) in Figure 13 proves that PUF responses (m) in higher temperature (300 K) are slightly shorter (smaller) than in the case of lower temperature (260 K). These results are in accordance with Monte Carlo analysis discussed in Section 3.

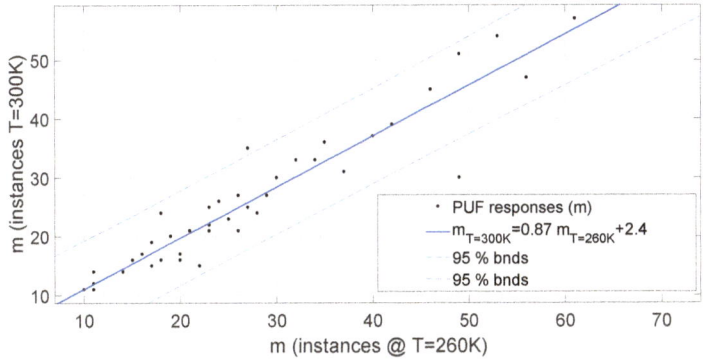

Figure 13. Regression of generated *m* values (PUF) keys in the same instances under different operating temperature.

In order to evaluate the PUF performance using common metrics (see, for example, [7,49]), we have combined three *m*-values (each represented by 8 bits) into one 24-bit PUF key. The differences between such keys (bit sequences) were measured with the use of Hamming distance—the number of different bits between two keys. Figure 14 shows normalized probability of occurrence of particular Hamming distances between keys in percentages for both inter-class (keys compared between different chips) and intra-class (keys repeatedly sampled within the same chip) distributions. The influence of the temperature change on such keys can be observed in Figure 15. Based on these results (and following common PUF metrics [7,49]), we were able to estimate the basic PUF keys parameters:

- Uniqueness: 41.16%,
- Reliability: 91.33%,
- Uniformity: 36.50%.

It is worthwhile to mention that such a chaotic circuit amplifies microscopic differences between instances in the way that even the device manufacturer cannot predict the results. Moreover, since every change in the implementation results in different keys, the number of PUF challenges within even a simple device strives for infinity. An invasive attempt to measure transistors and paths geometry, material heterogeneity as well as many other device parameters most likely would change the unique trajectory and destroy the keys. On the other hand, even if someone had succeeded, it would have been impossible to reconstruct an instance or to create simulation model that would have resulted in exactly the same chaotic trajectory and the same cryptographic keys.

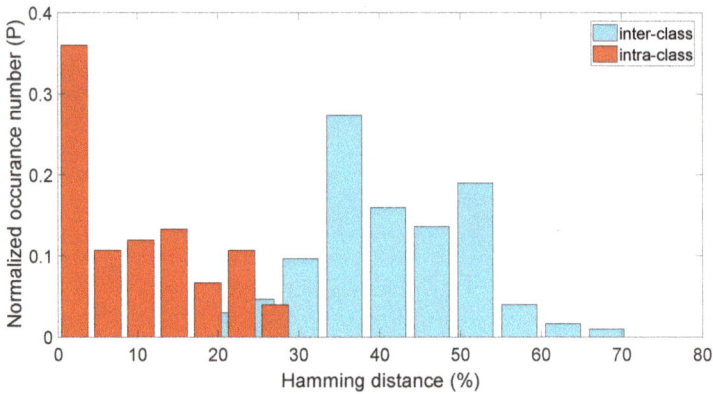

Figure 14. Deterministic density functions for intra- and inter-class Hamming distances of 24-bit PUF keys.

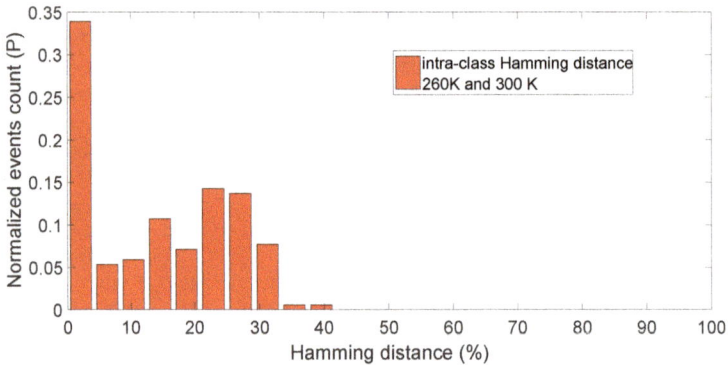

Figure 15. Deterministic density function for Hamming distances at two different temperatures.

5. Conclusions

The paper introduced an novel idea of generating unique PUF/POK cryptographic keys being derived from a chaotic circuit implemented in programmable devices. A new design of a chaotic circuit was also adapted—it utilizes time as an analog continuous state variable, which is very uncommon in purely digital systems, and as a result joins the advantages of analog chaotic signals with digital simplicity of implementation. The PUF keys were derived from the length of the deterministic part of a circuit's chaotic behavior. Both the simulations and physical measurements proved that the chaos theory should be explored for the sake of PUFs as a natural mechanism of amplifying random process variations of digital circuits (simple as well as advanced). The design was successfully tested in cheap CPLDs as well as in state-of-the-art FPGAs. The results showed significantly different keys derived from different instances programmed with precisely the same bit-stream as well as from slightly different implementations within one instance. The solution fits into the modern trends of developing highly secured hardware resistant to side-channel attacks, but not expensive and with universal application at the same time.

6. Patents

The presented solution is a patent pending technology. On 7 August 2018, an international patent application was filled under number PCT/IB2018/055943 based on three Polish patent applications (PL422486, PL422487 and PL425581 filed in the Polish patent office) with the earliest claimed priority date of 8 August 2017. The patent was internationally published by the World Intellectual Property Organization on 14 February 2019 under the number WO 2019/030670.

Author Contributions: conceptualization, K.G.; data curation, P.Z.W.; formal analysis, K.G. and P.Z.W.; funding acquisition, K.G. and P.Z.W.; investigation, P.Z.W.; methodology, K.G.; project administration, K.G.; resources, K.G. and P.Z.W.; software, P.Z.W.; supervision, K.G.; validation, P.Z.W.; visualization, K.G. and P.Z.W.; writing—original draft, K.G. and P.Z.W.; writing—review & editing, K.G. and P.Z.W.

Funding: This research received no external funding.

Conflicts of Interest: The authors declare no conflict of interest.

Abbreviations

The following abbreviations are used in this manuscript:

BR-PUF	bistable ring physical unclonable function
CPLD	complex programmable logic device
DL	delay line
FB	feedback loop
FPGA	field-programmable gate array
LUT	look-up table
MUX	multiplexer
PD	phase detector
POK	physical obfuscated key
PUF	physical unclonable function
RAM	random-access memory
ROM	read-only memory
SCA	side-channel attack
SCRO	switchable chain ring oscillator

References

1. Wang, A.; Chen, M.; Wang, Z.; Wang, X. Fault Rate Analysis: Breaking Masked AES Hardware Implementations Efficiently. *IEEE Trans. Circuits Syst. II Express Briefs* **2013**, *60*, 517–521. [CrossRef]
2. Skorobogatov, S.P. *Semi-Invasive Attacks: A New Approach to Hardware Security Analysis*; Technical report, UCAM-CL-TR-630; University of Cambridge: Cambridge, UK, 2005.
3. Torrance, R.; James, D. The State-of-the-Art in IC Reverse Engineering. In *Cryptographic Hardware and Embedded Systems—CHES 2009*; Clavier, C., Gaj, K., Eds.; Springer: Berlin/Heidelberg, Germany, 2009; pp. 363–381.
4. Xie, Y.; Xue, X.; Yang, J.; Lin, Y.; Zou, Q.; Huang, R.; Wu, J. A Logic Resistive Memory Chip for Embedded Key Storage with Physical Security. *IEEE Trans. Circuits Syst. II Express Briefs* **2016**, *63*, 336–340. [CrossRef]
5. Aziz, B.; Arenas, A.; Crispo, B. *Engineering Secure Internet of Things Systems*; Institution of Engineering and Technology: London, UK, 2016.
6. Bayat-Sarmadi, S.; Kermani, M.M.; Azarderakhsh, R.; Lee, C.Y. Dual-Basis Superserial Multipliers for Secure Applications and Lightweight Cryptographic Architectures. *IEEE Trans. Circuits Syst. II Express Briefs* **2014**, *61*, 125–129. [CrossRef]
7. Tanamoto, T.; Yasuda, S.; Takaya, S.; Fujita, S. Physically Unclonable Function Using an Initial Waveform of Ring Oscillators. *IEEE Trans. Circuits Syst. II Express Briefs* **2017**, *64*, 827–831. [CrossRef]
8. Tanaka, Y.; Bian, S.; Hiromoto, M.; Sato, T. Coin Flipping PUF: A Novel PUF with Improved Resistance against Machine Learning Attacks. *IEEE Trans. Circuits Syst. II Express Briefs* **2018**, *65*, 602–606. [CrossRef]

9. Barbareschi, M.; Natale, G.D.; Torres, L.; Mazzeo, A. A Ring Oscillator-Based Identification Mechanism Immune to Aging and External Working Conditions. *IEEE Trans. Circuits Syst. I Regul. Pap.* **2018**, *65*, 700–711. [CrossRef]

10. Marchand, C.; Bossuet, L.; Mureddu, U.; Bochard, N.; Cherkaoui, A.; Fischer, V. Implementation and Characterization of a Physical Unclonable Function for IoT: A Case Study With the TERO-PUF. *IEEE Trans. Comp. Aided Des. Integr. Circuits Syst.* **2018**, *37*, 97–109. [CrossRef]

11. Amsaad, F.; Niamat, M.; Dawoud, A.; Kose, S. Reliable Delay Based Algorithm to Boost PUF Security Against Modeling Attacks. *Information* **2018**, *9*, 224. [CrossRef]

12. Kömürcü, G.; Pusane, A.E.; Dündar, G. Enhanced challenge-response set and secure usage scenarios for ordering-based ring oscillator-physical unclonable functions. *IET Circuits Dev. Syst.* **2015**, *9*, 87–95. [CrossRef]

13. Liu, R.; Chen, P.Y.; Peng, X.; Yu, S. X-Point PUF: Exploiting Sneak Paths for a Strong Physical Unclonable Function Design. *IEEE Trans. Circuits Syst. I Regul. Pap.* **2018**, *65*, 1–10. [CrossRef]

14. Kim, H.; Hong, S. AES Sbox GF(2^4) inversion functions based PUFs. In Proceedings of the International SoC Design Conference (ISOCC), Jeju, Korea, 3–6 November 2014; pp. 15–16.

15. Wieczorek, P.Z.; Golofit, K. Metastability occurrence based physical unclonable functions for FPGAs. *Electron. Lett.* **2014**, *50*, 281–283. [CrossRef]

16. Vijayakumar, A.; Patil, V.C.; Kundu, S. On Improving Reliability of SRAM-Based Physically Unclonable Functions. *J. Low Power Electron. Appl.* **2017**, *7*, 2. [CrossRef]

17. Gong, M.; Liu, H.; Min, R.; Liu, Z. Pitfall of the Strongest Cells in Static Random Access Memory Physical Unclonable Functions. *Sensors* **2018**, *18*, 1776. [CrossRef] [PubMed]

18. Anagnostopoulos, N.A.; Katzenbeisser, S.; Chandy, J.; Tehranipoor, F. An Overview of DRAM-Based Security Primitives. *Cryptography* **2018**, *2*, 7. [CrossRef]

19. Anagnostopoulos, N.A.; Arul, T.; Fan, Y.; Hatzfeld, C.; Schaller, A.; Xiong, W.; Jain, M.; Saleem, M.U.; Lotichius, J.; Gabmeyer, S.; et al. Intrinsic Run-Time Row Hammer PUFs: Leveraging the Row Hammer Effect for Run-Time Cryptography and Improved Security. *Cryptography* **2018**, *2*, 13. [CrossRef]

20. Kumar, A.; Sahay, S.; Suri, M. Switching-Time Dependent PUF Using STT-MRAM. In Proceedings of the 31st International Conference on VLSI Design and 17th International Conference on Embedded Systems (VLSID), Pune, India, 8–10 January 2018; pp. 434–438. [CrossRef]

21. Chen, A. Reconfigurable physical unclonable function based on probabilistic switching of RRAM. *Electron. Lett.* **2015**, *51*, 615–617. [CrossRef]

22. Kim, J.; Ahmed, T.; Nili, H.; Yang, J.; Jeong, D.S.; Beckett, P.; Sriram, S.; Ranasinghe, D.C.; Kavehei, O. A Physical Unclonable Function With Redox-Based Nanoionic Resistive Memory. *IEEE Trans. Inf. Forensics Secur.* **2018**, *13*, 437–448. [CrossRef]

23. Schaller, A.; Xiong, W.; Anagnostopoulos, N.A.; Saleem, M.U.; Gabmeyer, S.; Skoric, B.; Katzenbeisser, S.; Szefer, J. Decay-Based DRAM PUFs in Commodity Devices. *IEEE Trans. Dependable Secur. Comp.* **2018**. [CrossRef]

24. Wang, W.C.; Yona, Y.; Diggavi, S.N.; Gupta, P. Design and Analysis of Stability-Guaranteed PUFs. *IEEE Trans. Inf. Forensics Secur.* **2018**, *13*, 978–992. [CrossRef]

25. Sahoo, D.P.; Mukhopadhyay, D.; Chakraborty, R.S.; Nguyen, P.H. A Multiplexer-Based Arbiter PUF Composition with Enhanced Reliability and Security. *IEEE Trans. Comp.* **2018**, *67*, 403–417. [CrossRef]

26. Gao, Y.; Ma, H.; Abbott, D.; Al-Sarawi, S.F. PUF Sensor: Exploiting PUF Unreliability for Secure Wireless Sensing. *IEEE Trans. Circuits Syst. I Regul. Pap.* **2017**, *64*, 2532–2543. [CrossRef]

27. Cao, Y.; Zhang, L.; Zalivaka, S.S.; Chang, C.H.; Chen, S. CMOS Image Sensor Based Physical Unclonable Function for Coherent Sensor-Level Authentication. *IEEE Trans. Circuits Syst. I Regul. Pap.* **2015**, *62*, 2629–2640. [CrossRef]

28. Herkle, A.; Becker, J.; Ortmanns, M. Exploiting Weak PUFs From Data Converter Nonlinearity—E.g., A Multibit CT $\Delta\Sigma$ Modulator. *IEEE Trans. Circuits Syst. I Regul. Pap.* **2016**, *63*, 994–1004. [CrossRef]

29. Wan, M.; He, Z.; Han, S.; Dai, K.; Zou, X. An Invasive-Attack-Resistant PUF Based On Switched-Capacitor Circuit. *IEEE Trans. Circuits Syst. I Regul. Pap.* **2015**, *62*, 2024–2034. [CrossRef]

30. Guo, Y.; Dee, T.; Tyagi, A. Barrel Shifter Physical Unclonable Function Based Encryption. *Cryptography* **2018**, *2*, 22. [CrossRef]

31. Addabbo, T.; Fort, A.; Marco, M.D.; Pancioni, L.; Vignoli, V. Physically Unclonable Functions Derived from Cellular Neural Networks. *IEEE Trans. Circuits Syst. I Regul. Pap.* **2013**, *60*, 3205–3214. [CrossRef]

32. Tao, S.; Dubrova, E. Ultra-energy-efficient temperature-stable physical unclonable function in 65 nm CMOS. *Electron. Lett.* **2016**, *52*, 805–806. [CrossRef]
33. Chen, Q.; Csaba, G.; Lugli, P.; Schlichtmann, U.; Rührmair, U. The Bistable Ring PUF: A new architecture for strong Physical Unclonable Functions. In Proceedings of the IEEE International Symposium on Hardware-Oriented Security and Trust, San Diego, CA, USA, 5–6 June 2011; pp. 134–141. [CrossRef]
34. Chen, Q.; Csaba, G.; Lugli, P.; Schlichtmann, U.; Rührmair, U. Characterization of the bistable ring PUF. In Proceedings of the Design, Automation Test in Europe Conference Exhibition (DATE), Dresden, Germany, 12–16 March 2012; pp. 1459–1462.
35. Yamamoto, D.; Takenaka, M.; Sakiyama, K.; Torii, N. Security evaluation of bistable ring PUFs on FPGAs using differential and linear analysis. In Proceedings of the Federated Conference on Computer Science and Information Systems, Warsaw, Poland, 7–10 September 2014; pp. 911–918. [CrossRef]
36. Wiggins, S. *Introduction to Applied Nonlinear Dynamical Systems and Chaos*, 2nd ed.; Springer: New York, NY, USA, 2003.
37. Keuninckx, L.; der Sande, G.V.; Danckaert, J. Simple Two-Transistor Single-Supply Resistor-Capacitor Chaotic Oscillator. *IEEE Trans. Circuits Syst. II Express Briefs* **2015**, *62*, 891–895. [CrossRef]
38. Huang, Y.; Zhang, P.; Zhao, W. Novel Grid Multiwing Butterfly Chaotic Attractors and Their Circuit Design. *IEEE Trans. Circuits Syst. II Express Briefs* **2015**, *62*, 496–500. [CrossRef]
39. Sprott, J.C. A New Chaotic Jerk Circuit. *IEEE Trans. Circuits Syst. II Express Briefs* **2011**, *58*, 240–243. [CrossRef]
40. Vaidelys, M.; Ragulskiene, J.; Ziaukas, P.; Ragulskis, M. Image Hiding Scheme Based on the Atrial Fibrillation Model. *Appl. Sci.* **2015**, *5*, 1980–1991. [CrossRef]
41. Tao, S.; Ruli, W.; Yixun, Y. Clock-controlled chaotic keystream generators. *Electron. Lett.* **1998**, *34*, 1932–1934. [CrossRef]
42. François, M.; Defour, D.; Negre, C. A Fast Chaos-Based Pseudo-Random Bit Generator Using Binary64 Floating-Point Arithmetic. *Informatica* **2014**, *38*, 115–124.
43. Li, C.; Sprott, J.C.; Thio, W.; Zhu, H. A New Piecewise Linear Hyperchaotic Circuit. *IEEE Trans. Circuits Syst. II Express Briefs* **2014**, *61*, 977–981. [CrossRef]
44. Jin, P.; Wang, G.; Iu, H.H.C.; Fernando, T. A Locally Active Memristor and Its Application in a Chaotic Circuit. *IEEE Trans. Circuits Syst. II Express Briefs* **2018**, *65*, 246–250. [CrossRef]
45. Wieczorek, P.Z.; Gołofit, K. True Random Number Generator Based on Flip-Flop Resolve Time Instability Boosted by Random Chaotic Source. *IEEE Trans. Circuits Syst. I Regul. Papers* **2018**, *65*, 1279–1292. [CrossRef]
46. Beirami, A.; Nejati, H. A Framework for Investigating the Performance of Chaotic-Map Truly Random Number Generators. *IEEE Trans. Circuits Syst. II Express Briefs* **2013**, *60*, 446–450. [CrossRef]
47. Shannon, C.E. A mathematical theory of communication. *Bell Syst. Tech. J.* **1948**, *27*, 379–423. [CrossRef]
48. Wieczorek, P.Z. Lightweight TRNG Based on Multiphase Timing of Bistables. *IEEE Trans. Circuits Syst. I Regul. Papers* **2016**, *63*, 1043–1054. [CrossRef]
49. Hori, Y.; Yoshida, T.; Katashita, T.; Satoh, A. Quantitative and Statistical Performance Evaluation of Arbiter Physical Unclonable Functions on FPGAs. In Proceedings of the International Conference on Reconfigurable Computing and FPGAs, Washington, DC, USA, 13–15 December 2010; pp. 298–303. [CrossRef]

![applied sciences logo] *applied sciences*

MDPI

Article

Re-Keying Scheme Revisited: Security Model and Instantiations

Yuichi Komano [1,*] and Shoichi Hirose [2]

[1]   Toshiba Corporation, Kawasaki 212-8582, Japan
[2]   Faculty of Engineering, University of Fukui, Fukui 910-8507, Japan; hrs_shch@u-fukui.ac.jp
*   Correspondence: yuichi1.komano@toshiba.co.jp; Tel.: +81-44-549-2156

Received: 31 January 2019; Accepted: 4 March 2019; Published: 11 March 2019

Abstract: The re-keying scheme is a variant of the symmetric encryption scheme where a sender (respectively, receiver) encrypts (respectively, decrypts) plaintext with a temporal session key derived from a master secret key and publicly-shared randomness. It is one of the system-level countermeasures against the side channel attacks (SCAs), which make attackers unable to collect enough power consumption traces for their analyses by updating the randomness (i.e., session key) frequently. In 2015, Dobraunig et al. proposed two kinds of re-keying schemes. The first one is a scheme without the beyond birthday security, which fixes the security vulnerability of the previous re-keying scheme of Medwed et al. Their second scheme is an abstract scheme with the beyond birthday security, which, as a black-box, consists of two functions; a re-keying function to generate a session key and a tweakable block cipher to encrypt plaintext. They assumed that the tweakable block cipher was ideal (namely, secure against the related key, chosen plaintext, and chosen ciphertext attacks) and proved the security of their scheme as a secure tweakable block cipher. In this paper, we revisit the re-keying scheme. The previous works did not discuss security in considering the SCA well. They just considered that the re-keying scheme was SCA resistant when the temporal session key was always refreshed with randomness. In this paper, we point out that such a discussion is insufficient by showing a concrete attack. We then introduce the definition of an SCA-resistant re-keying scheme, which captures the security against such an attack. We also give concrete schemes and discuss their security and applications.

Keywords: side channel attack; re-keying; tweakable block cipher; provable security

---

1. Introduction

Side channel attacks (SCAs) recover a secret key from a cryptographic device by collecting leakage information, such as the power consumption traces or the electro-magnetic traces, and by analyzing them statistically. Since the proposal by Kocher et al. [1], differential power analysis (DPA) has been one of the serious threats in the real world.

1.1. Background on Re-Keying Schemes

Against DPA, many countermeasures have been reported. At the device level, *masking* and *hiding* are studied well [2]. Masking is a countermeasure that randomizes the internal variables inside the module to disallow adversaries from analyzing the variables correctly. Hiding unlinks the internal values from the measured leakage to make the statistics meaningless. These countermeasures can be implemented within the device itself and do not change the interface of the device. Therefore, they can be add-ons to existing systems.

Appl. Sci. **2019**, *9*, 1002; doi:10.3390/app9051002   www.mdpi.com/journal/applsci

Appl. Sci. **2019**, *9*, 1002

On the other hand, as a system-level countermeasure, the re-keying scheme [3] has been proposed. It updates an encryption key frequently to make it infeasible for adversaries to collect leakage information. Medwed et al. [4] introduced the concept of "separation of duties" and proposed a concrete scheme for tiny devices. Their scheme consists of a re-keying function F and a block cipher BC. Here, F is a function that takes, as inputs, a master secret key mk and a randomness r to compute a temporal session key tk. They assumed that F is easily protected from the SCAs. To encrypt a message m with the length of the block size of BC, the scheme first chooses a randomness r and computes the session key tk by $F(mk, r)$. The scheme then encrypts m with BC using the session key tk, without a countermeasure against the SCAs. Dobraunig et al. [5] provided an attack to recover the master secret key from Medwed et al.'s scheme [4]. In the next year, Dobraunig et al. [6] proposed two improvements. As for the first one, they reconsidered the property of F. In their attack against Medwed et al.'s scheme, they used the property that F is invertible. To make the attack infeasible, they gave another example of F, which was non-invertible and pseudo-random in the ideal cipher model [7]. As for the second improvement, they gave a generic scheme replacing the block cipher of [4] with an ideal (secure against the related key attack in addition to the chosen cipher attack) tweakable block cipher to achieve the beyond birthday security. They showed its security by proving that the second scheme is a secure tweakable block cipher.

1.2. Our Contribution

This paper revisits the re-keying scheme. First, we point out that the previous works [4–6] did not give a formal security model. For example, Dobraunig et al. [6] discussed the security of their schemes with different security models. In addition, their models did not take the SCA resistance into consideration. In fact, as we give a concrete SCA attack, their model did not capture the security against the SCAs. Hence, we introduce a security model in considering the SCAs.

Second, we give two concrete first-order SCA-resistant re-keying schemes and discuss their security in our model. The first scheme is naturally derived from the combination of Dobraunig et al.'s second scheme [6] and Liskov et al.'s tweakable block cipher [8]. Unlike Dobraunig et al.'s abstract scheme, it is possible to discuss the SCA resistance with our concrete scheme. Our second scheme is a modification of the first scheme, which cannot be SCA resistant if it is used in the decryption device, which reveals the plaintext. However, it is useful for some applications in IoT systems, where the decrypted data are not revealed. We also discuss another scheme that is secure against the higher order SCA.

The paper is organized as follows. In the following subsection, we explain a related work recently reported. Section 2 reviews the definitions of the block cipher and the tweakable block cipher. In Section 3, we provide a message recovery attack with the SCA to the previous re-keying schemes. In considering such an attack, we introduce a new security model for the re-keying scheme in Section 4. We then give our concrete schemes and discuss their security in Sections 5 and 6, respectively. We also discuss the application of the re-keying scheme and the component of the re-keying function in Section 7. Finally, Section 8 concludes this paper.

1.3. Related Work

Dziembowski et al. [9] reconsidered the model of re-keying schemes and gave a new concrete scheme based on the hard physical learning problem. They first modeled the re-keying scheme, which assumed that the master secret key is stored as shares, and these shares were updated when the encryption/decryption process was invoked. Divide a secret key into randomized shares is one of the well-known countermeasures against the SCAs. Then, they gave two security models: against the black-box adversaries and the gray-box ones. The first one denotes the security against adversaries who access the inputs and outputs of the re-keying scheme. On the other hand, the second one allows adversaries to see leakages of the scheme in addition to its inputs and outputs. They also proposed a concrete scheme based on the learning parity with leakage (LPL) problem, which can be reduced to the well-known problem, the learning parity with noise (LPN) problem.

Their models supposed the countermeasure using the shares. Moreover, their scheme is unsuitable for tiny devices since it requires a costly non-volatile write memory for updating shares. On the other hand, our schemes do not use the countermeasure with shares, which were not captured by their model. In addition, our schemes do not require such non-volatile write memory; hence, they are suitable for tiny devices.

Another device-level approach to enhance the security is also proposed. In [10], Chittamuru et al. proposed a framework that protects data from snooping attacks and improves hardware security. The re-keying scheme can be used as a module to improve the security of this framework.

2. Preliminaries

2.1. Block Cipher

The block cipher is a fundamental tool used for secure communication. It is also a building block of the re-keying scheme in the latter. Let us start with a review of a pseudo-random function, and then, we will review the definitions of the secure block cipher.

Definition 1. *Let Φ be a family of functions with ℓ_n-bit input and ℓ_m-bit output. We say that $g(\cdot, \cdot)$ is a pseudo-random function with ℓ_n-bit input and ℓ_m-bit output parameterized by an ℓ_k-bit key k, if the advantage Adv below is negligible for any polynomial time adversary $\mathcal{A}$ who makes oracle queries to either $g(k, \cdot)$ or $\varphi \in \Phi$ up to q times:*

$$\text{Adv} := \max_{\mathcal{A}} |\Pr[\textbf{Exp}_{g,\mathcal{A}}^{real} = 1] - \Pr[\textbf{Exp}_{g,\mathcal{A}}^{rand} = 1]|,$$

where experiments $\textbf{Exp}_{g,\mathcal{A}}^{real}$ and $\textbf{Exp}_{g,\mathcal{A}}^{rand}$ are as in Figure 1. In these experiments, $\mathcal{O}_g(k, \cdot)$ and $\mathcal{O}_\varphi(\cdot)$ are oracles, which, with ℓ_n-bit input x, return $g(k, x)$ and $\varphi(x)$, respectively.

$\textbf{Exp}_{g,\mathcal{A}}^{real}$:	$\textbf{Exp}_{g,\mathcal{A}}^{rand}$:
$k \xleftarrow{\$} \{0,1\}^{\ell_k}$;	$\varphi \xleftarrow{\$} \Phi$;
return 1 iff $\mathcal{A}^{\mathcal{O}_g(k,\cdot)} = 1$	return 1 iff $\mathcal{A}^{\mathcal{O}_\varphi(\cdot)} = 1$

Figure 1. Experiments for a pseudo-random function.

Definition 2. *Let Π be a family of ℓ_n-bit permutations. We say that $(\text{BC}, \text{BC}^{-1})$ is a pair of ℓ_n-bit pseudo-random permutations parameterized by an ℓ_k-bit key k, if the advantage Adv below is negligible for any polynomial time adversary $\mathcal{A}$ who makes oracle queries, with an ℓ_n-bit input (plaintext), to either BC or $\pi \in \Pi$ up to q times.*

$$\text{Adv} := \max_{\mathcal{A}} |\Pr[\textbf{Exp}_{BC,\mathcal{A}}^{real} = 1] - \Pr[\textbf{Exp}_{BC,\mathcal{A}}^{rand} = 1]|$$

where experiments $\textbf{Exp}_{BC,\mathcal{A}}^{real}$ and $\textbf{Exp}_{BC,\mathcal{A}}^{rand}$ are as in Figure 2. In these experiments, $\mathcal{O}_{BC}(k, \cdot)$ and $\mathcal{O}_\pi(\cdot)$ are oracles, which, with ℓ_n-bit input x, return the ciphertext $BC(k, x)$ and $\pi(x)$, respectively.

Similarly, we say that $(\text{BC}, \text{BC}^{-1})$ is a strong pseudo-random permutation, if the advantage is negligible even if $\mathcal{A}$ allows making oracle queries, with an ℓ_n bit input (plaintext or ciphertext), to either (π, π^{-1}) or $(\text{BC}, \text{BC}^{-1})$, up to q times in total.

$\mathbf{Exp}^{real}_{BC,\mathcal{A}}$:	$\mathbf{Exp}^{rand.}_{BC,\mathcal{A}}$:
$k \xleftarrow{\$} \{0,1\}^{\ell_k}$;	$\pi \xleftarrow{\$} \Pi(\cdot)$;
return 1 iff $\mathcal{A}^{\mathcal{O}_{BC}(k,\cdot)} = 1$	return 1 iff $\mathcal{A}^{\mathcal{O}_\pi(\cdot)} = 1$

Figure 2. Experiments for the block cipher.

Definition 3. *Let Π be a family of ℓ_n-bit permutations. We say that (BC, BC^{-1}) is a pair of ℓ_n-bit pseudo-random permutations against related key attacks parameterized by an ℓ_k-bit key k, if the advantage* Adv *below is negligible for any polynomial time adversary $\mathcal{A}$ who makes oracle queries to either $\mathcal{O}_{BC}$ or $\mathcal{O}_\Pi$ up to q times.*

$$\text{Adv} \quad := \quad \max_{\mathcal{A}} |\Pr[\mathbf{Exp}^{real}_{BC,\mathcal{A}_{rk}} = 1] - \Pr[\mathbf{Exp}^{rand}_{BC,\mathcal{A}_{rk}} = 1]|$$

where experiments $\mathbf{Exp}^{real}_{BC,\mathcal{A}_{rk}}, \mathbf{Exp}^{rand}_{BC,\mathcal{A}_{rk}}$ are as in Figure 3. Within them, $\mathcal{O}_{BC}(k,\cdot,\cdot)$ and $\mathcal{O}_\Pi(l,\cdot,\cdot)$, given a difference $\Delta \in \{0,1\}^{\ell_k}$ and a plaintext m from $\mathcal{A}_{rk}$, return $\mathcal{O}_{BC}(k,\Delta,m) = BC(k \oplus \Delta, m)$ and $\mathcal{O}_\Pi(l,\Delta,m) = \pi(m)$, respectively, where π is chosen from Π in accordance with $l \oplus \Delta$.

Similar to Definition 2, we have the notion of strong pseudo-random permutation against the related key attack.

$\mathbf{Exp}^{real}_{BC,\mathcal{A}_{rk}}$:	$\mathbf{Exp}^{rand}_{BC,\mathcal{A}_{rk}}$:
$k \xleftarrow{\$} \{0,1\}^{\ell_k}$;	$l \xleftarrow{\$} \{0,1\}^{\ell_k}$;
return 1 iff $\mathcal{A}_{rk}{}^{\mathcal{O}_{BC}(k,\cdot,\cdot)} = 1$	return 1 iff $\mathcal{A}_{rk}{}^{\mathcal{O}_\Pi(l,\cdot,\cdot)} = 1$

Figure 3. Experiments for the block cipher with related key attack.

2.2. Tweakable Block Cipher

The tweakable block cipher [8] is a variant of the block cipher. Besides the plaintext and the key, it takes an auxiliary input, called *a tweak*, which acts as an initial vector of the mode of operations [11]. By using different tweaks, ciphertexts differ even though the pair of the plaintext and the key is unique. This property makes the statistical analysis difficult.

Note that the tweak can be shared between a sender and a receiver in public. For example, the sender may select a tweak at random to send it with a ciphertext; or if both the sender and receiver are stateful and if they share a seed for tweaks, they synchronously compute a tweak without sending it. For simplicity, we assume that the sender sends a tweak along with a ciphertext over the public channel.

The tweakable block cipher consists of a pair of two algorithms (TBC, TBC^{-1}). The encryption algorithm TBC takes, as inputs, a key k, a tweak t, and a plaintext m with bit lengths ℓ_k, ℓ_t, and ℓ_n, respectively, to output an ℓ_n-bit ciphertext c. The decryption algorithm TBC^{-1} takes, as inputs, k, t, and c to recover m. They should satisfy the completeness; namely, $TBC^{-1}(k,t,TBC(k,t,m)) = m$ holds for arbitrary k, t, and m. We then review its security model.

Definition 4. *Let* Π *be a family of* ℓ_n-*bit permutations parameterized by an* ℓ_t-*bit tweak.* $(\mathsf{TBC}, \mathsf{TBC}^{-1})$ *is a pair of* ℓ_n-*bit pseudo-random permutations parameterized by an* ℓ_t-*bit tweak and an* ℓ_k-*bit key, if the advantage* Adv *below is negligible for any polynomial time adversary* $\mathcal{A}$ *who makes oracle queries, with a tweak and a plaintext, to either* TBC *or* Π *up to q times.*

$$\text{Adv} \quad := \quad \max_{\mathcal{A}} |\Pr[\mathbf{Exp}^{real}_{\mathsf{TBC},\mathcal{A}} = 1] - \Pr[\mathbf{Exp}^{rand}_{\mathsf{TBC},\mathcal{A}} = 1]|$$

where experiments $\mathbf{Exp}^{real}_{\mathsf{TBC},\mathcal{A}}$, $\mathbf{Exp}^{rand}_{\mathsf{TBC},\mathcal{A}}$ *are as in Figure* 4. *Within them,* $\mathcal{O}_{\mathsf{TBC}}(k, \cdot, \cdot)$ *and* $\mathcal{O}_{\pi}(\cdot, \cdot)$, *given a tweak t and a plaintext m from* $\mathcal{A}$, *return* $\mathsf{TBC}(k, t, m)$ *and* $\pi(t, m)$, *respectively.*

Similar to Definition 2, *we have the notion of strong pseudo-random permutation for the tweakable block cipher.*

Note that, unlike the key, the tweaks may be selected in an insecure manner (using the time information or sequential number, for example) and the adversary may control the manner. Especially, as for the (stateless) decryption, the adversary can make oracle queries with arbitrary tweaks of his/her choice. Therefore, the above definitions assume the adversary in the related tweak attack setting, i.e., the open-tweak model [12]. Moreover, similar to the block cipher, the security model can be extended against the related key attacks. The model is similar to Definition 3, and we omit it here.

$\mathbf{Exp}^{real}_{\mathsf{TBC},\mathcal{A}}$:	$\mathbf{Exp}^{rand}_{\mathsf{TBC},\mathcal{A}}$:
$k \xleftarrow{\$} \{0,1\}^{\ell_k}$;	$\pi \xleftarrow{\$} \Pi(\cdot, \cdot)$;
return 1 iff $\mathcal{A}^{\mathcal{O}_{\mathsf{TBC}}(k,\cdot,\cdot)} = 1$	return 1 iff $\mathcal{A}^{\mathcal{O}_\pi(\cdot,\cdot)} = 1$

Figure 4. Experiments for the tweakable block cipher.

3. SCA on the Previous Re-Keying Schemes

In this section, we review the previous re-keying schemes; Medwed et al.'s re-keying scheme and Dobraunig et al.'s first re-keying scheme. We then show the plaintext recovery attacks with the SCA against these schemes.

3.1. Previous Works

Medwed et al. [4] introduced the design concept of "separation of duties" and proposed a concrete re-keying scheme suitable for lightweight devices. Their scheme consists of two parts as in Figure 5. The first part is a re-keying function F, which takes a master secret key mk and randomness r as inputs and outputs a session key tk. The re-keying function is designed with simple operations, which is easily protected from the SCAs. The second part is a block cipher BC, which encrypts a plaintext m with the session key tk. Decryption consists of the above F and BC^{-1}, which is the inverse of BC. They assumed that F was multiplicative; precisely, $tk = mk \cdot r$ in a finite field. Against their scheme, Dobraunig et al. [5] showed an attack that first searches tk with the birthday attack and then recovers mk by $mk = tk \cdot r^{-1}$.

Dobraunig et al. [6] then gave two other schemes and discussed their security. Figure 6 depicts their first scheme. Their first scheme is an improvement of Medwed et al.'s scheme by replacing the re-keying function with a non-invertible function. Figure 6 depicts their construction, consisting of an SCA-resistant function g and a one-way function h.

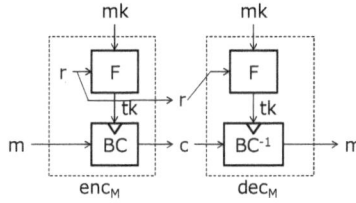

Figure 5. Medwed et al.'s scheme.

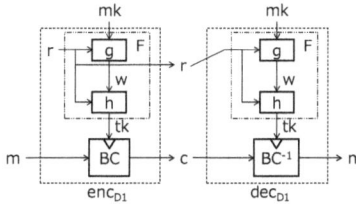

Figure 6. Dobraunig et al.'s first scheme.

3.2. The Attacks

The previous works supposed that the re-keying schemes are secure if the randomness r is fresh, and they discussed the security against the SCAs insufficiently. In fact, as they supposed, *the key recovery attack* on both the master secret key mk and the session key tk seems infeasible with the SCA if the randomness is fresh, because of the property (assumption) of g.

Let us discuss another attack, a message recovery attack, with the SCAs. Assume an adversary who, monitoring a ciphertext (r, c), aims to recover the plaintext m corresponding to c. Against Medwed et al.'s re-keying scheme and Dobraunig et al.'s first scheme, such an attack is feasible if the decryption device accepts decrypting an arbitrary input without checking the freshness of r. The attack proceeds as follows. The adversary repeats sending a decryption query (r, c') for randomly-chosen c'. The attacker measures the side channel information. For the fixed randomness r, the session key tk is also fixed; and therefore, the adversary can determine tk with the SCA. Once the session key tk is determined, the adversary can recover m by decrypting c with tk. The assumption, where the decryption device does not check the freshness of r, is realistic because Medwed et al. assumed stateless devices as the lightweight devices.

4. Security Model of the Re-Keying Scheme in Considering the SCA

In this section, we introduce a security model considering the SCA. Except the SCA resistance, the model is similar to that of the tweakable block cipher. Namely, the re-keying scheme is secure if it is indistinguishable from a random function. To take the SCA resistance into consideration, we assume that the encryption and decryption oracles leak the side channel information.

Unlike the tweakable block cipher, we assume that the encryption device chooses a randomness (which corresponds to the tweak in the tweakable block cipher) uniformly random for each encryption. Hence, we disallow the adversary from choosing it in the encryption oracle query. On the other hand, the decryption device receives the randomness as one of the inputs; therefore, we allow the adversary to choose it in the decryption query.

As for the side channel information, we introduce a leakage function $\mathcal{L}(\mathsf{BC})$, which returns the leakages (e.g., a power consumption trace) through the block cipher operation. In the real world, the SCAs on the key loading and XORing with the key have been reported. However, there are protection mechanisms to decrease the platform leakage (e.g., for the key loading), and the leakage is small in XORing compared to the complex operations in the block cipher. Hence, we omit them. Moreover, we assume that the re-keying function is properly designed not to leak the side channel information of the key.

Definition 5. *Let Π be a family of ℓ_n-bit permutations parameterized by an ℓ_r-bit randomness. (E, E^{-1}) is a pair of ℓ_n-bit pseudo-random permutations parameterized by an ℓ_r-bit randomness and an ℓ_k-bit key, if the advantage Adv below is negligible for any polynomial time adversary $\mathcal{A}$ who makes oracle queries, with a plaintext, to either E or Π up to q times.*

$$\mathrm{Adv} := \max_{\mathcal{A}} |\Pr[\mathbf{Exp}_{\mathsf{RK},\mathcal{A}}^{real} = 1] - \Pr[\mathbf{Exp}_{\mathsf{RK},\mathcal{A}}^{rand} = 1]|$$

where experiments $\mathbf{Exp}_{\mathsf{RK},\mathcal{A}}^{real}$, $\mathbf{Exp}_{\mathsf{RK},\mathcal{A}}^{rand}$ are as in Figure 7. Within them, $\mathcal{O}_{\mathsf{RK}}(k, R, \cdot)$, given a plaintext m from $\mathcal{A}$, returns both $\mathsf{RK}(k, r, m)$ for a fresh randomness r chosen by the oracle itself and $\mathcal{L}(\mathsf{BC}(tk, m))$, where tk is a session key derived from the re-keying function with k and r. On the other hand, $\mathcal{O}_{\Pi}(R, \cdot)$, given a plaintext m from $\mathcal{A}$, returns both $\pi(r, m)$ for a fresh randomness r chosen by the oracle itself and $\mathcal{L}(\mathsf{BC}(tk, m'))$, where tk and m' are a session key derived from the re-keying function with k and r and an ℓ_n-bit random plaintext chosen by the oracle itself, respectively.

Similar to Definition 2, we have the notion of strong pseudo-random permutation for the re-keying scheme. In this case, note that the adversary is allowed to choose the randomness r as an input of the decryption oracle.

$\mathbf{Exp}_{\mathsf{RK},\mathcal{A}}^{real}$:	$\mathbf{Exp}_{\mathsf{RK},\mathcal{A}}^{rand}$:
$k \xleftarrow{\$} \{0,1\}^{\ell_k}$;	$\pi \xleftarrow{\$} \Pi(\cdot, \cdot)$;
return 1 iff $\mathcal{A}^{\mathcal{O}_{\mathsf{RK}}(k,R,\cdot)} = 1$	return 1 iff $\mathcal{A}^{\mathcal{O}_{\pi}(R,\cdot)} = 1$

Figure 7. Experiments for the re-keying scheme.

5. New Concrete Re-Keying Schemes

We first recall our building blocks: Liskov et al.'s tweakable block cipher and Dobraunig et al.'s second re-keying schemes. We then give our concrete re-keying schemes.

5.1. Building Blocks

Liskov et al. [8] introduced the concept of the tweakable block cipher and gave several schemes. One of the schemes, known as LRW1, calls the block cipher with a secret key k twice as inner components where the input of one block cipher is the XOR of the tweak and the output of the other block cipher. Its latency is about twice as one of the block cipher's. Liskov et al. proved that LRW1 is a pseudo-random permutation.

The other scheme, LRW2, described in Figure 8, reduces the number of the block cipher operations by one. It, however, requires additional operation of a keyed hash function. The output of the hash function is XORed with both the input and output of the block cipher. If an encryption device computes the keyed hash function in advance and stores the output, its latency is half of LRW1's. Liskov et al. proved that LRW2 is a strong pseudo-random permutation, if h_l is chosen from the δ_h-AXU2 hash function family h_L. Here, h_L is the δ_h-AXU2 (δ_h-almost two-XOR-universal [8]) hash function family if $\Pr_l[h_l(x) \oplus h_l(y) = z] \leq \delta_h$ holds for all x, y, and z such that $x \neq y$.

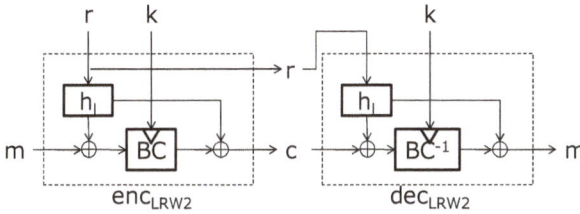

Figure 8. Liskov et al.'s tweakable block cipher: LRW2.

Dobraunig et al. [6] followed the concept of "separation of duties" to propose two other re-keying schemes. Their first scheme is identical to Medwed et al.'s scheme, but the requirement of F is different. They added the one-wayness property (precisely, Dobraunig et al. divided F into two parts: the first and second parts are assumed to have SCA resistance and one-wayness, respectively) to F and proved its security in the ideal cipher model. Note that their first scheme avoids the attack to recover mk from tk; however, it is possible to search for the session key tk by the birthday attack.

Their second scheme, depicted in Figure 9, replaces BC with a tweakable block cipher to achieve the beyond birthday security. The birthday attack above succeeds because the re-keying function and the block cipher work independently, and adversaries are able to collect the input-output pairs of the block cipher and the re-keying scheme, step by step. Their second scheme, however, binds the re-keying function with the (tweakable) block cipher by inputting the randomness into the tweakable block cipher as a tweak. It makes it difficult for adversaries to collect the input-output pairs independently.

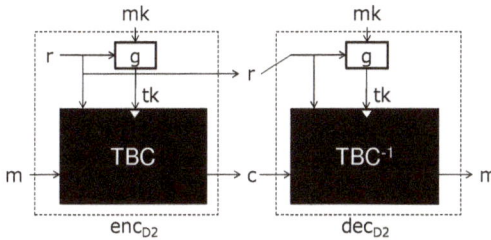

Figure 9. Dobraunig et al.'s second re-keying scheme (generic construction).

5.2. Our Concrete Schemes

5.2.1. First-Order SCA-Resistant Re-Keying Scheme

Our first scheme is based on the combination of LRW2 and Dobraunig et al.'s second scheme. From Theorem 3 of [6], the combination of these schemes is a concrete re-keying scheme; however, the keyed hash function in LRW2 may leak the side channel information. Hence, we modified it as depicted in Figure 10 to achieve the security against the first-order SCAs.

This scheme uses three functions. The first function is a re-keying function g_1, which, given the master secret key mk_1 and the randomness r, returns the session key tk. We assume that g_1 is a pseudo-random permutation when one of mk_1 and r is fixed and that g_1 leaks no side channel information on inputs. The second function is a pseudo-random function g_2, which, given another master secret key mk_2 and the randomness r, returns the ℓ_n-bit string n. We assume that g_2 also leaks no side channel information on the inputs. The third function is a block cipher BC. As discussed later, the re-keying scheme is secure against the first-order SCA because the input and output of BC are XORed (masked, as in the masking countermeasure) by n, which is unknown to the adversary. The procedures of this scheme (enc_1, dec_1) are described in Figure 10 and Algorithms 1 and 2.

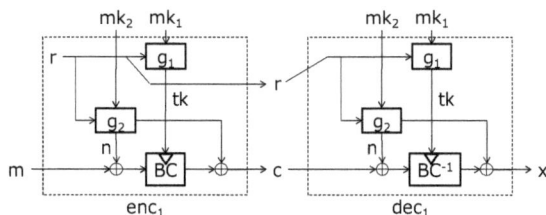

Figure 10. First-order SCA-resistant re-keying scheme.

Algorithm 1 First-order SCA-resistant re-keying scheme: enc_1

Input: master secret key (mk_1, mk_2) and plaintext m
Output: ciphertext (r, c)

---

1. Choose a randomness r
2. Compute $tk = g_1(mk_1, r)$
3. Compute $n = g_2(mk_2, r)$
4. Compute $c = BC(tk, m \oplus n) \oplus n$
5. Return (r, c)

Algorithm 2 First-order SCA-resistant re-keying scheme: dec_1

Input: master secret key (mk_1, mk_2) and ciphertext (r, c)
Output: plaintext m

---

1. Compute $tk = g_1(mk_1, r)$
2. Compute $n = g_2(mk_2, r)$
3. Compute $m = BC^{-1}(tk, c \oplus n) \oplus n$
4. return m

5.2.2. SCA-Resistant Re-Keying Encryption Scheme

Our second scheme is a modification of the first scheme, by removing the XOR operation before the block cipher operation. The second scheme lacks the resistance against the message recovery attack with the SCA, if it is used in the decryption device as follows. Assume that an SCA adversary, given a target (r, c), is allowed to use the decryption device as an oracle. The attacker queries (r, c') for randomly-chosen c' and receives m' with the leakage $\mathcal{L}(BC^{-1}(tk, c))$ where $tk = g_1(mk_1, r)$. Although the input of BC^{-1} is protected with a mask $g_2(mk_2, r)$, the output is unprotected; and hence, the adversary can recover tk with the first-order SCA and decrypt m with the recovered tk.

Note that, if the decryption device does not output the decrypted data, i.e., the decryption oracle returns only $\mathcal{L}(BC^{-1}(g_1(mk_1, r), c))$ without m, the above message recovery attack with the SCA is infeasible.

Furthermore, note that if the encryption device (i.e., the encryption oracle) chooses a randomness r properly, the encryption device is secure against the first-order SCA without masking the input of BC, since the encryption key tk is a fresh randomness. The procedures of this scheme (enc_2, dec_2) are described in Figure 11 and Algorithms 3 and 4.

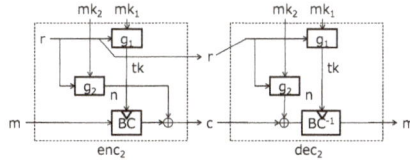

Figure 11. First-order SCA-resistant re-keying encryption scheme.

Algorithm 3 First-order SCA-resistant re-keying encryption scheme: enc_2

Input: master secret key (mk_1, mk_2) and plaintext m
Output: ciphertext (r, c)

---

1. Choose a randomness r
2. Compute $tk = g_1(mk_1, r)$
3. Compute $n = g_2(mk_2, r)$
4. Compute $c = BC(tk, m) \oplus n$
5. Return (r, c)

Algorithm 4 First-order SCA-resistant re-keying encryption scheme: dec_2

Input: master secret key (mk_1, mk_2) and ciphertext (r, c)
Output: plaintext m

---

1. Compute $tk = g_1(mk_1, r)$
2. Compute $n = g_2(mk_2, r)$
3. Compute $m = BC^{-1}(tk, c \oplus n)$
4. Return m

---

6. Security Considerations

In this section, let us discuss the security of our schemes. For each scheme, we first show that the adversary has no meaningful information from the SCA. Then, we show that the adversary's advantage for the distinguishing game in Definition 5 is negligible.

6.1. Security of the SCA-Resistant Re-Keying Scheme

In the scheme in Figure 10, the input and the output of BC are masked with $g_2(mk_2, r)$, which is unknown to the adversary. Hence, the adversary cannot guess the internal value of BC, nor retrieve the meaningful side channel information. Therefore, it is obvious that the scheme is resistant against the first-order SCA.

The scheme is a modification of the combinatorial scheme of LRW2 and Dobraunig et al.'s second scheme, restricting the keyed hash function in LRW2 not to leak the side channel information. The security model of the re-keying scheme in Definition 5 is a subset of the tweakable block cipher in Definition 4. From Theorem 3 of [6] and the above discussions, the scheme naturally satisfies Definition 5. More precisely, we have the following theorem for the scheme.

Theorem 1. *Assume that BC is a (strong) pseudo-random permutation and that g_1 is a pseudo-random permutation. Then, the scheme of Figure 10 is a (strong) pseudo-random re-keying scheme such that:*

$$\text{Adv} \le \epsilon_{g_1} + q\epsilon_B,$$

where ϵ_B and ϵ_{g_1} are upper bounds on the advantage of the adversaries against BC and g_1, respectively.

Proof. For simplicity, this proof is for a pseudo-random permutation BC. The proof for a strong pseudo-random permutation BC is very similar.

Let $\mathcal{A}$ be an adversary against the re-keying scheme. $\mathcal{A}$ has oracle access to $\mathcal{O}_{RK}$.

Let $\mathcal{D}_1$ be an adversary against g_1. $\mathcal{D}_1$ runs $\mathcal{A}$ and simulates $\mathcal{O}_{RK}$ for $\mathcal{A}$. Then,

$$\Pr[\textbf{Exp}^{real}_{RK,\mathcal{A}} = 1] = \Pr[\textbf{Exp}^{real}_{g_1,\mathcal{D}_1} = 1] \le \epsilon_{g_1} + \Pr[\textbf{Exp}^{rand}_{g_1,\mathcal{D}_1} = 1].$$

Let $\mathcal{D}_2$ be an adversary against BC. $\mathcal{D}_2$ is given q different oracles, which are either $(BC(k_1, \cdot), \ldots, BC(k_q, \cdot))$ or $(\pi_1(\cdot), \ldots, \pi_q(\cdot))$, where each k_i is chosen uniformly at random from $\{0,1\}^{\ell_k}$ and each π_i is chosen uniformly at random from the set of all permutations over $\{0,1\}^{\ell_n}$. $\mathcal{D}_2$ runs $\mathcal{A}$ and simulates $\mathcal{O}_{RK}$ for $\mathcal{A}$. For a query (r, m) made by $\mathcal{A}$, if there exists a previous query (r', m') such that $r = r'$, then $\mathcal{D}_2$ asks m from the oracle from which $\mathcal{D}_2$ asked m'. Otherwise, $\mathcal{D}_2$ asks m from a new oracle. Then,

$$\Pr[\textbf{Exp}^{rand}_{g_1,\mathcal{D}_1} = 1] = \Pr[\mathcal{D}_2^{BC(k_1,\cdot),\ldots,BC(k_q,\cdot)} = 1] \le q\epsilon_B + \Pr[\mathcal{D}_2^{\pi_1(\cdot),\ldots,\pi_q(\cdot)} = 1].$$

It is not difficult to see that:

$$\Pr[\mathcal{D}_2^{\pi_1(\cdot),\ldots,\pi_q(\cdot)} = 1] = \Pr[\textbf{Exp}^{rand}_{RK,\mathcal{A}} = 1].$$

Thus,

$$\left| \Pr[\textbf{Exp}^{real}_{RK,\mathcal{A}} = 1] - \Pr[\textbf{Exp}^{rand}_{RK,\mathcal{A}} = 1] \right| \le \epsilon_{g_1} + q\epsilon_B.$$

□

The above theorem holds whether BC is secure against the related key attack or not. Although Dobraunig et al.'s second scheme requires the ideal tweakable block cipher (namely, secure against the related key attack), our scheme can relax the requirement for its building block.

6.2. Security of the SCA-Resistant Re-Keying Encryption Scheme

Let us discuss the security of the scheme in Figure 11. Assume that an SCA adversary mounts the attacks on the encryption device, whereas the SCA on the decryption is restricted. Similar to the previous scheme, this scheme is also SCA resistant if it is used in the encryption device.

Let us discuss the SCA resistance of the decryption device. Without restriction, it is vulnerable to the message recovery attack with the SCA, because the output of BC^{-1} is unprotected. If we restrict the adversary not to receive m, which is the output of BC^{-1}, the attack is difficult only with the leakage information.

The security of our second scheme, by ignoring the side channel information, can be proven in a similar way as Theorem 1.

6.3. Toward the Higher Oder SCA-Resistant Schemes

The schemes in Figures 10 and 11 are secure against the first-order SCAs. To resist the higher order SCA, we need more random freshnesses in general. Assume that, in addition to r, the randomnesses $r_2, r_3, \cdots$ are randomly generated in the encryption device to compute $n_2, n_3, \cdots$, which mask the input and the output of BC, and sent to the decryption device as elements of $(r, r_2, r_3, \cdots, c)$. It seems to lead the higher order SCA-resistant scheme. However, it is insecure because the message recovery attack, by reusing $(r, r_2, r_3, \cdots)$, is possible.

Let us consider another scheme in Figure 12 and Algorithms 5 and 6. The idea to achieve the higher order SCA resistance is to generate a fresh randomness for a pair (m, r). In the SCA, the adversary queries m or (r, c) as the encryption query or the decryption query, respectively. By letting the mask depend on the pair, the masks for the input and the output of BC should be always fresh for each query. This prevents the higher order SCA.

The scheme in Figure 12 is complicated, unlike the previous two schemes, which shows the trade-off between the security and efficiency. Another construction, where the session key also depends on both (r, m), can be considered.

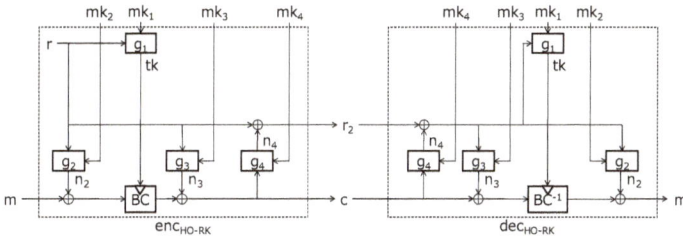

Figure 12. Higher order SCA-resistant re-keying scheme.

Algorithm 5 Higher order SCA-resistant re-keying encryption scheme: enc_{HO-RK}

Input: master secret key (mk_1, mk_2, mk_3) and plaintext m
Output: ciphertext (r, c)

---

1. Choose a randomness r
2. Compute $tk = g_1(mk_1, r)$
3. Compute $n_2 = g_2(mk_2, r)$
4. Compute $c = BC(tk, m \oplus n_2) \oplus g_3(mk_3, r)$
5. Compute $r_2 = r \oplus g_4(mk_4, c)$
6. Return (r_2, c)

Algorithm 6 Higher-order SCA-resistant re-keying encryption scheme: dec_{HO-RK}

Input: master secret key (mk_1, mk_2, mk_3) and ciphertext (r_2, c)
Output: plaintext m

---

1. Compute $r = r_2 \oplus g_4(mk_4, c)$
2. Compute $n_4 = g_4(mk_4, r)$
3. Compute $tk = g_1(mk_1, r)$
4. Compute $m = BC^{-1}(tk, c \oplus n_4) \oplus g_2(mk_2, r)$
5. Return m

---

7. Discussion

The following subsections discuss two applications for the first-order SCA-resistant encryption re-keying scheme and the first-order SCA-resistant decryption re-keying scheme.

7.1. Application to Sensor Network Devices

Let us assume sensor devices that collect the sensor data and send them to a server via an edge device. The sensor devices and the edge device are supposed to be resource-constrained. Since these devices are located in the field, there is a fear of the SCA on these devices.

Furthermore, assume that the sensor data are confidential. For example, the sensor devices are medical ones, and all the data include sensitive information of the patients. Alternatively, the sensor devices are located at an agricultural field, and all the data include key information, such as the temperature or the humidity, which is useful for optimal cultivation. Furthermore, let us assume that the data sent from the server to the devices are less sensitive; for example, the instructions for collecting and sending sensor data. Hence, let us consider that the sensor device encrypts its sensor data and sends them to the edge device; and then, the edge device decrypts the sensor data, re-encrypts them with a key shared between the edge device and the server, and sends the ciphertext to the server. Note that the decrypted sensor data are not revealed by the edge device.

The first-order SCA-resistant encryption re-keying scheme is suitable for such a situation. Let us regard the sensor devices and the edge device as the encryption device and the decryption one in the scheme, respectively. As we discussed in Section 5.2.2, the SCA against the sensor device is difficult because of the fresh randomness r. As for the edge device, it does not reveal the decrypted sensor data, but the re-encrypted ciphertext; therefore, the SCA against the edge device is also difficult, as we discussed in Section 5.2.2.

Note that the previous schemes, Medwed et al.'s scheme and Dobraunig et al.'s first scheme, are unsuitable for this application. This is because, in these schemes, the input of BC^{-1} in the edge device is unprotected and the message recovery attack with the SCA against the edge device is possible.

7.2. Construction of g

If g is resistant to the SCAs, our schemes are theoretically resistant to the SCAs. Therefore, our aim is to construct g, which consists of operations with less side channel leakage and/or easily-added countermeasures against the SCAs.

Generally speaking, the non-linear operations, such as SubByte in AES [13], tend to leak the side channel information. The implementations of these operations tend to be complicated circuits, which require much power consumption, including the meaningful information. Moreover, from the implementability, the inputs and outputs of these operations are restricted by a small bit length. This enables adversaries to guess the inputs and outputs to succeed in the statistics of the SCAs. Therefore, it is better to construct the re-keying function with the simple (linear) operations, rather than the non-linear one, to mix the master secret key and the randomness.

Note that Theorem 1 requires that g is a (strong) pseudo-random permutation; namely, it is a one-to-one pseudo-random function if one of mk and r is fixed. An example of such g is a composition of an SCA-free function g_{MIX} and a pseudo-random permutation g_{PRP}. The SCA-free function is, for example, $g_{\text{MIX}}(mk, r) = mk \cdot T \oplus r$, where T is a regular 128×128 matrix consisting of zero or one, and $mk \cdot T$ is a multiplication of a 128×1 vector mk and T. Other examples can be obtained by dividing the above T into 64×64 or 32×32 matrices. In addition, the masking countermeasures are easily applicable to linear functions including such g_{MIX}.

Appl. Sci. **2019**, *9*, 1002

8. Conclusions

In this paper, we reconsidered the re-keying scheme. We pointed out that the previous works lacked the security consideration on the SCA and introduced a security model of the re-keying scheme considering the SCA. We then gave concrete schemes and discussed their security and applications. In the IoT era, whereas the SCAs are a serious threat to the IoT devices, these devices are resource-constrained, and it is difficult to implement the existing countermeasures on them. In addition to the device-level approach, systematic countermeasures such as the re-keying scheme are some of the promising countermeasures.

Author Contributions: conceptualization, Y.K.; methodology, Y.K. and S.H.; formal analysis, Y.K. and S.H.; writing—original draft preparation, Y.K.; writing—review and editing, Y.K. and S.H.

Funding: This work was supported by JSPS KAKENHI Grant Number JP18H05289.

Conflicts of Interest: The authors declare no conflict of interest.

References

1. Kocher, P.C.; Jaffe, J.; Jun, B. Differential Power Analysis. In *Advances in Cryptology, 19th Annual International Cryptology Conference, Santa Barbara, CA, USA, 15–19 August 1999*; Wiener, M.J., Ed.; Lecture Notes in Computer Science; Springer: Cham, Switzerland, 1999; Volume 1666, pp. 388–397.
2. Mangard, S.; Oswald, E.; Popp, T. *Power Analysis Attacks: Revealing the Secrets of Smart Cards*; Springer: Cham, Switzerland, 2007.
3. Kocher, P.C. Leak-Resistant Cryptographic Indexed Key Update. U.S. Patent 6,539,092, 25 March 2003.
4. Medwed, M.; Standaert, F.; Großschädl, J.; Regazzoni, F. Fresh Re-keying: Security against Side-Channel and Fault Attacks for Low-Cost Devices. In *Progress in Cryptology, Proceedings of the Third International Conference on Cryptology in Africa, Stellenbosch, South Africa, 3–6 May 2010*; Bernstein, D.J., Lange, T., Eds.; Lecture Notes in Computer Science; Springer: Cham, Switzerland, 2010; Volume 6055, pp. 233–244.
5. Dobraunig, C.; Eichlseder, M.; Mangard, S.; Mendel, F. On the Security of Fresh Re-keying to Counteract Side-Channel and Fault Attacks. In *Smart Card Research and Advanced Applications, Proceedings of the 13th International Conference, Paris, France, 5–7 November 2014*; Joye, M., Moradi, A., Eds.; Lecture Notes in Computer Science; Springer: Cham, Switzerland, 2015; Volume 8968, pp. 233–244.
6. Dobraunig, C.; Koeune, F.; Mangard, S.; Mendel, F.; Standaert, F. Towards Fresh and Hybrid Re-Keying Schemes with Beyond Birthday Security. In *Smart Card Research and Advanced Applications, Proceedings of the 14th International Conference, CARDIS 2015, Bochum, Germany, 4–6 November 2015*; Homma, N., Medwed, M., Eds.; Lecture Notes in Computer Science; Springer: Cham, Switzerland, 2016; Volume 9514, pp. 225–241.
7. Coron, J.S.; Patarin, J.; Seurin, Y. The Random Oracle Model and the Ideal Cipher Model are Equivalent. In Proceedings of the Annual International Cryptology Conference, Santa Barbara, CA, USA, 17–21 August 2008.
8. Liskov, M.; Rivest, R.L.; Wagner, D. Tweakable Block Ciphers. *J. Cryptol.* **2011**, *24*, 588–613. [CrossRef]
9. Dziembowski, S.; Faust, S.; Herold, G.; Journault, A.; Masny, D.; Standaert, F. Towards Sound Fresh Re-keying with Hard (Physical) Learning Problems. In *Advances in Cryptology-CRYPTO 2016, Proceedings of the 36th Annual International Cryptology Conference, Santa Barbara, CA, USA, 14–18 August 2016*; Robshaw, M., Katz, J., Eds.; Lecture Notes in Computer Science; Springer: Cham, Switzerland, 2016; Volume 9815, pp. 272–301.
10. Chittamuru, S.V.R.; Thakkar, I.G.; Bhat, V.; Pasricha, S. SOTERIA: Exploiting process variations to enhance hardware security with photonic NoC architectures. In Proceedings of the 55th Annual Design Automation Conference, DAC 2018, San Francisco, CA, USA, 24–29 June 2018; p. 81.
11. National Institute of Standards and Technology (NIST). NIST Special Publication 800-38A: Recommendation for Block Cipher Modes of Operation: Methods and Techniques. 2001. Available online: https://csrc.nist.gov/publications/detail/sp/800-38a/final (accessed on 11 March 2019).

12. Jean, J.; Nikolic, I.; Peyrin, T. Tweaks and Keys for Block Ciphers: The TWEAKEY Framework. In *Advances in Cryptology-ASIACRYPT 2014, Proceedings of the 20th International Conference on the Theory and Application of Cryptology and Information Security, Kaoshiung, Taiwan, 7–11 December 2014*; Sarkar, P., Iwata, T., Eds.; Lecture Notes in Computer Science; Springer: Cham, Switzerland, 2014; Volume 8874, pp. 274–288.

13. National Institute of Standards and Technology (NIST). Federal Information Processing Standards Publication (FIPS) 197, Specification for the ADVANCED ENCRYPTION STANDARD (AES). 2001. Available online: https://csrc.nist.gov/publications/detail/fips/197/final (accessed on 11 March 2019).

applied
sciences

MDPI

Article

Machine-Learning-Based Side-Channel Evaluation of Elliptic-Curve Cryptographic FPGA Processor

Naila Mukhtar [1,*], Mohamad Ali Mehrabi [1], Yinan Kong [1] and Ashiq Anjum [2]

[1] School of Engineering, Macquarie University, Sydney 2109, Australia;
mohamadali.mehrabi@hdr.mq.edu.au (M.A.M.); yinan.kong@mq.edu.au (Y.K.)
[2] Department of Computing and Mathematics, University of Derby, Derby DE22 1GB, UK;
ashiq.anjum@cern.ch
* Correspondence: naila.mukhtar@students.mq.edu.au

Received: 6 November 2018; Accepted: 18 December 2018; Published: 25 December 2018

Abstract: Security of embedded systems is the need of the hour. A mathematically secure algorithm runs on a cryptographic chip on these systems, but secret private data can be at risk due to side-channel leakage information. This research focuses on retrieving secret-key information, by performing machine-learning-based analysis on leaked power-consumption signals, from Field Programmable Gate Array (FPGA) implementation of the elliptic-curve algorithm captured from a Kintex-7 FPGA chip while the elliptic-curve cryptography (ECC) algorithm is running on it. This paper formalizes the methodology for preparing an input dataset for further analysis using machine-learning-based techniques to classify the secret-key bits. Research results reveal how pre-processing filters improve the classification accuracy in certain cases, and show how various signal properties can provide accurate secret classification with a smaller feature dataset. The results further show the parameter tuning and the amount of time required for building the machine-learning models.

Keywords: side-channel analysis; power-analysis attack; embedded system security; machine-learning classification

1. Introduction

Security is the core requirement in embedded systems nowadays and is ensured by using secure cryptographic algorithms on the embedded chips inside these systems. When designing and standardizing cryptographic algorithms, it is ensured that no mathematical relationship can be found between the key, the plain-text, and the ciphertext. However, side-channel attacks are still a threat to the embedded system. In side-channel attacks, physical leakages of the system are exploited to recover the private secret key. Side-channel attacks were introduced by Paul Kocher in the 90s [1,2], which was followed by the discovery of more side-channel attacks on hardware implementation of popular algorithms like AES, DES, RSA and ECC [3–6]. All these algorithms are proven to be prone to various kinds of side-channel attacks including power-analysis attack (PA), electromagnetic-analysis attack (EMA), timing attacks (TA). In 2003, Standaert et al. presented a practical PA attack on a Field Programmable Gate Array (FPGA) implementation of AES (symmetric algorithm) [7], and during the same year Siddika et al. presented a power-analysis attack on an FPGA (Virtex 800) implementation of an elliptic-curve cryptosystem [8]. Mulder et al. have presented techniques of key recovery by capturing, processing, and analyzing EM radiations using statistical models [9]. Based on similar techniques, the authors in [10] performed side-channel analysis for retrieving secret information. To perform the side-channel-based key-recovery analysis, various statistical and mathematical methods are used [11–16]. However, noise in leaked signals is one of the main hurdles to the success of side-channel attacks, which leads to the need for huge data sets for accurate key recovery. To cater to this issue, researchers have proposed to use statistical tools like principal-component analysis (PCA). PCA is used as the pre-processing step to eliminate the noise from side-channel leaked data, hence enhancing the success of differential power analysis (DPA) [17]. PCA can also act as a distinguisher [18].

Recently, researchers have performed machine-learning and neural-network-based analysis to improve side-channel attack efficiency, using various classifiers, to recover the key from the DES, AES and RSA hardware implementations [19–23]. Some of the major challenges of machine learning are over-fitting and the curse of dimensionality, in which a model trains itself to the specific data so well that it fails to predict accurately with new unseen data. To solve this problem, various feature extraction and selection techniques are used [24,25]. Some of them have been tested for AES data classification as well [26]. There is a very limited literature on machine-learning-based side-channel analysis of elliptic-curve cryptosystems, which is the standard for public-key cryptosystems and is ideally used for resource-constrained environments like IoT-based systems. The focus of this research is to analyze the resistance of the elliptic-curve cryptosystem algorithm, double-and-add-always (which is designed to be resistant against DPA), against a machine-learning-based power analysis attack. To check the immunity of an elliptic-curve cryptosystem against this attack; a hardware system was set up which is capable of capturing the power being consumed by the FPGA Kintex-7 chip while the elliptic-curve cryptography (ECC) algorithm is running on it (as ECC power-signal data does not exist for a Kintex-7) and then analyzing it against various machine-learning-based algorithms with a specifically designed feature dataset. We have chosen an FPGA for analysis because of its popularity for rapid prototyping. Our contributions in detail are listed below.

Our Contributions. Our contributions are threefold. Firstly, analysis of captured power signals is carried out using three machine-learning and neural-network-based classification techniques to classify and recover the secret-key bits of the ECC algorithm. The complete methodology is formulated for formation of the input datasets for further machine-learning analysis. In the existing literature, machine-learning-based side-channel attacks are launched on the raw samples, but no clear information is provided about the attack methodology and input feature datasets. Moreover, the ECC double-and-add-always (public-key) algorithm is selected for attack, as not much analysis is done on key recovery in the public-key domain.

Secondly, we propose to use signal properties as features for efficient analysis instead of using raw samples. This ensures the elimination of redundant data during processing, hence increasing the computational power and reducing the time to train and to test the network. The same proposed hypothesis has been tested for the AES (symmetric cryptography) algorithm in our previous work [26]. Based on the findings of our previous work for the symmetric key algorithm, we have tested the public-key algorithm for five particular signal properties only. Please note that this study is conducted for the ECC (public-key cryptography) algorithm.

Thirdly, our contribution is the application of filters to datasets, before processing them using the classification process. We have used Principal Components and Chi-square filters for pre-processing. The purpose of applying filters is to avoid the problem of wrong classification and to check if the accuracy can be improved by selecting/extracting important features. Again, we have selected one feature-selection and one feature-extraction algorithm, based on our previous findings from the research related to the symmetric algorithm AES. This double layer of pre-processing is applied just to verify if this can improve the accuracy further.

The rest of the paper is organized as follows. Section 2 describes related work, the ECC algorithm under test and the classification algorithms used for this analysis, Section 3 describes the implementation design of ECC (algorithm under analysis), Section 4 explains our attack methodology for key recovery using machine-learning-based classification techniques and describes the feature formation procedure, Section 5 outlines the hardware and software experimental setup, Section 6 gives an analysis of the results while Section 7 concludes the paper.

2. Background and Related Terminologies

2.1. Power-Analysis Attacks

The PA is a strong passive attack, meaning that the attacker does not need to manipulate the device in any way to extract the secret key. In fact, whenever a command is executed by the device, the consumed power is measured by putting a resistor between V_{ss} or V_{dd} and the true V_{dd}, for processors implemented in CMOS technology. The voltage drop by the current through the resistor is recorded. The voltage measurements are then analyzed using statistical methods to recover the secret key. The details of CMOS leakage can be found in [27].

PAs can be categorized into simple (SPA) and DPA. The feasibility of a simple PA depends upon the assumption that each instruction will have a unique power trace, which is normally caused by key-dependent branching. For scenarios where traces are not related to the key and instructions but are related to the data key, such attacks are categorized as differential power-analysis attacks. In DPA, the results of hypothetical models are compared with the actual experimental results.

2.2. Classification Algorithms

For the analysis in this paper, four main classification algorithms are used—three machine-learning and one simple neural-network-based algorithm. These algorithms have been tested for similar nonlinear data, having independent features, for other symmetric and asymmetric algorithms.

2.2.1. Random Forest (RF)

RF belongs to the class of supervised machine-learning algorithm which is based on decision trees [28]. The outcome of each tree contributes towards the prediction which makes is more reliable and accurate. RF helps in overcoming the problem of over-fitting by using feature-bagging technique. It produces better results even without hyper-parameter tuning which we will verify for our leaked data as well.

2.2.2. Support Vector Machine (SVM)

The support vector machine is another supervised-learning algorithm, which maps and represents data points in n-dimensional spaces to create a clear hyper-plane to separate classes. High-dimensionality can be an issue with SVM which can be handled using feature-extraction methods like PCA.

2.2.3. Naive Bayes (NB)

NB is also a supervised-learning algorithm. It is based on Bayes theorem, in which a probability model is created for the possible outcomes. It is useful for large datasets and is based on the assumption that predictors are independent, i.e., the features present in a sample are completely uncorrelated with each other, which is true for our key classification problem feature set as well.

2.2.4. Multilayer Perceptron (MLP)

A multilayer Perceptron is a type of feed-forward neural network, which uses backpropagation for training. This supervised-learning algorithm is used for solving complex problems stochastically. It is a fully connected network with layers having specific weights 'w' and neurons having a linear activation function which maps the weighted inputs to outputs. These weight values are adjusted based on the output error as compared to the expected value and is achieved through backpropagation.

2.3. Validation

It is important to validate the model against the existence of bias, after training with a machine-learning classification algorithm. For our analysis, the k-fold cross-validation mechanism is applied for validation. In the k-fold cross-validation, a hold-out method is used in which the model is trained k times, using k-1 subsets of the training data, and an error is estimated for the testing portion (which is one subset of the data) to analyze the performance of the model. The process is repeated k times to get better validation accuracy.

2.4. Feature/Attribute Selection and Extraction

In a feature-selection procedure, several features/attributes are selected, from the existing feature dataset, which are then used in classification-model construction. However, in feature-extraction methods, a new feature/attribute dataset is formed based on the existing features. Both techniques help in reducing the features which helps in better classification. We have selected one feature-selection (Chi-Square) and one feature-extraction (PCA) method for our analysis. As mentioned before, PCA has proven to be the best choice for pre-processing if a support vector machine (SVM) algorithm is used before classification. One of the purposes of this research is to analyze the effect of this best-performing feature-extraction technique on our reduced proposed feature data set (which is formed based on signal properties). Chi-square is randomly selected from the list of feature-extraction techniques. The reason for this selection is that our previous machine-learning-based power analysis on AES data, showed that all feature-selection give almost similar results [26,29]. We just picked one feature selection as the scope of analysis is wider than just analyzing the feature pre-processing.

3. Design and Implementation of Elliptic-Curve Cryptosystem F256 on FPGA

This section explains FPGA design of the elliptic-curve double-and-add-always algorithm (1) used for this analysis. The understanding of the implementation of the algorithm is important for re-launching the attacks for achieving the same results.

3.1. Power Analysis and ECC

ECC, introduced by Koblitz and Millers in the early 80s, is a preferred powerful public-key cryptosystem, especially for resource-constrained environments like smart cards, mobile phones, IoT-based devices, and RFIDs. In ECC, point multiplication is the resource-expensive operation in which a point on an elliptic-curve is added to itself successively. Let ′*P*′ be the point and ′*k*′ be the number of times ′*P*′ is required to be added, then output ′*Q*′ will be ′*k*′ times point ′*P*′ multiplication and is given by (1). Elliptic-curve point multiplication is also referred to as Elliptic-curve scalar multiplication (ECSM). Security of an elliptic-curve cryptosystem is based on the elliptic-curve discrete-logarithm problem, which relies on the fact that for an elliptic curve E and given points $P(x,y,z)$ and $Q(x,y,z)$, it is hard to find the integer k such that $Q = kxP$.

$$Q = kxP \tag{1}$$

To compute ECSM, double-and-add is the simplest straightforward algorithm, in which operations are performed depending upon the ′k′ key bits. If the key bit is ′0′ then only the point-double operation is performed. However, point-double and point-addition both are performed if the key bit is ′1′. The simple double-and-add algorithm is susceptible to a simple power-analysis (SPA) attack; simply by analyzing the power consumption of the chip, scalar key ′k′ can be resolved, by merely looking at the oscilloscope, without using any advanced processing. Countermeasures are proposed in the literature to help safeguard against SPA attacks. The simplest of all is to add an extra operation so that the double-and-add operations are performed always irrespective of the scalar k bit as can be seen from Algorithm 1. Double-and-add-always seems to be resistant against PA but is not secure against the safe-error attack, where an attacker introduces an error and examines if the output will show an error or not. Depending upon the output, the scalar key bit k is determined. However, double-and-add-always still seems to be feasible due to the low cost. Further details of the algorithm can be found in [30].

Algorithm 1 double-and-add-always

1: Input: $P, k[n]$
2:
3: Output: $Q = kP$
4:
5: $R0 = P, R1 = 0$
6:
7: **for** $i = 1$ to $n - 2$ **do**
8:
9:     $R0 = 2R0$
10:
11:     $R1 = R0 + P$
12:
13:     **if** $k_i = 1$ **then**
14:
15:         $R0 = R1$
16:
17:     **end if**
18:
19: **end for**
20:
21: return $Q = R0$
22:

3.2. Nist Standard for 256-Bit Koblitz Curve

The NIST curve (SECP256K1), used in this analysis, over prime fields F_p, is defined as E: $y^2 = x^3 + ax + b$ mod p, where $a = 0$ and $b = 7$ and $p = 2^{256} - 2^{32} - 2^9 - 2^8 - 2^7 - 2^6 - 2^4 - 1$ [31]. The two main field operations in the double-and-add-always algorithm, point doubling and point addition in Jacobian coordinates over curve E, used for this study, are described in [32]. Jacobian coordinates are preferred over affine coordinates because inversions can be avoided while performing the addition or doubling operation, which is not the case in the affine coordinate system.

3.3. Point Doubling in Jacobian Coordinates

This section gives the formulas used for implementing point doubling.
Suppose: $P(X_1, Y_1, Z_1)$ and

$$\alpha = 3X_1^2 + aZ_1^4, \beta = 4X_1Y_1^2, \tag{2}$$

Point Q on curve E is defined as: $Q(X_2, Y_2, Z_2) = 2.P(X_1, Y_1, Z_1)$

$$X_2 = \alpha^2 - 2\beta, \tag{3}$$

$$Y_2 = \alpha(\beta - X_2) - 8Y_1^4, \tag{4}$$

$$Z_2 = 2Y_1Z_1, \tag{5}$$

3.4. Point Addition in Jacobian Coordinates

This section gives the formulas used for implementing point addition. Suppose: $P_1(X_1, Y_1, Z_1)$ and $P_2(X_2, Y_2, Z_2)$ are two points on curve (E) and

$$\gamma = Y_1Z_2^3, \lambda = X_1Z_2^2, \mu = Y_2Z_1^3 - Y_1Z_2^3, \zeta = X_2Z_1^2 - X_1Z_2^2, \tag{6}$$

The new point P3 on Curve (E) such that: $P_3(X_3, Y_3, Z_3) = P_1(X_1, Y_1, Z_1) + P_2(X_2, Y_2, Z_2)$ is:

$$X_3 = \mu^2 - \zeta^3 - 2\lambda\zeta^2 \tag{7}$$

$$Y_3 = \mu(\lambda\zeta^2 - X_3) - \gamma\zeta^3, \tag{8}$$

$$Z_3 = Z_1Z_2\zeta, \tag{9}$$

All calculations are to be done in finite field F_p, meaning that *mod p* reduction is applied to Formulas (2)–(9).

A modular reduction unit is designed based on an interleaved modular multiplier architecture similar to the one proposed in [33,34]. Based on the implementation results in [33,34], an interleaved modular multiplier has more efficient area and timing characteristics. For fast and area-efficient implementation of such a multiplier, we use just one CSA adder and a look-up table. The structure of our design is depicted in Figure 1.

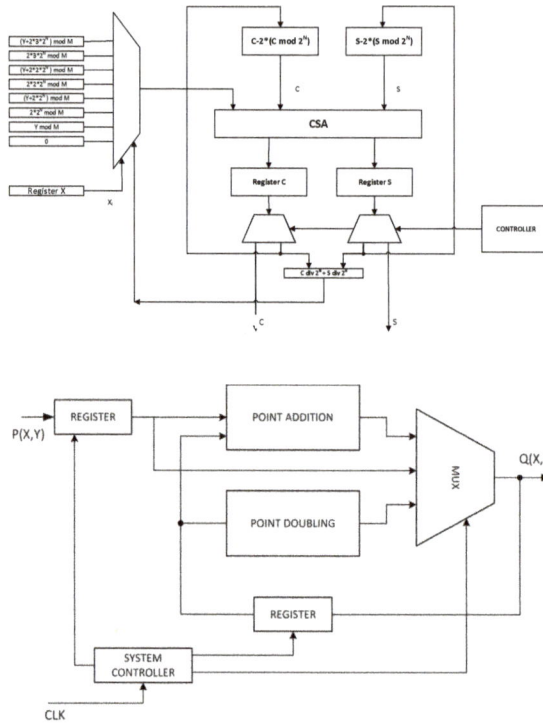

Figure 1. Interleaved Modular Multiplier (top) and ECC core Diagram (bottom).

The look-up table code is given by;

$$LUT(0) = 0;$$
$$LUT(1) = Y;$$
$$LUT(2) = 2 * 2^n \bmod M;$$
$$LUT(3) = (2 * 2^n + Y) \bmod M;$$
$$LUT(4) = 4 * 2^n \bmod M;$$
$$LUT(5) = (Y + 4 * 2^n) \bmod M;$$
$$LUT(6) = 6 * 2^n \bmod M;$$
$$LUT(7) = (Y + 7 * 2^n) \bmod M;$$
$$\text{and}$$
$$(S,C) = CSA(A,B,C) \text{ is}$$
$$S_i = A_i \oplus B_i \oplus C_i,$$
$$C_{i+1} = (A_i.B_i)V(A_i.C_i)V(B_i.C_i), C_0 = 0$$

The modular multipliers use one clock cycle to register inputs, 256 (n) clock cycles in the loop, one clock cycle to calculate the CSA addition and one clock to register output, so the calculation is done in just 259 n + 3 clock cycles.

3.5. ECC Core Design

The ECC core design gets a point on the ECC curve in Jacobian coordinates **P(X,Y,Z)** and calculates point **Q = kxP** within the same coordinate system. Figure 1 illustrates the ECC core design.

3.6. Elliptic-Curve Point Doubling—ECPD

Point doubling uses three modular multiplier units to calculate (2)–(5) in parallel. Ten modular multiplications are done in five stages that reduce the point-doubling calculation time to $5(n + 3) + 4$ clock cycles.

For curve SECP256K1, as $a = 0$, the logic can be reduced. Using just one modular reduction unit, ECPD can be performed at 7 logic levels or $7(n + 3) + 2$ clock cycles by the optimized-area ECPD. Figure 2 shows the data-flow diagram of the ECPD doubling with and without optimized area.

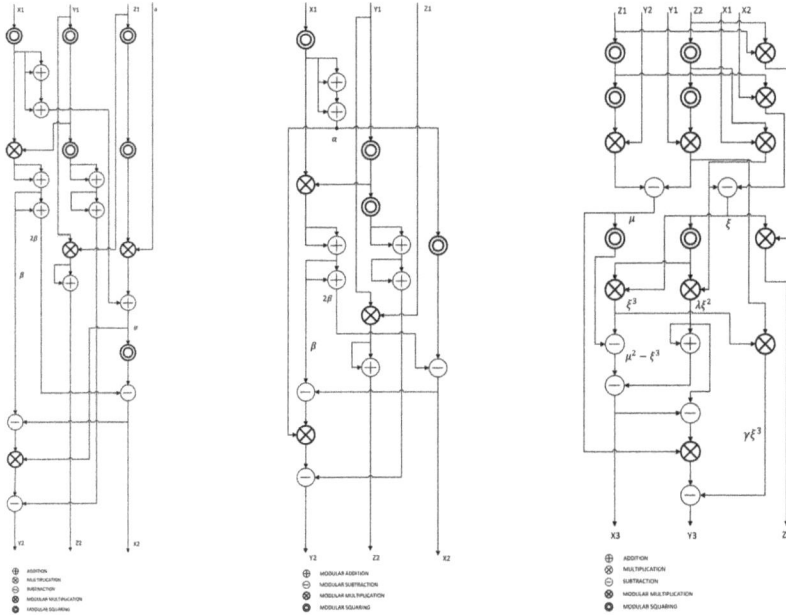

Figure 2. Data Flows (Left to Right) 1. Point-doubling ECPD for SEC2P256K1 curve, 2. Optimized area for point doubling, 3. Point Addition.

3.7. Elliptic-Curve Point Addition

Point addition uses three modular multiplier units to calculate point $Q + P$ on the elliptic curve in parallel. Sixteen modular multiplications are done in seven stages as shown in Figure 2, so the latency of point addition will reduce to $7(n + 3) + 5$ clock cycles.

3.8. Scalar Factor (Private Key) k

The scalar factor 'k' is stored in internal RAM and can be changed via software command. To implement a point multiplication, the double-and-add-always algorithm is used as given in Algorithm 1. A point doubling is done followed by a point addition at every stage i, but the result of the point addition is used only when the *i*th bit of the scalar k is '1'. Otherwise, the result of point addition will not be used.

In this method, N times PD and PA are required (here N = 256). This algorithm uses the same hardware resources for the zero and one bits of the key k, so the power consumption during calculations is homogeneous. The resources consumed by the design of the interleaved multiplier are given in Table 1.

Table 1. Implementation results on Xilinx FPGA XC7k160tfbg676-1.

Resource	Utilization
CORE AREA (LUT)	26,570
ECPA (LUT)	14,382
ECPD (LUT)	11,760
CLK frequency	100.00 MHz
DYNAMIC POWER	0.20 W
TOTAL POWER	0.313 W

4. Attack Methodology

The purpose of this research is to capture and analyze the power-consumed signals of the FPGA (Kintex-7) while the ECC double-and-add-always algorithm is encrypting data with a secret key. The idea is to attack one bit at a time. For our analysis, we will attack the least-significant three bits of the nibble i.e., bit 2, bit 3 and bit 4. The bit at location one does not need to be attacked as it does not contribute to the encryption. To achieve this purpose, a random 31-bytes (which are the most significant 248 bits) fixed key is selected and the value of the last byte is changed in ascending order, from 2^1 till $2^4 - 1$. For further simplification, in this paper we have attacked bit locations 2, 3 and 4 only, and bit locations 5 to 8 are set to "0000". From now on, 'key' refers to the last nibble of the key as shown in Figure 3.

For the analysis, machine-learning classification will be used. For classification using machine learning, the data samples should consist of the properly labeled features. We propose to use a different set of features as opposed to the raw samples' amplitude, which leads to a division of our attack into two main steps:

- Step 1—Training dataset preparation
- Step 2—Classification using machine learning

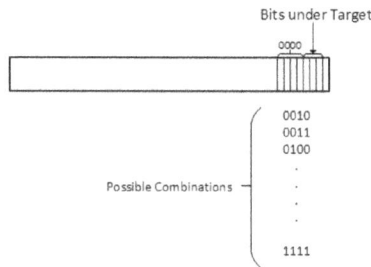

Figure 3. Key Bits Under Target.

4.1. Step 1—Training Dataset Preparation

Let N be the number of randomly selected ECC points, in the Jacobian coordinate system, from the elliptic curve E, and M represent the set of ECC points, then each ECC point in M, over curve E, can be represented as follows:

$$M = \{(X_i, Y_i, Z_i) \text{ where } i = 1, 2, ...N\} \tag{10}$$

Let K be the least-significant four bits of the 256-bit key. Out of the 4 LSBs, the three bits at locations 2nd, 3rd and 4th, are the target of this analysis. The first bit is not considered, as the double-and-add-always algorithm's implementation starts encryption using a second bit of the key. For each bit location, raw traces of length Len_{Trace} are collected and then processed to form samples. $S_{BitLoc} = N * S_p t$ samples are collected for N ECC points from the set M, where $S_p t$ represents the number of samples for each ECC point from the pool of N ECC points. As the number of possible combinations for the last nibble is 2^4 and there is no point in attacking the first bit, so in total $S_N = S_{BitLoc} * (2^n - 2)$ samples are collected. For creating a training dataset for machine-learning classification, data samples need to be labeled. After data sample collection, labeling is an important task. To ease the process of attacking and labeling, we have divided the attack into three levels according to the bit location under attack and have categorized the samples into two groups. Each is further explained below.

4.1.1. Group Labeling

All data samples are divided into two groups 'GB0' and 'GB1'. GB0 means that the sample represents a bit '0' and GB1 means that the sample represents a bit '1'. Each attack level will have different samples marked as GB0 or GB1 according to the bit location.

4.1.2. Attack Levels

Attack levels are designed based on the bit locations under attack, called 'LB{b}'. LB stands for the 'Location bit' and 'b' represents the actual location of the bit. Based on this information, three bit levels are defined as follows.

- LB2—At this attack level, LSB '2' is targeted and each sample for the key having '0' at the second location is marked as 'GB0' and all samples for the key byte having '1' at the second location are marked as 'GB1'.
- LB3—At this attack level, LSB '3' is targeted and each sample for the key having '0' at the third location is marked as 'GB0' and all samples for the key byte having '1' at the third location are marked as 'GB1'.
- LB4—At this attack level, LSB '4' is targeted and each sample for the key having '0' at the fourth location is marked as 'GB0' and all samples for the key byte having '1' at the fourth location are marked as 'GB1'.

The attack levels along with the labeling of the samples are shown in Figure 4.

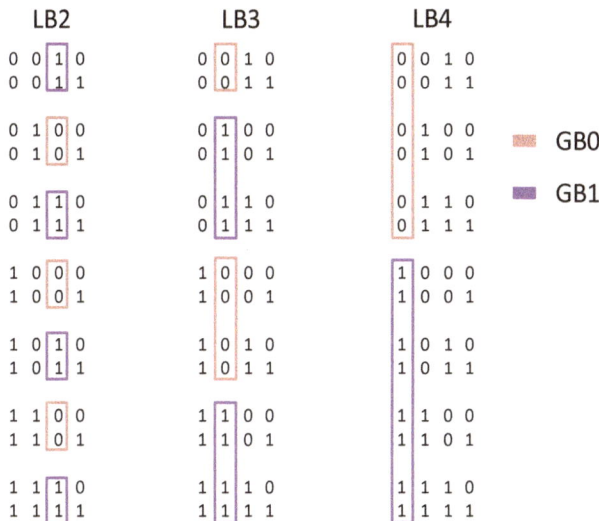

Figure 4. Sample Labeling.

4.1.3. Features Dataset Formation

As all the raw samples have been labeled, the next step is to calculate the features. For our analysis, we have used time-domain and frequency-domain signal properties as features. The reason for selecting these particular signal properties is based on our previous analysis on AES leaked data. We selected and analyzed more than six signal properties and concluded that a combination of time-domain and frequency-domain signal properties leads to better classification [26,29]. An explanation of each signal property (used in this work) is given below:

- Mean of Absolute Value (MAV)—Mean of the signal is calculated.
- Kurtosis (Kur)—For Kurtosis, Frequency-distribution curve peak's sharpness is noted.
- Median PSD (FMD)—For Median PST, median is calculated in frequency domain.
- Frequency Ratio (FR)—For Frequency ratio, frequency ratio of the frequencies is recorded.
- Median Amplitude Spectrum (MFMD)—For MFMD, the median amplitude spectrum of signals is calculated.

For all the captured S_N samples, the above-mentioned features are calculated, returning one sample value for each instead of Len_{Trace}, hence reducing the data sample size, which is the advantage of using the above-proposed features.

The overall training dataset preparation process is shown in Figure 5.

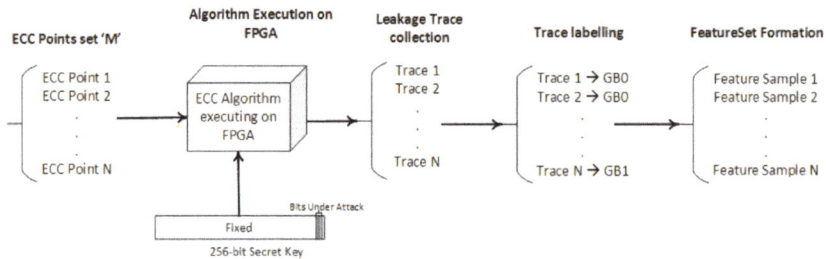

Figure 5. Dataset Preparation.

4.2. Step 2—Classification Using Machine Learning

Traditionally, statistical methods are used to perform the analysis but for this research machine-learning and neural-network-based classifiers are applied on the feature datasets, formed in Section 4.1. The classification algorithms selected for analysis are Support Vector Machines (SVM), Naive Bayes (NB), Random Forest (RF) and Multilayer Perceptron (MLP). An explanation of each is given in Section 2.2. According to author's knowledge, there is very little work done in the field of machine-learning-based power analysis on elliptic curves which includes analysis of ECC leaked data (from a FPGA) using PCA-SVM. Hence, in our analysis, the comparison is provided with respect to the machine-learning-based analysis only.

There are two parts of the analysis as given below.

4.2.1. Analysis without Pre-Processing

In the first phase of analysis, classification is performed on the feature datasets without any pre-processing. This analysis will help in identifying the impact of using signal properties as features.

4.2.2. Analysis with Pre-Processing

In the second phase of analysis, the feature dataset is first processed through a feature selection and extraction mechanism before training the model and is then subjected to the classification. The feature selection and extraction techniques used for pre-processing are PCA and Chi-Square (Chi-Sq). Details are given in Section 2.4. The signals' noise makes the side-channel attacks harder to launch. The evaluators/selectors are used to filter out the features to overcome the problem of noisy signals, hence reducing the training time and computational complexity. Another benefit of using these extractors/selectors is to reduce the possibility of a wrong classification. For testing the trained model, another feature dataset is formed based on the same methodology. This is done to gain more confidence in the results, as it ensures that the model has never seen the test data before. The process of classification on the training feature dataset is shown in Figure 6. Moreover, the effect of changing of various variables/parameters was observed. The time required to build the model has also been recorded.

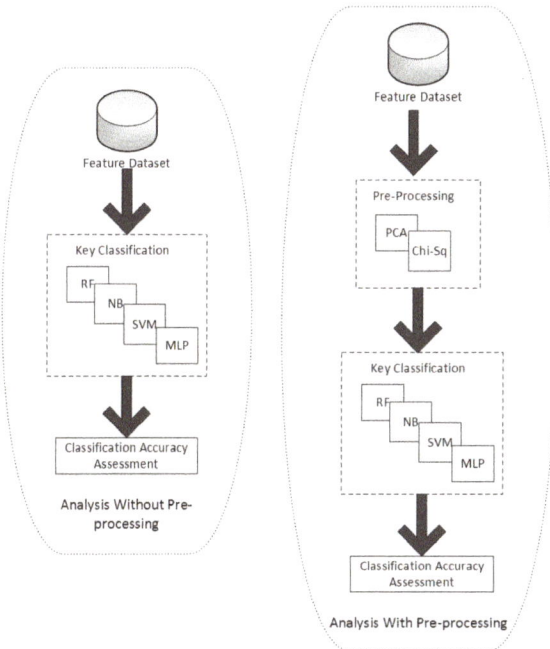

Figure 6. Classification process without pre-processing (**left**) and with pre-processing (**right**).

5. Experimental Setup

This section explains the hardware and software setup for testing the methodology explained in the above sections.

5.1. Step 1—Data Capture

To conduct our experiments, we must capture the leakage traces, as a power signals database for ECC does not exist. For the hardware setup, we captured the power signals for ECC FPGA (Kintex-7) implementation, operating at 24 MHz. For this research, specialized side-channel analysis board, named as SAKURA-X, is used [35]. On SAKURA-X, for calculating the power being consumed, a resistor is connected in series and a voltage is measured across that. The user does not need to tweak the board, as the connector is available to get the power signal directly. Traces are captured using a Tektronix oscilloscope having a 5 GS/sec sampling frequency and a 1 GHz bandwidth. We have acquired N = 100 traces for randomly selected ECC points from set M, and for each point S_pt = 10 traces were captured. Thus, in total S_N = 14,000 traces are collected where each trace has 10 k sampling points.

For the software side of the data-collection process, we have developed bespoke codes using C# and the MATLAB library to form an automated standalone application which requires little or no intervention from the user. The hardware setup and the application GUI is shown in Figure 7. A few modules of the C# application provided by SAKURA are used to achieve the purpose [35]. The new bespoke C# application consists of three main units: control unit, data unit, and configuration unit as shown in Figure 8, and an explanation of each is given below.

- Configuration Unit—The configuration unit uses MATLAB library support for C# and configures the oscilloscope through the C# application. This eliminates setting up the oscilloscope on every start up; the application automatically restores it to the settings required for the data capturing. The configuration unit communicates with the oscilloscope only.
- Control Unit—The control unit has the role of sending the ECC points to the FPGA after taking them from the data unit. When the FPGA receives an ECC point, it starts the process of encryption and sends a trigger signal to the oscilloscope. As soon as the trigger signal is received at the oscilloscope, it will start collecting the leaked information from the FPGA and will transmit it to the control unit. The control unit then stores the information by communicating with the data unit. The control unit communicates with both the oscilloscope and the FPGA.
- Data Unit—The data unit handles the data. It is responsible for storing and retrieving the data in files. The data unit communicates with the control unit only.

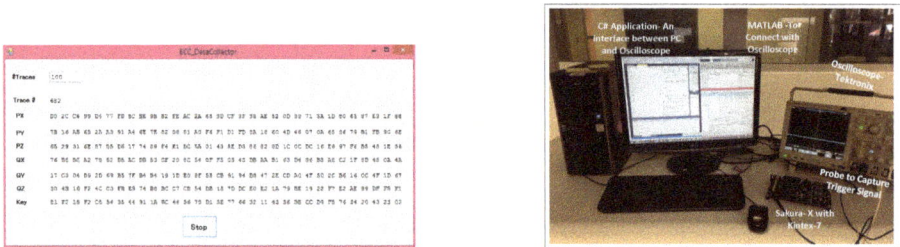

Figure 7. GUI for raw Sample Collection Application and hardware setup for power analysis data capture.

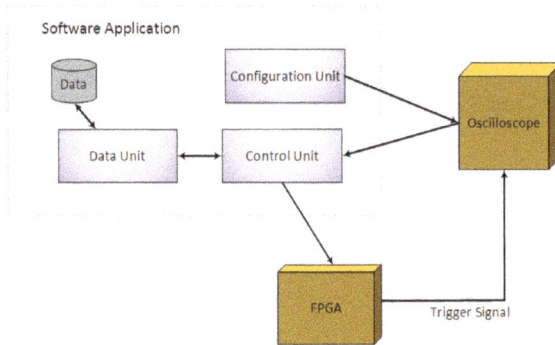

Figure 8. Software Setup Design.

5.2. Step 2—Feature Datasets Formation

After collection of the raw traces, samples are labeled according to the description given in Section 4.1.1, using a bespoke java snippet. After labeling, features (properties) are calculated using bespoke MATLAB code, and act as features for further classification.

5.3. Step 3—Analysis

Classification models are then trained, using the proposed feature datasets. Feature datasets are trained and tested with and without applying the pre-processing filters. For training and testing, tools like weka and organe3 are used [36]. Parameters settings for each classification algorithm are discussed in the results.

6. Results and Discussion

Results and discussion are divided into four sections. In each section, results are discussed with reference to classification algorithms.

6.1. Analysis Phase 1—Accuracy without Pre-Processing

In the first part of phase-1 analysis, the classification accuracy is calculated on the raw-signal feature data set. It is observed that RF gives an accuracy of 79% for LB4. For NB, SVM and MLP, the accuracy is even lower i.e., 52%, 55% and 71%. For LB2 and LB3, the accuracy is less than LB4, as shown in Table 2. These results clearly show that the data cannot be correctly classified due to the large number of features in the dataset.

Table 2. Accuracy for Raw sample analysis.

Algorithm	LB4	LB3	LB2
RF	79.2%	56.9%	58.0%
SVM	55.4%	49.3%	45.7%
MLP	71.9%	58.7%	55.6%
NB	52.5%s	57.0%	55.7%

In the second part of phase-1 analysis, the classification accuracy is calculated on the proposed processed feature datasets without any pre-processing (i.e., feature selector/extractor) for all three levels of attack (LB2-LB4), as given in Figure 9. Models are trained and tested using the four classifiers SVM, RF, NB, and MLP. It can be seen that, without the pre-processing step, SVM does not perform well for any level of bit classification. However, for the fourth-bit classification, RF gives an accuracy of approximately 90% while NB and MLP give an accuracy of 85% and 88%, respectively. RF and NB perform well for the datasets in which the features are completely independent of each other. These results prove that the features in the signals feature datasets (for fourth-bit location) are independent of each other.

For bit 2 and bit 3 classification, the maximum accuracy achieved is 71–73% with RF. Both SVM and MLP perform poorly in these cases. The reason for MLP's low performance could be the feature dataset size. For neural-network algorithms, the training data should be huge, roughly a hundred times more than the number of features in each trace/row. It is worth exploring if MLP or any other neural network can behave better if the number of samples is increased for better training. This analysis is out of the scope of this paper and is a future prospect of this particular research.

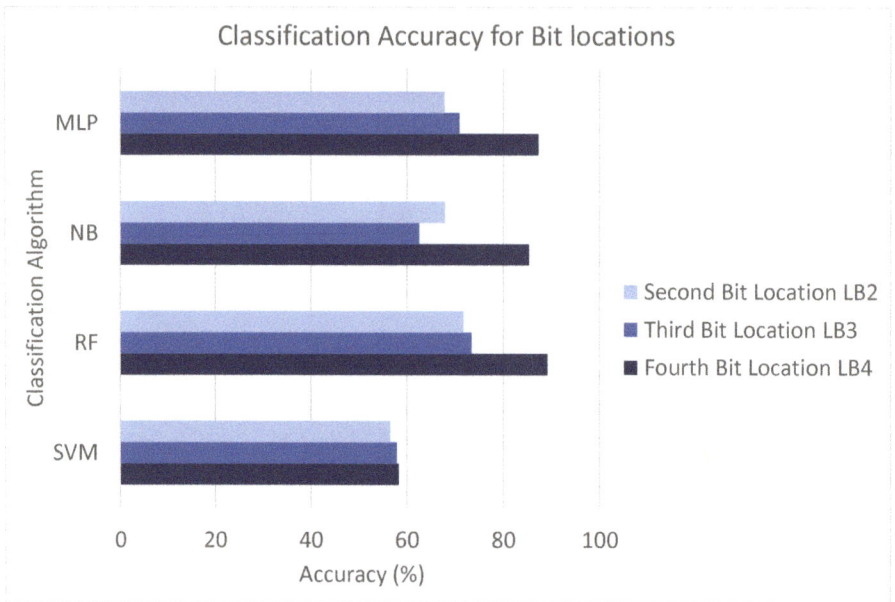

Figure 9. Classification Accuracy without any pre-processing.

6.2. Analysis Phase 2—Accuracy with Pre-Processing

In the second phase of the analysis, the classification accuracy is calculated on the feature datasets after pre-processing them using the feature selectors/extractors. The results of 'LB2', 'LB3' and 'LB4' are given in Figure 10. The results show that if PCA is applied then, for SVM, the accuracy improves for all three cases. This happens because the obtained traces are noisy, having redundant information. PCA extracts the important features/components so when SVM is applied on the reduced feature set then the accuracy is improved. The maximum accuracy attained is 87% for 'LB4'. However, Chi-square did not show any improvement in any of the LBs. It is worth noting that the accuracy of RF got worst after pre-processing with PCA, because RF works on the assumption that there is no dependence between features. PCA reduces the number of features and at the same time removes the col-linear features from the feature dataset. For MLP, accuracy increases after applying filters in case of LB4 but strange behavior is observed in case of LB3 and LB2, which requires further analysis.

The authors in [37] have obtained 96% accuracy after applying SVM on 4-bit implementation of ECC leaked data. Our results of classification algorithms are obtained after applying SVM on 256-bit key (out of which first 31 bytes of the key are fixed random numbers). Our results show that, with PCA-SVM, accuracy of around 86% can be achieved to recover the least-significant nibble from a 256-bit key.

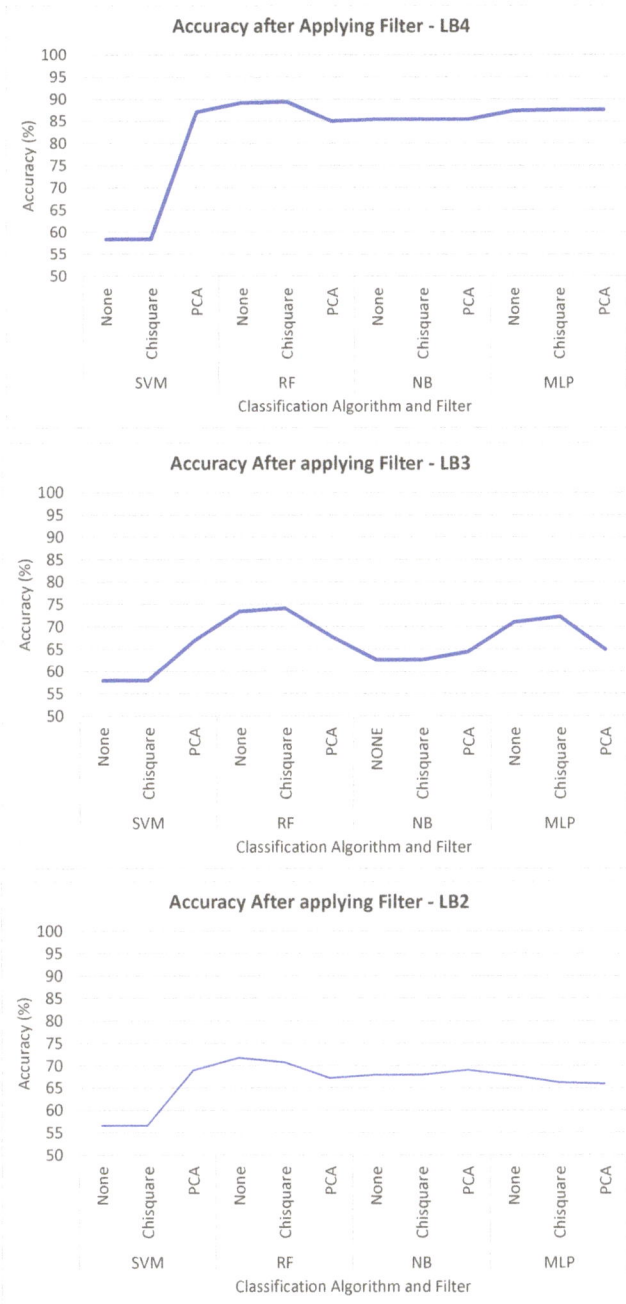

Figure 10. Classification Accuracy with pre-processing—LB2, LB3 and LB4.

6.3. Analysis Phase 3—Time to Build Models

It has been seen that the time taken to build the model with raw signals varies from 90–150 s for classifiers. However, the time taken to build the model on the proposed processed feature dataset is less. The reason is obviously that, with raw signals, the number of features per trace is 10 k times more than for the proposed processed feature datasets. In particular it was observed that the time taken to build the model for the MLP, with proposed processed feature dataset, is more than the time taken by SVM, RF and NB. After applying PCA, the time taken to build the model is reduced in all cases (LB2–LB4) for all algorithms, as can be seen from Figure 11. The reason for the decrease in the time required to build the model is that the number of features is reduced after applying filters.

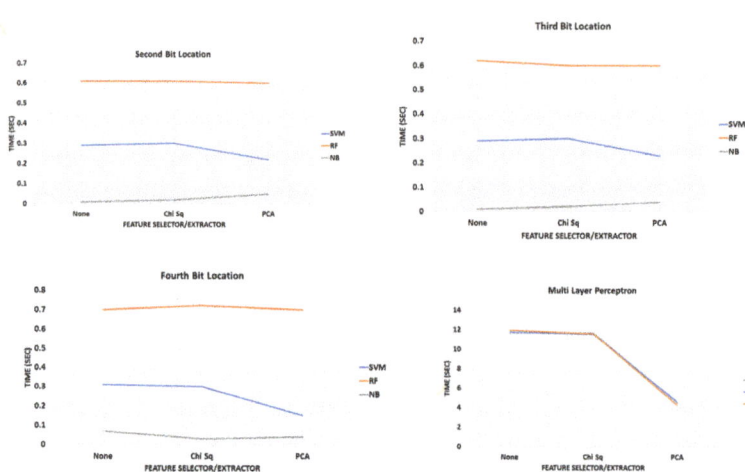

Figure 11. Timing Results for model building for LB2, LB3 and LB4.

6.4. Hyper-Parameter Tuning

Based on the analysis so far, the best-performing combination of filters and classifier is selected, and different parameters are tuned for LB4. The parameters used for analysis, using four classification algorithms, are discussed below.

The parameters tuned for RF are the number of trees and the number of features per tree. Experimental results show that with 90 trees within the forest, the highest accuracy of approximately 89.4%, is achieved. The accuracy decreases if the number of trees is increased or decreased beyond 90. Therefore, 90 trees are selected for further analysis, and number of features per tree are changed, when it is observed that maximum accuracy is obtained when the number of features per tree is 30, as shown in Figure 12.

In NB, two important parameters are kernel estimator and supervised discretization. It was observed that turning on kernel estimator gives an accuracy of 86.77%. However, accuracy is increased if supervised discretization mode is turned on.

For SVM, gamma is changed to see the effect on accuracy. As SVM uses nonlinear kernel functions, so a lower gamma value means low bias and high variance. It is seen that with higher values of gamma, the accuracy decreases as can be seen from Figure 13.

For any neural network, the two most important parameters to analyze are the change of learning rate and the batch size. For this analysis, the number of neurons is fixed, and one hidden layer is used. Learning rate is the rate of training with which the model is trained, and the batch size is the number of samples that are given to the model for one training period. It was observed that the batch size does not have any effect on the accuracy. However, the model is trained best with learning rate of 0.01, as can be seen from Figure 13.

Parameter Tuning for RF

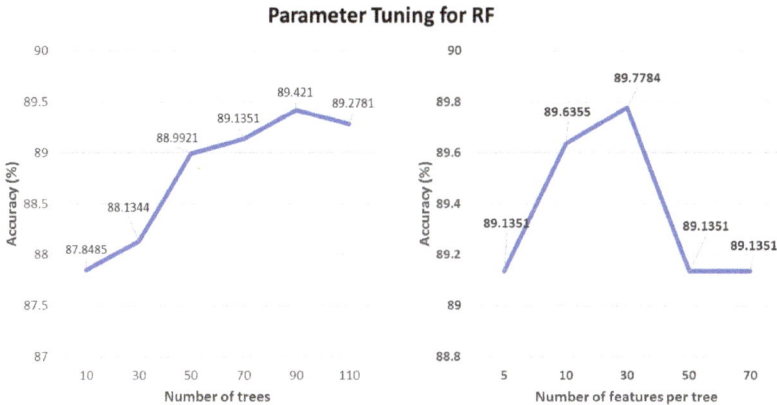

Figure 12. Parameter Tuning for RF.

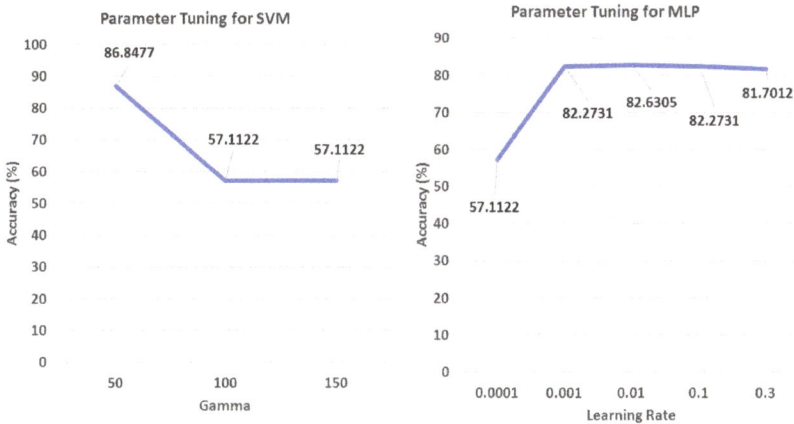

Figure 13. Parameter Tuning for SVM and MLP.

7. Conclusions

After analyzing the results on the power-consumed signals obtained from the Kintex-7, we can conclude that signal properties of the captured leaked power signals can be used as features, as they give an accuracy of approx. 90% with RF. This means that we can recover the secret key from the leaked signals with approx. 90% accuracy. For classification algorithms like RF, NB, and MLP, pre-processing did not show any improvement at all, because these classifiers already perform well for noisy data having redundant information. However, for SVM, using PCA as a pre-processing step improved the accuracy to 86%, as PCA extracted the important relevant features. SVM still shows less accuracy than the others. Moreover, the time taken for building the model has been analyzed and it is observed that the time for training the model is more for raw signals and for the neural-network-based MLP model. The parameters for all four classification algorithms have been tuned and the best recommendations are put forward in the paper.

8. Future Work

Future work will be based on the findings of this research. We aim to recover the middle and initial most significant bits of the key. As RF performs the best, the first preference for all future analysis would be RF. As the data samples would be different for the key bits according to their locations, which might introduce non-linearity in the system, thus increasing the possibility of improved attack accuracy using neural networks. We would like to analyze the data using deep-learning algorithms like Convolutional Neural Networks and Long-Short-Term-Memory networks, to explore these avenues.

Author Contributions: Conceptualization, N.M. and Y.K.; methodology, N.M.; software, N.M. and M.A.M.; validation, N.M., M.A.M. and Y.K.; formal analysis, N.M.; investigation, N.M.; resources, N.M.; data curation, N.M.; writing—original draft preparation, N.M. and M.A.M.; writing—review and editing, A.A.; visualization, N.M.; supervision, Y.K.; project administration, N.M. and Y.K.; funding acquisition, N.M. and Y.K.

Funding: This research received no external funding.

Acknowledgments: The work in this paper was supported by School of Engineering, Macquarie University, Sydney, Australia.

Conflicts of Interest: The authors declare no conflict of interest.

Abbreviations

The following abbreviations are used in this manuscript:

ML	Machine Learning
SCA	Side-Channel Analysis
PCA	Principal-Component Analysis
Chi-Sq	Chi-Square
RF	Random Forest
NB	Naive Bayes
MLP	Multilayer Perceptron
SVM	Support Vector Machines
AES	Advanced Encryption Standard
ECC	Elliptic-curve Cryptography

References

1. Kocher, P.C. Timing Attacks on Implementations of Diffie-Hellman, RSA, DSS, and Other Systems. In Proceedings of the Advances in Cryptology—CRYPTO '96: 16th Annual International Cryptology Conference, Santa Barbara, CA, USA, 18–22 August 1996; Springer: Berlin/Heidelberg, Germany, 1996; pp. 104–113.
2. Kocher, P.C.; Jaffe, J.; Jun, B. Differential Power Analysis. In Proceedings of the Advances in Cryptology—CRYPTO' 99: 19th Annual International Cryptology Conference, Santa Barbara, CA, USA, 15–19 August 1999; Springer: Berlin/Heidelberg, Germany, 1999; pp. 388–397.
3. Rivest, R.L. Cryptography and machine-learning. In Proceedings of the Advances in Cryptology—ASIACRYPT '91: International Conference on the Theory and Application of Cryptology, Fuji Yoshida, Japan, 11–14 November 1991; Springer: Berlin/Heidelberg, Germany, 1993; pp. 427–439.
4. Genkin, D.; Pachmanov, L.; Pipman, I.; Tromer E.; Yarom, Y. ECDSA Key Extraction from Mobile Devices via Nonintrusive Physical Side Channels. In Proceedings of the 2016 ACM SIGSAC Conference on Computer and Communications Security, Vienna, Austria, 2016; ACM: New York, NY, USA; pp. 1626–1638.
5. Kadir, S.A.; Sasongko, A.; Zulkifli, M. Simple power analysis attack against elliptic curve cryptography processor on FPGA implementation. In Proceedings of the 2011 International Conference on Electrical Engineering and Informatics, Bandung, Indonesia, 17–19 July 2011; pp. 1–4.
6. Genkin, D.; Shamir, A.; Tromer, E. RSA Key Extraction via Low-Bandwidth Acoustic Cryptanalysis. In Proceedings of the Advances in Cryptology—CRYPTO 2014: 34th Annual Cryptology Conference, Santa Barbara, CA, USA, 17–21 August 2014; pp. 444–461.
7. Standaert, F.X.; tot Oldenzeel, L.V.O.; Samyde, D.; Quisquater, J.J. Power Analysis of FPGAs: How Practical Is the Attack? In Proceedings of the Field Programmable Logic and Application, Lisbon, Portugal, 1–3 September 2003; Springer: Berlin/Heidelberg, Germany, 2003; pp. 701–710.

8. Yalcin Ors, S.B.; Oswald, E.; Preneel, B. Power-Analysis Attacks on an FPGA—First Experimental Results. In Proceedings of the Cryptographic Hardware and Embedded Systems (CHES), Cologne, Germany, 8–10 September 2003; Springer: Berlin/Heidelberg, Germany, 2003; pp. 35–50.

9. De Mulder, E.; Ors, S.B.; Preneel, B.; Verbauwhede, I. Differential Electromagnetic Attack on an FPGA Implementation of Elliptic Curve Cryptosystems. In Proceedings of the Automation Congres, Budapest, Hungary, 24–26 July 2006; pp. 1–6.

10. Longo, J.; De Mulder, E.; Page, D.; Tunstall, M. *SoC it to EM: Electromagnetic Side-Channel Attacks on a Complex System-on-Chip*; Cryptographic Hardware and Embedded Systems—CHES; Lecture Notes in Computer Science; Springer: Berlin, Germany, 2015; Volume 9293, pp. 620–640.

11. Gierlichs, B.; Batina, L.; Tuyls, P.; Preneel, B. Mutual information analysis. In Proceedings of the Cryptographic Hardware and Embedded Systems—CHES, Washington, DC, USA, 10–13 August 2008.

12. Renauld, M.; Standaert, F.; Veyrat-Charvillon, N. Algebraic Side-Channel Attacks on the AES: Why Time also Matters in DPA. In Proceedings of the Cryptographic Hardware and Embedded Systems—CHES 2009, Lausanne, Switzerland, 6–9 September 2009; Springer: Berlin/Heidelberg, Germany, 2009; pp. 97–111.

13. Bhasin, S.; Danger, J.; Guilley, S.; Najm, Z. Side-Channel Leakage and Trace Compression using Normalized Inter-Class Variance. In Proceedings of the 3rd International Workshop on Hardware and Architectural Support for Security and Privacy, HASP, Portland, OR, USA, 14 June 2015; p. 7.

14. Oswald, D.; Paar, C. Improving Side-Channel Analysis with Optimal Linear Transforms. In Proceedings of the 11th International Conference on Smart Card Research, and Advanced Applications, CARDIS, Graz, Austria, 28–30 November 2012; Springer: Berlin/Heidelberg, Germany, 2013; pp. 219–233.

15. Hospodar, G.; Mulder, E.D.; Gierlichs, B.; Verbauwhede, I.; Vandewalle, J. Least Squares Support Vector Machines for Side-Channel Analysis. In Proceedings of the 2nd Workshop on Constructive Side-Channel Analysis and Secure Design (COSADE), Darmstadt, Germany, 24–25 February 2011.

16. Willi, R.; Curiger, A.; Zbinden, P. On Power-Analysis Resistant Hardware Implementations of ECC-Based Cryptosystems. In Proceedings of the 2016 Euromicro Conference on Digital System Design (DSD), Limassol, Cyprus, 31 August–2 September 2016; pp. 665–669. [CrossRef]

17. Batina, L.; Hogenboom, J.; van Woudenberg, J.G.J. Getting More from PCA: First Results of Using Principal Component Analysis for Extensive Power Analysis. In Proceedings of the Cryptographers? Track at the RSA Conference, San Francisco, CA, USA, 27 February–2 March 2012; Springer: Berlin/Heidelberg, Germany, 2012; pp. 383–397.

18. Souissi, Y.; Nassar, M.; Guilley, S.; Danger, J.L.; Flament, F. First Principal Components Analysis: A New Side Channel Distinguisher. In Proceedings of the International Conference on Information Security and Cryptology, Seoul, Korea, 1–3 December 2010; pp. 407–419. [CrossRef]

19. Lerman, L.; Bontempi, G.; Markowitch, O. A machine learning approach against a masked AES. *J. Cryptogr. Eng.* **2013**, *5*, 123–139. [CrossRef]

20. Lerman, L.; Bontempi, G.; Markowitch, O. Power analysis attack: An approach based on machine learning. *Int. J. Appl. Cryptogr. (IJACT)* **2014**, *3*, 97–115. [CrossRef]

21. Maghrebi, H.; Portigliatti, T.; Prouff, E. Breaking Cryptographic Implementations Using Deep Learning Techniques. In Proceedings of the SPACE, Hyderabad, India, 14–18 December 2016; pp. 3–26.

22. Gilmore, R.; Hanley, N.; O'Neill, M. Neural network based attack on a masked implementation of AES. In Proceedings of the Hardware Oriented Security and Trust (HOST), Washington, DC, USA, 5–7 May 2015; pp. 106–111. [CrossRef]

23. Levina, A.; Sleptsova, D.; Zaitsev, O. Side-channel attacks and machine learning approach. In Proceedings of the FRUCT, Saint-Petersburg, Russia, 18–22 April 2016; pp. 181–186.

24. Kira, K.; Rendell, L.A. A Practical Approach to Feature Selection. In Proceedings of the Ninth International Workshop on Machine Learning, Aberdeen, Scotland, UK, 1–3 July 1992; Morgan Kaufmann Publishers Inc.: San Francisco, CA, USA; pp. 249–256.

25. Yun, C.; Shin, D.; Jo, H.; Yang, J.; Kim, S. An Experimental Study on Feature Subset Selection Methods. In Proceedings of the Seventh International Conference on Computer and Information Technology, Fukushima, Japan, 16–19 October 2007.

26. Mukhtar, N.; Kong, Y. On features suitable for power analysis?Filtering the contributing features for symmetric key recovery. In Proceedings of the 2018 6th International Symposium on Digital Forensic and Security (ISDFS), Antalya, Turkey, 22–25 March 2018; pp. 1–6.

27. Messerges, T.S.; Dabbish, E.A.; Sloan, R.H. Investigations of Power Analysis Attacks on Smartcards. In Proceedings of the USENIX Workshop on Smartc ard Technology, Chicago, IL, USA, 11–14 May 1999; pp. 151–162.

28. Breiman, L. Random Forests. *Mach. Learn.* **2001**, *45*, 5–32. [CrossRef]

29. Mukhtar, N.; Kong, Y. Secret key classification based on electromagnetic analysis and feature extraction using machine-learning approach. In Proceedings of the Future Network Systems and Security: 4th International Conference, FNSS 2018, Paris, France, 9–11 July 2018; Doss, R., Piramuthu, S., Zhou, W., Eds.; Communications in Computer and Information Science; Springer: Cham, Switzerland, 2018; Volume 878, pp. 80–92. [CrossRef]

30. Blake, I.; Seroussi, G.; Seroussi, G.; Smart, N. *Elliptic Curves in Cryptography*; Cambridge University Press: Cambridge, UK, 1999; Volume 265.

31. Standards for Efficient Cryptography (SEC): Recommended Elliptic Curve Domain Parameters. 2000. Available online: http://www.secg.org/ (accessed on 25 December 2018).

32. Hankerson, D.; Menezes A.J.; Vanstone, S. *Guide to Elliptic Curve Cryptography*, 1st ed.; Springer: Berlin/Heidelberg, Germany, 2004.

33. Amanor, D.N.; Paar, C.; Pelzl, J.; Bunimov, V.; Schimmler, M. Efficient Hardware Architectures for Modular multiplication on FPGA. In Proceedings of the International Conference on Field Programmable Logic and Applications, Tampere, Finland, 24–26 August 2005.

34. AbdelFattah, A.M.; El-Din, A.M.B.; Fahmy, H.M. An Efficient Architecture for Interleaved Modular Multiplication. In Proceedings of the WCSET 2009: World Congress on Science, Engineering and Technology, Singapore, 26–28 August 2009.

35. Available online: http://satoh.cs.uec.ac.jp/SAKURA/index.html (accessed on 6 December 2018).

36. Available online: http://www.cs.waikato.ac.nz/ml/weka/ (accessed on 6 December 2018).

37. Saeedi, E.; Hossain, M.S.; Kong, Y. Multi-class SVMs analysis of side-channel information of elliptic curve cryptosystem. In Proceedings of the 2015 International Symposium on Performance Evaluation of Computer and Telecommunication Systems (SPECTS), Chicago, IL, USA, 26–29 July 2015; pp. 1–6. [CrossRef]

applied
sciences

MDPI

Article

Improving Security and Reliability in Merkle Tree-Based Online Data Authentication with Leakage Resilience

Dongyoung Koo [1,†], Youngjoo Shin [2], Joobeom Yun [3,*] and Junbeom Hur [4,*]

[1] Department of Electronics and Information Engineering, Hansung University, 116 Samseongyo-ro 16-gil, Seongbuk-gu, Seoul 02876, Korea; dykoo@hansung.ac.kr
[2] Department of Computer and Information Engineering, Kwangwoon University, 20 Kwangwoon-ro, Nowon-gu, Seoul 01897, Korea; yjshin@kw.ac.kr
[3] Department of Computer and Information Security, Sejong University, 209 Neungdong-ro, Gwangjin-gu, Seoul 05006, Korea
[4] Department of Computer Science and Engineering, Korea University, 145 Anam-ro, Seongbuk-gu, Seoul 02841, Korea
* Correspondence: jbyun@sejong.ac.kr (J.Y.); jbhur@korea.ac.kr (J.H.);
  Tel.: +82-2-6935-2425 (J.Y.); +82-2-3290-4603 (J.H.)
† Current address: Rm. #508, Research Bldg., 116 Samseongyo-ro 16-gil, Seongbuk-gu, Seoul 02876, Korea.

Received: 30 September 2018; Accepted: 3 December 2018; Published: 7 December 2018

Abstract: With the successful proliferation of data outsourcing services, security and privacy issues have drawn significant attention. Data authentication in particular plays an essential role in the storage of outsourced digital content and keeping it safe from modifications by inside or outside adversaries. In this paper, we focus on online data authentication using a Merkle (hash) tree to guarantee data integrity. By conducting in-depth diagnostics of the side channels of the Merkle tree-based approach, we explore novel solutions to improve the security and reliability of the maintenance of outsourced data. Based on a thorough review of previous solutions, we present a new method of inserting auxiliary random sources into the integrity verification proof on the prover side. This prevents the exposure of partial information within the tree structure and consequently releases restrictions on the number of verification execution, while maintaining desirable security and reliability of authentication for the long run. Based on a rigorous proof, we show that the proposed scheme maintains consistent reliability without being affected by continuous information leakage caused by repetitions of the authentication process. In addition, experimental results comparing with the proposed scheme with other state-of-the-art studies demonstrate its efficiency and practicality.

Keywords: data outsourcing; integrity; online authentication; Merkle (hash) tree; data loss; information leakage; reliability

1. Introduction

In accordance with the dramatic increase in data volume, advances in information and communication technology (ICT) have facilitated the move from local data management to remote data outsourcing services. Although data outsourcing has several benefits in terms of its low cost, agility, scalability, and ease of maintenance, it also has potential problems that users may overlook. Outsourcing data to third-party storage means that control of the data is delegated to the authority managing the remote repository. Unintended data breaches or losses are possible because third-party storage service may be less vigilant than the data owner. Data breaches and losses may lead to serious financial damage as well as wasteful efforts, and can happen for various reasons such as negligent management [1], improper operations [2], and poor resource utilization [3], among other reasons [4,5].

Nonetheless, some remote storage service providers may even attempt to hide losses to protect their reputation [6]. Given the issues surrounding data outsourcing, there is an increasing need for a method of effectively and efficiently verifying the integrity of the data stored in a remote repository [7–11].

Of the various methods available for data integrity verification, Merkle tree [12] is a particularly well-known authenticated data structure. The verification of data integrity based on Merkle tree is implemented by a challenge-response protocol, where a prover provides a series of node values on the path from a leaf node (randomly chosen by a verifier) to the root node of the tree. The prover needs to generate the entire Merkle tree for the attested data, while the verifier can store the single value for the root node in the tree. In other words, this approach only requires the verifier to store and operate on a few number of hash values, and the prover to generate the proof by using the entire data blocks. Thanks to its lightweight computation and low memory requirements, Merkle tree-based authentication has been widely adopted to various systems including blockchain technologies [13,14] such as Bitcoin [15]. Of particular note is that its efficiency increases significantly as the amount of data to be verified increases.

When it comes to online authentication, however, adversarial entities can gather meaningful information from transcripts by repeatedly conducting integrity verification for the same data. In an extreme case, an eavesdropper who collects authentication information can illicitly obtain ownership of the data stored in a remote repository if this method is directly used for authenticating ownership [16–18]. To minimize this risk, we examine possible information leakage from Merkle tree-based online authetication. Based on an in-depth analysis, we present a new Merkle tree-based protocol which inserts random sources every time the proof generation is executed. This eliminates information leakage we have found, thus providing consistently reliable online authentication.

The main contributions of this paper can be summarized as follows: (This work is an extended version of a conference paper published in ICWS 2017 [19] . In the earlier version, the proof generated by the prover was allowed to be extended but not shortened while this version provides both options for better flexibility. In addition, rigorous security analysis and a complete comparison with related work are provided in this version.)

- We analyze potential information leakage during the online verification process. It includes partial information of the Merkle tree and size information, which weaken the security and reliability of authentication (Section 3).
- We propose a leakage-resilient integrity verification protocol (Section 4). Through a rigorous security proof, we illustrate its effectiveness regardless of the number of executions without requiring additional trusted third-party (Section 5).
- We evaluate efficiency of the proposed scheme by implementing it in a real-world application. It shows that our approach can flexibly be adjusted to required system resources with minimal overhead. Nonetheless, it still supports leakage resilience that was not guaranteed in previous research (Section 6).

This paper is organized as follows. Merkle tree-based authentication is described in Section 2. Possible information leakage of Merkle tree-based authentication is analyzed and then vulnerabilities of the previous schemes are analyzed in Section 3. In Section 4, a leakage-resilient online data integrity verification protocol is proposed. The security and efficiency of the propssed scheme are then analyzed in Sections 5 and 6, respectively. Finally, the paper concludes in Section 7.

2. Merkle Tree-Based Authentication

A Merkle tree [20] is constructed from a series of data blocks, where the value of an internal node is assigned based on the hash value of its children, while the value of a leaf node is assigned the direct hash value of the corresponding data block (Figure 1). In the tree construction procedure, the hash function satisfies the preimage-resistance property, which implies it is computationally infeasible to find the preimage of the given hash value. Also, since this forms a binary tree, the maximum depth from leaf to root is at most $\lceil \log_2 n \rceil$ for n data blocks. Thus, the Merkle tree acts as an authenticated data structure for efficient verification of the online content.

In Merkle tree-based online authentication, there are two entities, prover $\mathcal{P}$ and verifier $\mathcal{V}$:

- **Prover** $\mathcal{P}$ is an entity who attempts to convince the other party (i.e., the verifier $\mathcal{V}$) that it owns all of the data. To converve network bandwidth, the prover sends a small piece of verifiable information instead of all of the content.
- **Verifier** $\mathcal{V}$ is another entity who tries to determine whether prover $\mathcal{P}$'s claim is correct or not. To reduce storage requirements, the verifier usually stores only the value of the root node of the Merkle tree instead of all nodes of the tree.

It is notable that the Merkle tree-based online authentication is a protocol that verifies that the prover and verifier own the same data. Unlike public verification, therefore, it assumes that the verifier has some secret (i.e., not publicly available) information about the data to be validated. This issue is dealt with in detail in Section 5.

Based on the hardness assumption that it is infeasible to find a preimage of a given hash value within a computationally reasonable time [21], it can be guaranteed that only entities possessing the same data can obtain the same Merkle tree. In brief, the security of authentication based on Merkle tree is based on the security of hash function in use. Therefore, the verifier $\mathcal{V}$ only stores the value of the root node of the tree and removes the rest of the metadata once the tree is constructed. On the other hand, the prover $\mathcal{P}$ is required to generate a series of (different) hash values leading to a value of the root node that is identical to the one held by the verifier with each authentication cycle.

In the example shown in Figure 1, the verifier $\mathcal{V}$ chooses a random block index (e.g., 1) as a challenge. The prover $\mathcal{P}$ then constructs a Merkle tree from its local data, followed by sending the corresponding unique sibling paths from the leaves to the root node (i.e., (H_1, H_2, H_{3-4})) to the verifier. Upon receiving the proof response, the verifier $\mathcal{V}$ derives the root value of the Merkle tree (i.e., $H(H(H_1, H_2), H_{3-4})$) and determines whether the result is identical to the value of the root node held in local storage.

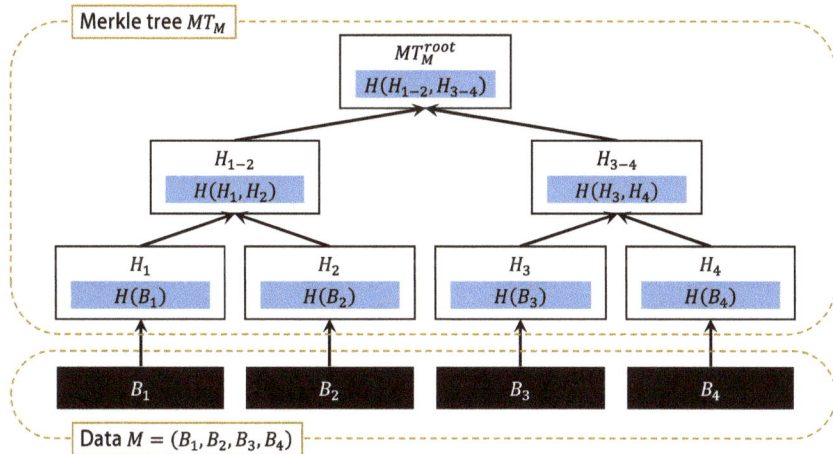

Figure 1. Merkle Tree of data M composed of four blocks.

In the above protocol, adversaries may not be able to uncover the underlying plain data from the communication as long as a secure (preimage resistant) hash function is used. However, we observed that it is vulnerable to a side-channel attack, which allows deducing meaningful information from communications during the authentication process and narrowing down the scope of the attack vector, thereby weakening the reliability of the authentication and possibly nullyfying its effectiveness of authentication completely. Thus, we first analyze the weakness of Merkle tree-based authentication method to side-channel attacks on the same data in Section 3. After this, we present a simple method for improving security and reliability, of Merkle tree-based authentication with minimal overhead in Section 4.

3. Information Leakage Analysis of Merkle Tree-Based Authentication Schemes

In this section, we investigate the vulnerabilities against side-channel attacks of Merkle tree-based authentication method (Section 3.1). Then, we demonstrate the previous authentication schemes are not secure against the side-channel attacks we found (Section 3.2). For the rest of this paper, we assume that there is an adversary eavesdropping communication between the prover $\mathcal{P}$ and verifier $\mathcal{V}$.

Exposure of structural information in Merkle tree-based authentication and its potential risks were previously analyzed by Kundu et al. [22] and Buldas and Laur [23]. Kundu et al. [22], especially, developed a notion called secure name to prevent information leakage about the correlation between nodes in the tree and the graph. However, to the best of our knowledge, detailed diagnosis of information leakage has not yet been conducted in the research.

3.1. Analysis of Merkle Tree-Based Authentication

Prior to authentication, the prover and verifier need to agree on the hash function to be used in Merkle tree construction, the size of the data blocks, and the rules for identifying specific data. In the authentication process, the prover $\mathcal{P}$ first sends an identifier of the data to the verifier $\mathcal{V}$ and proves complete possession of the data in question.

3.1.1. Leakage of Data Size Information

Looking at the communication between $\mathcal{P}$ and $\mathcal{V}$, an eavesdropper can figure out the approximate size of the underlying data from a single authentication proof. Specifically, the adversary can determine the length of sibling path(s) from the knowledge of the hash function in use, data block size, and the size of the proof transmitted by $\mathcal{P}$. The minimum and maximum number of leaf nodes can be easily determined from the height of the tree (i.e., the length of the sibling path -1) when the Merkle tree is constructed in a left-to-right and bottom-up manner.

Let us assume that the size of a single data block is $|B|$, the size of hash value is $|\mathcal{H}|$, and the size of the proof is $|P|$. The length of the sibling path L can then be derived from $L = |P|/|\mathcal{H}|$. When the length of the sibling path is acquired, $(L-1)$ becomes the height of the constructed Merkle tree, and the total size of target data S can be approximated as

$$(2^{(L-2)} + 1) \cdot |B| \leq |S| \leq 2^{(L-1)} \cdot |B|, \tag{1}$$

where the right-hand-side of the inequality is the full and complete binary tree [24].

This information about size obtained by eavesdropping can be used to narrow the attack space for the range of target sizes and filter out unnecessary data. It gives the attacker the powerful option to select target data of an appropriate size. Therefore, it is more desirable for an authentication method to hide size information of data.

3.1.2. Leakage of Merkle Tree Hash Values

Typically, data authentication is expected to operate reliably, regardless of the number of times it occurs. Contrary to this expectation, however, the maximum number of effective authentication is bounded by the size of the Merkle tree due to the leakage of hash values of it. Specifically, authentication can be conducted only as many times as the logarithmic number of leaf nodes in the tree, because responded hash values of the tree are leaked to the adversaries. After the limited number of authentications, the attacker can construct the entire Merkle tree through eavesdropping even though it has no information about the underlying content.

Observing Figure 1, the two sibling paths which are proof for the challenged leaf nodes (1 or 2) and (3 or 4) provide all of the information required to construct the entire Merkle tree (i.e., $\{H_1, H_2, H_{3-4}, (H_{1-2}, MT_M^{root})\} \cup \{H_3, H_4, H_{1-2}, (H_{3-4}, MT_M^{root})\} = MT_M = \{H_1, H_2, H_3, H_4, H_{1-2}, H_{3-4}, MT_M^{root}\}$, where the values inside the parentheses can be derived from the other values given as part of the proof).

Therefore, it can be seen that the authentication range of the data is reduced at an exponential rate in the presence of an adversary exploiting information accumulated about the tree gained during repeated authentication attempts. Once the prover $\mathcal{P}$ passes the authentication proof for a challenged block (e.g., B_1), an eavesdropper can obtain infomation about the subtree rooted at the child of the root node (e.g., (H_1, H_2, H_{1-2}) as a subtree rooted at H_{1-2} in Figure 1). The subsequent authentication attempt guarantees the integrity of at most half of the entire data (e.g., $\{B_3, B_4\}$ in Figure 1). Otherwise, the attacker can reuse the other half of the tree already known from the previous authentication attempt. Therefore, the authentication coverage reduces further or the adversary becomes able to bypass the verification process with overwhelming probability even when it does not know the corresponding data by exploiting the obtained hash values.

The typical coverage pattern is illustrated in Figure 2. When data is composed of 2^i blocks, the maximum authentication coverage $C(\cdot)$ of the data (M) at the j-th execution attempt can be defined as

$$C(M)^j = \begin{cases} 1, & \text{if } j = 0 \\ \frac{1}{2}^{\lfloor \log_2 j \rfloor + 1}, & \text{if } 0 < j < 2^i \\ 0, & \text{otherwise.} \end{cases} \tag{2}$$

As the number of demanding sibling paths for challenged leaf nodes in a single verification increases, the number of allowable verification attempts decreases sharply. (According to Ateniese et al. [25], data composed of 10,000 blocks requires 460 samples of leaf nodes to be verified in order to achieve 99% confidence. When it comes to Merkle tree-based authentication [16], effectiveness is only guaranteed for 21 times. After this, the entire Merkle tree can be reconstructed via eavesdropping so that the attacker can successfully pass authentication.)

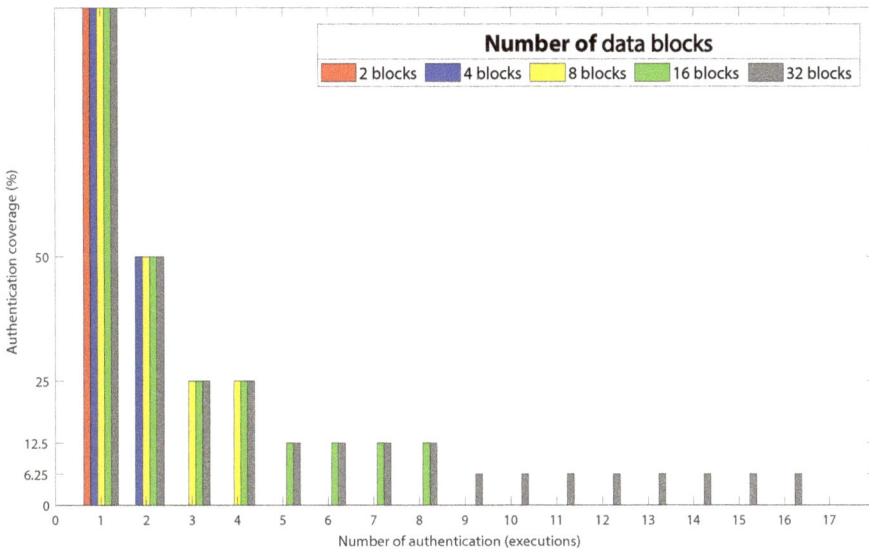

Figure 2. Authentication coverage for data by the number of blocks.

3.2. Previous Schemes and Their Vulnerabilities

Taking aforementioned side channels into account, we analyze weakness of the previous authentication schemes.

3.2.1. Generic Merkle Tree-Based Authentication

In generic Merkle tree-based authentication [12], the original data block is used together with the tree, rather than using the tree only, which was further adopted to proof-of-ownership process in the cloud data deduplication literature [16].

As shown in Figure 3, the proof in authentication includes the content of the challenged block along with its sibling path. In this example, the challenged data block B_1 and partial information about the Merkle tree rooted at H_{1-4} (i.e., $\{H_1, H_2, H_{1-2}, H_{3-4}, H_{1-4}\} = MT_{(B_1, B_2, B_3, B_4)} \setminus \{H_3, H_4\}$) can be exposed to the public after the first authentication request. Therefore, in the second authentication attempt, the challenging block might be randomly chosen in $\{B_5, B_6, B_7, B_8\}$ for maximal authentication coverage, and this covers only half of the entire data set (because it excludes the blocks $\{B_1, B_2, B_3, B_4\}$). The next challenging block can be chosen from $\{B_3, B_4\}$ or $\{B_5, B_6\}$, covering a quarter of the data in a similar way.

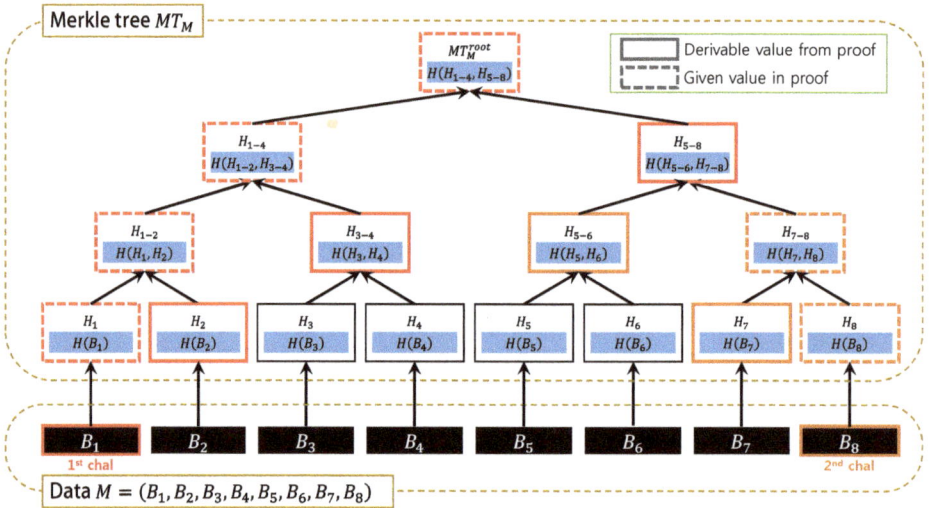

Figure 3. Merkle tree-based authentication for data M with eight blocks.

Thus, generic Merkle tree-based authentication method does not guarantee consistent authentication coverage as authentication is done repeatedly, and the maximum number of challenge-responses is limited to the number of data blocks.

However, the adversary is still able to guess the size of the authenticated data by using Equation (1), given publicly accessible data block size $|B|$, proof size $|P|$, hash value size $|\mathcal{H}|$, and the size of sibling path $L = (|P| - |B|)/|\mathcal{H}| + 1$.

3.2.2. Authentication without a Merkle Tree

Zhao and Chow [26] pointed out the possibility of a replay attack on Merkle tree-based data authentication and proposed a probabilistic protocol inserting randomness based on hardcore function to achieve resilience against the replay attack (the protocol is summarized in Algorithm 1).

Algorithm 1 Randomized online authentication exploiting a hardcore function

Public parameters: Hardcore function $G : \{0,1\}^l \to \{0,1\}^{|M|+\log|M|-1}$ where $l \ll |M|$
i-th bit in bitstring S, S_i

<table>
<tr><td align="center">Verifier $\mathcal{V}$</td><td align="center">Prover $\mathcal{P}$</td></tr>
</table>

1. Choose uniform random bitstring $s \in \{0,1\}^l$

2-1. Compute $r \leftarrow G(s) \in \{0,1\}^{|M|+\log|M|-1}$ $\xrightarrow{\;s\;}$ 2-2. Compute $r \leftarrow G(s) \in \{0,1\}^{|M|+\log|M|-1}$

3-1. With possessing data M, generate proof $h(M,r)$ such that   3-2. Possessing data M', generate proof prf such that

$$h(M,r) = (b_1(M,r), \ldots, b_{\log|M|}(M,r)),$$

$$prf = (b_1(M',r), \ldots, b_{\log|M|}(M',r)),$$

where $b_j(M,r) = (\sum_{i=1}^{|M|} M_i \cdot r_{i+j-1}) \mod 2$ for $1 \leq j \leq$ $\xleftarrow{\;prf\;}$ where $b_j(M',r) = (\sum_{i=1}^{|M'|} M'_i \cdot r_{i+j-1}) \mod 2$ $(1 \leq j \leq \log|M|)$
$\log|M|$

4. Check the integrity of M by comparing $h(M,r)$ and prf

Unfortunately, in their scheme, the size of data to be verified is known to the public (because the hash function G in Algorithm 1 specifies the size of the output dependent on the data). Due to this assumption, this approach does not prevent the leakage of size information.

The protocol requires the verifier to perform the same computation as the prover unless the same random seed s is used repeatedly. If the same seed value is used repeatedly, the proof becomes always the same, thus an adversary can bypass authentication for the entire data with a single eavesdropping on the proof. Consequently, as long as a newly chosen seed is used for every authentication attempt, the efficiency on the verifier side would be reduced because it uses all of the data in the verification process and pre-computation cannot occur.

In addition, according to their analysis based on the Goldreich-Levin theorem, the probability that an adversary deceives the verifier is at most $1/|M|$, which is not negligible on the data size. Because the bitstring r (**Step** 2 in Algorithm 1) can become known to the adversary, the adversary can extract at most $\log(|M|)$ bits from the transferred proof. To overcome this problem and prevent further information leakage, they recommended encrypting all traffic using a session key generated by additional protocols for secure connection establishment (e.g., SSL/TLS, IPSed), which requires non-negligible overhead in practice. This is dealt with in Sections 3.2.3 and 3.2.4.

There are also other approaches that do not rely on Merkle tree structure in data authentication. For example, Atallah et al. [27] proposed a technique for efficient integrity verification of 2-dimensional range data such as image and GIS data and suggested a method to maintain the communication overhead constant. Atallah et al. [28] and Benjamin and Atallah [29] also proposed several novel integrity verification techniques without Merkle tree.

3.2.3. Merkle Tree-Based Authentication of Encrypted Data

Bellare et al. [30] and Xu et al. [31] independently investigated data authentication of encrypted data in the context of proof-of-ownership (PoW). Their strategy is to encrypt the data first, then to perform authentication over the encrypted data based on a Merkle tree to guarantee data confidentiality and verify the complete possesion of the underlying plain data. In combination with a secure encryption algorithm, the underlying plain content can be hidden from unauthorized users, including malicious service providers.

Even if their schemes preserve data confidentiality by encrypting data itself, however, they are vulnerable to the side-channel attacks we found due to the inherent property of Merkle tree. In other words, even if the plaintext corresponding to the encrypted data is not known, the information about the Merkle tree generated from the ciphertext can be obtained by eavesdropping, so the authentication coverage falls with repeated authentication attempts. As with the Merkle tree-based authentication of unencrypted data, the size of the underlying (encrypted) data can still be inferred.

3.2.4. Merkle Tree-Based Authentication with Transmission in Encrypted Form

Li et al. [32] employed a Merkle tree-based approach for online authentication in a smart grid system. To avoid information leakage, they combined it with a secure encryption algorithm. Specifically, the prover (i.e., a smart meter) and verifier (i.e., a neighborhood gateway) engage in a Diffie-Hellman key agreement protocol, followed by AES encryption to preserve the privacy of the authentication proof generated by the prover. (Inspired by Li et al.'s work [32], we can employ secure encryption algorithms to obfuscate communication between the prover and verifier. In [32], a Merkle tree is used for sender identification instead of data authentication, in which the Merkle tree is constructed from random elements chosen by $\mathcal{P}$ and then $\mathcal{V}$ validates the origin of the received power measurement (in a way that guarantees only one who can construct the valid Merkle tree is $\mathcal{P}$). However, in their research, the main purpose of exploiting secure key agreement and encryption algorithms was to minimize side effects caused by side channels during Merkle tree-based online authentication.) The overall data authentication process for Merkle tree-based data authentication using encrypted channels is described in Algorithm 2. (This algorithm requires key agreement and encryption of communications. Compared to adopting full SSL/TLS, it is more efficient since it requires Diffie-Hellman key agreement and efficient encryption algorithms such as AES, which will be analyzed in Section 6.)

Algorithm 2 Merkle tree-based online authentication with encrypted communication

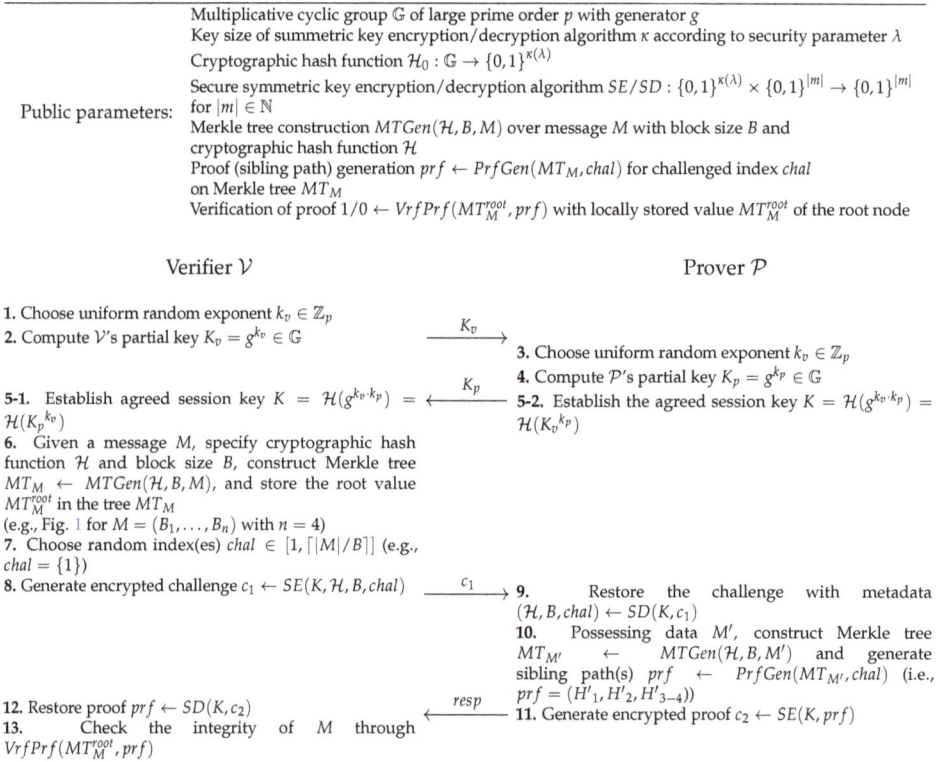

Public parameters:
Multiplicative cyclic group $\mathbb{G}$ of large prime order p with generator g
Key size of symmetric key encryption/decryption algorithm κ according to security parameter λ
Cryptographic hash function $\mathcal{H}_0 : \mathbb{G} \rightarrow \{0,1\}^{\kappa(\lambda)}$
Secure symmetric key encryption/decryption algorithm $SE/SD : \{0,1\}^{\kappa(\lambda)} \times \{0,1\}^{|m|} \rightarrow \{0,1\}^{|m|}$ for $|m| \in \mathbb{N}$
Merkle tree construction $MTGen(\mathcal{H}, B, M)$ over message M with block size B and cryptographic hash function $\mathcal{H}$
Proof (sibling path) generation $prf \leftarrow PrfGen(MT_M, chal)$ for challenged index $chal$ on Merkle tree MT_M
Verification of proof $1/0 \leftarrow VrfPrf(MT_M^{root}, prf)$ with locally stored value MT_M^{root} of the root node

Verifier $\mathcal{V}$		Prover $\mathcal{P}$
1. Choose uniform random exponent $k_v \in \mathbb{Z}_p$		
2. Compute $\mathcal{V}$'s partial key $K_v = g^{k_v} \in \mathbb{G}$	$\xrightarrow{\quad K_v \quad}$	
		3. Choose uniform random exponent $k_p \in \mathbb{Z}_p$
		4. Compute $\mathcal{P}$'s partial key $K_p = g^{k_p} \in \mathbb{G}$
5-1. Establish agreed session key $K = \mathcal{H}(g^{k_v \cdot k_p}) = \mathcal{H}(K_p^{k_v})$	$\xleftarrow{\quad K_p \quad}$	5-2. Establish the agreed session key $K = \mathcal{H}(g^{k_v \cdot k_p}) = \mathcal{H}(K_v^{k_p})$
6. Given a message M, specify cryptographic hash function $\mathcal{H}$ and block size B, construct Merkle tree $MT_M \leftarrow MTGen(\mathcal{H}, B, M)$, and store the root value MT_M^{root} in the tree MT_M (e.g., Fig. 1 for $M = (B_1, \ldots, B_n)$ with $n = 4$)		
7. Choose random index(es) $chal \in [1, \lceil |M|/B \rceil]$ (e.g., $chal = \{1\}$)		
8. Generate encrypted challenge $c_1 \leftarrow SE(K, \mathcal{H}, B, chal)$	$\xrightarrow{\quad c_1 \quad}$	9. Restore the challenge with metadata $(\mathcal{H}, B, chal) \leftarrow SD(K, c_1)$
		10. Possessing data M', construct Merkle tree $MT_{M'} \leftarrow MTGen(\mathcal{H}, B, M')$ and generate sibling path(s) $prf \leftarrow PrfGen(MT_{M'}, chal)$ (i.e., $prf = (H'_1, H'_2, H'_{3-4})$)
12. Restore proof $prf \leftarrow SD(K, c_2)$	$\xleftarrow{\quad resp \quad}$	11. Generate encrypted proof $c_2 \leftarrow SE(K, prf)$
13. Check the integrity of M through $VrfPrf(MT_M^{root}, prf)$		

Although the exact proof becomes indistinguishable when the transmitted data is encrypted, size information for the underlying data can be deduced from the size of the transferred ciphertext. One approach to prevent size information leakage is to dynamically change the size of the data blocks and the hash function used for each authentication attempt (**Step** 6 in Algorithm 2). However, this approach requires the verifier to construct a new Merkle tree for every authentication request, rendering pre-computation of the Merkle tree and its reuse on the verifier side impossible. Therefore, dynamically changing the size of the data blocks and hash functions for each authentication attempt significantly reduces efficiency from the verifier's perspective. Another approach is to insert dummy data into the ciphertext. While this can obfuscate information by increasing the size of data, it also increases the computational and communication overhead for both sides.

With regard to efficiency, it requires higher computation cost for data encryption and decryption during data verification than in Merkle tree-based authentication, which is another practical drawback of this approach. (Detailed analysis can be found in Section 6.)

4. Randomized Online Authentication

In this section, we present a probabilistic authentication protocol by exploiting Merkle tree without a reduction in verification coverage. Before describing the proposed protocol, the adversarial model and its goals are summarized in the following subsections.

4.1. Adversarial Model

We consider adversaries who are able to collect valid proofs of data authentication from public channels. This adversary can be either (1) a passive attacker eavesdropping on communications between valid prover $\mathcal{P}$ and verifier $\mathcal{V}$ without intervention, or (2) an adaptive online adversary. In the latter case, the adversary acts as a more active attacker by passing a set of random challenges of its choosing to an oracle and collecting a valid proof set for the upload-requested data. In other words, valid prover $\mathcal{P}$ can be an oracle for the target data and the adversary attempts to circumvent the authentication process by manipulating the obtained proofs.

Without loss of generality, we assume that the adversary has no prior knowledge about the data to be challenged (proved). Specifically, we assume that the adversary is unable to extract size information from eavesdropping on interactions during the authentication process, except for initial upload.

The goal of these adversaries is to weaken the reliability of the authentication process by exploiting information gathered through wiretapping.

4.2. Goal

In order to minimize information leakage when a Merkle tree is used for online data authentication, the proposed scheme needs to satisfy the following requirements:

- Prevention of size information leakage: The authentication mechanism should block the outflow of information about the size of the target data, which can be used by adversaries to select and predict the required number of authentication proofs.
- Prevention of replay attacks: The protocol should not allow adversaries to launch replay attacks, in which a collected valid set of authentication proofs are used in subsequent authentication requests. In other words, the adversary cannot learn any information from the disclosed information via public channels during the authentication process.
- Minimal requctions in efficiency: The effective handling of side channels should be achieved with acceptable computation and communication overhead, maintaining the advantages of the Merkle tree-based approach.
- Compatability: Given that the Merkle tree-based approach is widely deployed in industry and academia due to its intuitive nature and ease of utilization, the proposed approach should be applicable to existing uses. This includes adaptability to lightweight devices with limited resources and restrictions on the installation of additional libraries depending on the system architecture, such as IoT terminal devices and sensors.

4.3. Construction

To be resistant against information leakage regardless of the number of authentication attempts, the transmitted authentication proof (i.e., sibling paths) needs to be randomized so that an eavesdropping adversary cannot gather any valuable information from the transcript. In this section, we present a simple amendment that inserts random inputs and significantly increases the reliability of online authentication. The overall process is illustrated in Algorithm 3, with example data composed of four blocks. Associated notations are summarized in Table 1.

Algorithm 3 Merkle tree-based online authentication with randomized input

Verifier $\mathcal{V}$	Prover $\mathcal{P}$

1. Given data D, invoke **ConstructMerkleTree**(D) to obtain number of blocks n, Merkle tree MT_D for D, and its height h
   Keep n, h, MT_D^{root} in local storage privately, where MT_D^{root} is the value of the root node extracted from the Merkle tree MT_D
2. Choose random index(es) $chal \in \mathbb{N}$ and the length of sibling path(s) $L \in \mathbb{N}$ (e.g., $chal = 100, L = 7$ for $D = (B_1, \ldots, B_4)$)

$$\xrightarrow{\ \substack{chal(=100),\\ L(=5)}\ }$$

3. Calculate challenged index $chal \leftarrow (chal \bmod n) + 1$ (i.e., $(100 \bmod 4) + 1 = 1$)
4. Possessing data D', invoke **ConstructMerkleTree** to obtain n', h', and $MT_{D'}$ followed by **GenerateProof**($MT_{D'}, chal, L$) to obtain authentication proof prf of $chal$, in sequence (e.g., $prf = (H'_1, H'_2, H'_{3-4}, MT_{D'}^{root})$)
5. Make prf look random by invoking **ObfuscateProof**(prf, L)
   (i.e., $prf = (R, H'_1 \oplus mask, H'_2 \oplus mask, H'_{3-4} \oplus mask, MT_{D'}^{root} \oplus mask)$)

$$\xleftarrow{\ prf\ }$$

6. Invoke **RestoreProof**(prf, h, MT_D^{root}) to obtain the restored original sibling path of $chal$ as proof prf (e.g., $prf = (H'_1, H'_2, H'_{3-4})$)
7. Invoke **VerifyProof**(prf, MT_F^{root}) to validate the proof.
   The return value *True* indicates successful authentication. Otherwise, verification has failed

---

$(n, h, MT_D) \leftarrow$ **ConstructMerkleTree**(D):
  Split D into $|B|$-bit blocks as $\tilde{D} = (B_1, B_2, \ldots, B_n)$, where n is the number of blocks that make up D
  Construct Merkle tree MT_D with $\tilde{D}$ as leaf nodes

$prf \leftarrow$ **GenerateProof**($MT_D, chal, L$):
  Put the values for the siblings of the nodes that lie on the path from the $chal$-th leaf node to the root node in MT_D (including the root) into prf
  while $prf > L$:
    $prf \leftarrow (\mathcal{H}(prf_1, prf_2), prf_3, \ldots, prf_{|prf|})$
  endif

$res \leftarrow$ **VerifyProof**(prf, MT_D^{root}):
  Set vt to be the first element in prf
  Compute verification term vt by evaluating hash function $\mathcal{H}$ of vt and an element in prf from second element to the last one .5emin a recursive manner
  if $vt = MT_D^{root}$:
    $res \leftarrow True$
  else:
    $res \leftarrow False$

$prf \xleftarrow{\$}$ **ObfuscateProof**($prf, reqLen$):
  Let $prf = (prf_1, \ldots, prf_{|prf|})$
  $mask \leftarrow s \in_R \{0,1\}^{|\mathcal{H}|}$
  for 1 to $i = |prf|$:
    $prf_i \leftarrow prf_i \oplus mask$
  endfor
  if $|prf| < reqLen$:
    $R \in_R \{0,1\}^{(reqLen - |prf|) \cdot ||\mathcal{H}||)}$
    $prf \leftarrow (R, prf)$
  endif

$prf \leftarrow$ **RestoreProof**(prf, h, MT_D^{root}):
  # confirm that $|prf| = reqLen$
  Let $prf = (prf_1, \ldots, prf_{reqLen})$
  $mask \leftarrow prf_{reqLen} \oplus MT_D^{root}$
  for $i = reqLen - 1$ **down to** $reqLen - h - 1$:
    $prf_i \leftarrow prf_i \oplus mask$
  endfor
  $prf \leftarrow (prf_{reqLen-h-1}, \ldots, prf_{reqLen-1})$

Table 1. Notations.

Notation	Description
$\mathcal{H}(\cdot)$	Cryptographic hash function
$||\mathcal{H}||$	Size of the hash value (in bits)
n	Number of data blocks for the entire data D
MT_D	Merkle tree constructed from data D
MT_D^{root}	Root node in the Merkle tree MT_D
h	Height of the Merkle tree
L	Required length of the sibling path in the authentication proof
$valN$	Value of the given node N
$sib(N)$	Sibling node in the sibling path for the given node N
A_i	i-th element in sequence A (cardinality)
$|A|$	Number of elements in set A
$a \in_R A$	Random selection of element a in set A
$b \leftarrow B$	Assignment of the result of the deterministic algorithm (operation) B to b
$b \xleftarrow{\$} B$	Assignment of the result of the probabilistic algorithm (operation) B to b

4.3.1. Authentication Initiation

First of all, the verifier $\mathcal{V}$ constructs a Merkle tree for data D to be verified following the generic Merkle tree construction process (**Step** 1). Notice that $\mathcal{V}$ needs to construct this tree only once (usually before verification) to store the number of leaf nodes and the root node value in the tree, and then discards all remaining information about the tree. As for specifying the data to be verified, the verifier and the prover can use $\mathcal{H}(\mathcal{H}(D))$ as an identifier. Due to the collision-resistant property of the cryptographic hash function, we assume that the probability that different data files produce the same hash tree is negligible.

4.3.2. Randomized Challenge Generation

In this phase, the verifier $\mathcal{V}$ generates a random challenge for the claim that prover $\mathcal{P}$ manages the data properly as $\mathcal{V}$ desires. Unlike previous Merkle tree-based authentication approaches in which the challenge is selected from within a limited range (i.e., $\{1, 2, \ldots, n\}$), the verifier $\mathcal{V}$ selects a random integer without restriction. $\mathcal{V}$ also specifies the length of the proof to be received and sends this value and the challenge to the prover (**Step** 2). The length of the proof can be an arbitrary number when it is greater than 2. In our approach, the proof length does not depend on the Merkle tree structure, unlike the original Merkle tree-based verification process. Specifically, the value of the sibling nodes in the Merkle tree may not be used in the proposed scheme when the requested proof length is shorter than the length of the sibling path from the leaf (challenged) node to the root. For a detailed description, see Section 4.3.4.

4.3.3. Original Challenge Restoration

Upon receipt of the challenge and proof length specification, the prover $\mathcal{P}$ restores the intended challenge index *chal* (**Step** 3). The prover $\mathcal{P}$ can specify the data block to which the challenge points, on the assumption that the prover $\mathcal{P}$ and the verifier $\mathcal{V}$ have common knowledge about the number of data blocks constituting the data D. The index of the challenged block becomes the remainder after dividing the challenge by the total number of data blocks.

4.3.4. Proof Generation

Using the restored challenge and proof length specification, the prover $\mathcal{P}$ generates the corresponding proof *prf*. Unlike the typical Merkle tree-based approach, the proposed protocol requires the value of root node to be appended to the end of the proof (**Step** 4).

It is worth noting that the proposed protocol does not depend on the Merkle tree structure for concealing size information. In other words, a proof that is shorter than the length of the sibling nodes is possible, thus making the data look smaller than its actual size. (In this case, the values of the nodes closest to the leaf node in the certificate must be removed in order, but the value remaining at the end after this removal (corresponding to the leaf node in the certificate) must be derived from the removed values.) In addition, a proof longer than the length of the sibling nodes is also possible, making the data look larger than its actual size, as described in the following subsection.

4.3.5. Proof Obfuscation

In this phase, the prover $\mathcal{P}$ obfuscates the proof, prepends a random bitstring to the proof if necessary for the purpose of concealing the size information (making the data appear larger than the actual size), and then passes the resulting proof to the verifier $\mathcal{V}$ (**Step** 5).

Based on the algorithm **ObfuscateProof**, the prover $\mathcal{P}$ first selects a random bitstring s with a length equal to the hash value. The bit string s is then masked iteratively by applying a bitwise XOR operation to each element of the sibling path *sib*.

To obtain the bit-length of the resuting proof $L \cdot |\mathcal{H}|$ when the requested proof length is longer than the obfuscated proof, a randomly selected bitstring R of length $(L - h - 2) \cdot |\mathcal{H}|$ is prepended to the proof *prf* (when $L > h + 2$). Note that, before generating the proof, the prover $\mathcal{P}$ can derive the height h of the Merkle tree that is to be constructed because $\mathcal{P}$ knows the total number of leaf nodes (i.e., data blocks). Thus, $\mathcal{P}$ calculates the length of hash values to be appended as $L - h - 2$ (h for the number of siblings on the path from the challenged node to the root, 1 for the challenged node, and another 1 for the root node). From this calculation, prover $\mathcal{P}$ generates an arbitrary bitstring R with a length equal to $(L - h - 2)$ hash values .

The key to this phase is allowing individual provers to insert random sources into the verification proof in a non-deterministic manner.

4.3.6. (Original) Proof Restoration

When the verifier $\mathcal{V}$ receives the masked authentication proof from the prover $\mathcal{P}$ on the challenge *chal* with a bit-length equal to $L \cdot |\mathcal{H}|$, $\mathcal{V}$ restores it to the generic form of a sibling path (**Step** 6).

First, the unnecessary heading $(L - h - 2) \cdot |\mathcal{H}|$-bit bitstring is removed from the obfuscated proof *prf*. The last element in *prf*, corresponding to the masked value of the root node $MT_{D'}^{root}$ is then XORed with $\mathcal{V}$'s value of MT_{D}^{root} to obtain the masking factor *mask*. For the remaining elements, *mask* is recursively XORed for each one in reverse order.

4.3.7. Proof Verification

In this phase, the restored proof *prf* is validated by the verifier $\mathcal{V}$ in the same way as in the typical approach (**Step** 7).

The hash value corresponding to the root node of the tree is obtained by repeating the process of re-hashing two neighboring hash values from the first hash value of the proof. If the calculated hash value is the same as the value stored by the verifier $\mathcal{V}$, the authentication succeeds. Otherwise, validation is considered a failure.

5. Security Analysis

In this section, the security of the proposed scheme is analyzed in detail. First, the security of Merkle tree-based online authentication, which is assumed to be conducted only once, is discussed. Using this as a baseline, the security of the proposed method and its ability to improve reliability are then examined.

5.1. Security of Merkle Tree-based Authentication

The primitive used to construct a Merkle tree is a cryptographic hash function that satisfies preimage resistance, second preimage resistance, and collision resistance properties.

Definition 1. *(Preimage-resistant hash function) Given image y of a hash function h, for all pre-defined outputs, the function is preimage-resistant if it is computationally infeasible to find any preimage x such that* $y = h(x)$ *[33].*

In the verification of data integrity stored in remote storage, the Merkle tree-based approach begins with the assumption that the verifier and the prover share the same information (i.e., the value of the root node and the number of leaf nodes in the tree). Otherwise, the verifier can neither generate a valid challenge nor validate the correctness of the proof. Under this assumption, when online authentication is performed only once, its security can be summarized as Theorem 1.

Theorem 1. *(Security of Merkle tree-based authentication) Given a randomly chosen leaf index, the probability that an adversary without knowledge of the entire tree (data) can forge a valid sibling path is negligible if a cryptographic (specifically, preimage-resistant) hash function is used to construct the tree.*

Proof. Suppose that there is an adversary who knows the number of leaf nodes and the value of the root node in the Merkle tree generated from the target data. For the adversary to pass validation, it has to find a preimage of the root node with a bit-length twice that of the hash value. Regarding each half of the discovered preimage as children of the target node, the adversary has to repeatedly search for preimage of each half until the preimage corresponds to the leaves (i.e., $\lceil \log n \rceil$ times, where n is the number of leaf nodes in the tree). However, this contradicts the assumption that each hash value is an output of the cryptographic hash function. Therefore, the probability of the adversary forging a valid proof is negligible as long as a cryptographic (specifically, preimage-resistant) hash function is used to build the Merkle tree. Formal security model and proof of unforgeability for Merkle tree-based authentication can be found in [20,34–36]. □

5.2. Security of the Proposed Scheme

The data authentication process can be completely bypassed if eavesdropping is performed on the initial transmission of the underlying data. As noted in [35], the reliability of online authentication is weakened through extra information gathered by eavesdropping unless the Merkle tree is combined with private keys. In short, one-time online authentication is reliable only when a Merkle tree is used without modification. However, since this data transmission is performed at most once and subsequent data authentication can be conducted several times, the general assumption that the initial data is transmitted through a secure channel if necessary is reasonable.

5.2.1. Security of One-time Secret Delivery

One simple way to improve security and reliability is to have the prover $\mathcal{P}$ and the verifier $\mathcal{V}$ agree on an extra shared secret additional to the Merkle tree itself. To achieve this, we devise a one-way secret delivery mechanism following Definition 2.

Definition 2. *(One-way secret delivery) Let two parties, say $\mathcal{A}$ and $\mathcal{B}$, share secret information shared of bit-length λ. $\mathcal{A}$ can send another secret value toShare to $\mathcal{B}$ by embedding toShare in shared such that*

$$transmitted = shared \oplus toShare$$

where $\oplus$ represents a bitwise exclusive-or (XOR) operation and transmitted is data transmitted via a public channel. The recipient $\mathcal{B}$ can then recover the secret key such that

$$toShare = transmitted \oplus shared.$$

Specifically, every time the prover $\mathcal{P}$ tries to convince the verifier $\mathcal{V}$, $\mathcal{P}$ can choose a uniform random mask and securely send it to $\mathcal{V}$ by exploiting the one-way secret delivery mechanism in the proposed scheme. This can be achieved by embedding the mask in the shared value, which is the value of the root node in the Merkle tree such that $transmitted = mask \oplus MT_D^{root}$, where $mask$ and MT_D^{root} are a mask randomly chosen by $\mathcal{P}$ and the value of root node shared between $\mathcal{P}$ and $\mathcal{V}$, respectively. The one-way secret delivery mechanism can be thought of as a one directional password-authenticated key exchange (PAKE), in which the previously shared information is considered to be a password [37].

Prior to examining the security of the proposed scheme, notice that the result of the bitwise-exclusive (XOR) operation of a random value is also random regardless of other operands.

Lemma 1. *Let* $X, Y \in \{0, 1\}$ *be random variables, where* $Pr[X = 0] = Pr[X = 1] = 1/2$ *and Y is drawn from any distribution. The distribution for* $X \oplus Y$ *is also random as long as Y is independent of X such that*

$$Pr[Y = y | X = x] = Pr[Y = y]$$

for any fixed bits $x, y \in \{0, 1\}$.

Proof. Let $b \in \{0, 1\}$ be a fixed bit. Then,

$$
\begin{aligned}
Pr[X \oplus Y = b] &= Pr[X \oplus Y = b | X = 0] \cdot Pr[X = 0] + Pr[X \oplus Y = b | X = 1] \cdot Pr[X = 1] \\
&= Pr[0 \oplus Y = b | X = 0] \cdot (1/2) + Pr[1 \oplus Y = b | X = 1] \cdot (1/2) \\
&= Pr[Y = b | X = 0] \cdot (1/2) + Pr[Y = b \oplus 1 | X = 1] \cdot (1/2) \\
&= Pr[Y = b] \cdot (1/2) + Pr[Y = b \oplus 1] \cdot (1/2) \\
&= (Pr[Y = b] + Pr[Y = b \oplus 1]) \cdot (1/2) \\
&= 1/2.
\end{aligned}
$$

Therefore, the result of XORing a certain bit with a random bit is also random. □

Lemma 2. *Let* $X' = (X_1, X_2, \ldots, X_r) \in \{0, 1\}^r$ *be a random variable where* $Pr[X_i = 0] = Pr[X_i = 1] = 1/2$ *and* X_i *and* X_j *are independent of each other for any positive integer r,* $1 \le i, j \le r$*, and* $i \ne j$*. The distribution for* $X' \oplus Y'$ *is also random as long as* Y' *is independent of* X' *regardless of the distribution Y is drawn from.*

Proof. Let each bit of X' and Y' be X_i and Y_i for $1 \le i \le r$, respectively. The probability that the XOR result of X_i and Y_i becomes any of $\{0, 1\}$ is thus $1/2$ according to Lemma 1. Because X_i and X_j for $1 \le i, j \le r$ and $i \ne j$ are independent variables, the probability of $Pr[X' \oplus Y' = bs]$ is $(1/2)^r$ for any fixed bitstring $bs \in \{0, 1\}^r$. Therefore, the result of XORing a certain bitstring with a random bitstring is also random. □

Using Lemma 2, the one-time security of the one-way secret delivery mechanism can be proven.

Theorem 2. *(One-time Security of One-way Secret Delivery) The one-way secret delivery protocol in the proposed scheme is one-time secure against adversaries as long as the mask value is drawn independently and uniformly at random.*

Proof. Following the definition of entropy [38], the random mask has maximum uncertainty because it is chosen independently and uniformly at random from $\{0, 1\}^\lambda$, where λ is the bit-length of the hash value. According to Lemma 2, the XORed value with this random mask is also unpredictable (i.e., indistinguishable from other random bitstrings). □

5.2.2. Security of the Proposed Scheme

In the proposed approach, the size of the proof generated by the prover $\mathcal{P}$ can be either shorter than, exactly equal to, or longer than that of the typical Merkle tree-based approach. Typical types of proof according to proof size are presented in Figure 4.

First, consider a passive adversary who does not affect the designated protocol but collects proof information leaked by eavesdropping on a public channel. (Cases in which the prover and the verifier collude are not considered in this paper because this action invalidates the effectiveness of the authentication process and is beyond the scope of our discussion.) Because this kind of adversary has no knowledge of the underlying data used to construct the Merkle tree, it can be assumed not to have all of the necessary information in advance before the attack.

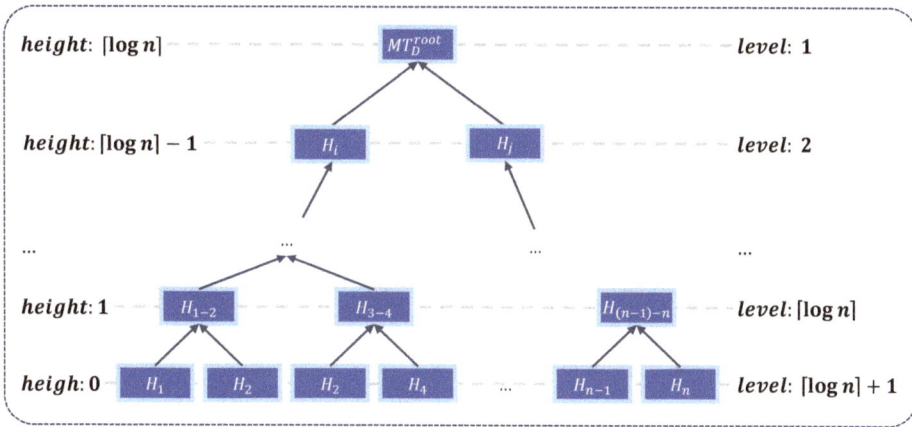

(**a**) Merkle tree MT with n leaf nodes

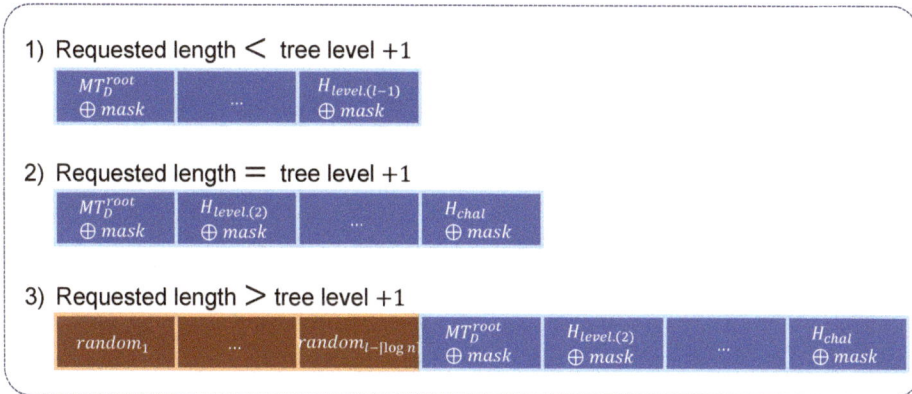

(**b**) Three possible types of proof

Figure 4. Classification of the proofs in the proposed scheme.

Theorem 3. *(Resilience to the leakage of size information) An adversary who knows neither the number of blocks nor the value of the root node in the Merkle tree (nor the original data used to construct the tree) cannot obtain any meaningful information through eavesdropping on the proposed authentication protocol.*

Proof. The proposed protocol allows the verifier to randomly select the proof length regardless of the Merkle tree structure. In the proposed scheme, the verifier sends a uniform random value as part of the challenge and the prover obtains the index value of the leaf node by taking the remainder after dividing the value by the number of blocks already known. This prevents the adversary from inferring the upper bound of the leaf node index during a repeated verification process as long as the verifier chooses different uniform random challenges with each execution, unlike typical Merkle tree-based authentication. Further, the sibling path generated using the Merkle tree is reduced or enlarged according to the requested proof length, which is also chosen uniformly at random by the verifier as another component of the challenge. As a result, there is no relationship between the size of the final proof and the size of the actual Merkle tree, making it impossible for the adversary to infer the size of the underlying data by calculating how the proof size corresponds to the number of leaf nodes in the tree. □

Upon closer inspection, the proof generated in the proposed scheme includes the value of the root node in the tree in masked form, which differs from typical Merkle tree-base authentication. Using this feature, it may be possible to uncover the mask when the number of data blocks is known to the adversary. In this case, the adversary can identify the starting position of the meaningful portion (i.e., the location of MT_D^{root} in Figure 4b) of the proof sent by the prover to the verifier. All they need to do is to uncover this mask value.

Theorem 4. *(Reliability of the proposed authentication) An adversary who knows the number of blocks within the underlying data cannot obtain any meaningful information from the proposed authentication protocol.*

Proof. Even if it were possible to track the location of the beginning portion (excluding bitstrings filled with random values) of the meaningful proof from the total number of blocks and the requested proof length, the adversary cannot recover the value of the root node in the tree from the masked proof in Theorem 2. Consequently, the probability of recovering the value of a root node is identical to finding the mask value, which is $(1/2)^\lambda$, where λ is the bit-length of the security parameter (or the hash value). It is notable that the prover chooses different random mask values on every proof generation. Therefore, the reliability of the challenge-response protocol remains the same as long as the adversaries cannot uncover the value of the root node. □

In this context, the increased security of the proposed scheme exploiting Merkle tree is dependent on the secrecy of the value of the root node of the tree. We define an experiment to show the formal security of the proposed scheme in the presence of eavesdropping adversary.

Definition 3. *The experiment is defined for the proposed Merkle tree-based authentication Π for the security parameter λ and an adaptive adversary $\mathcal{A}$ who only receives oracle accesses to the prover. The oracle access to the prover is again divided into access to the proof $\mathcal{O}_P^{proof}$ and access to the corresponding data $\mathcal{O}_P^{data}$.*

The indistinguishability experiment $PrivK_{\mathcal{A},\Pi}(\lambda)$:

1. (Init) The adversary $\mathcal{A}$ is given input 1^λ and the hash function h used to construct a Merkle tree.

2. (QueryI-I) $\mathcal{A}$ requests oracle acces to $\mathcal{O}_P^{proof}$, with challenging index $chal_i$ and required proof length L_i each chosen randomly and independently, polynomially many times. The oracle randomly creates and stores data D_i locally, generates obfuscated proof $proof_i$ (by performing $(n_i, h_i, MT_{D_i}) \leftarrow \Pi.\textbf{ContructMerkleTree}(D_i)$, $proof'_i \leftarrow \Pi.\textbf{GenerateProof}(MT_{D_i}, chal_i, L_i)$, and $proof_i \leftarrow \Pi.\textbf{ObfuscateProof}(MT_{D_i}, L_i)$ successively), and sends it to $\mathcal{A}$.

3. (QueryI-II) $\mathcal{A}$ requests oracle access to $\mathcal{O}_P^{data}$ by sending the received proof $proof_i$ to obtain the corresponding underlying data. The oracle retrieves the underlying data D_i used to generate the proof $proof_i$ and sends it to $\mathcal{A}$.

4. (Challenge) The challenger creates random data D that is not already queried and generate an obfuscated proof $proof_1$ following the specified protocol in Π. It also generates an arbitrary bitstring $proof_2$ of the same length as $proof_1$. Then, according to the result of the fair coin toss $b \in \{0,1\}$, $proof_b$ is delivered to $\mathcal{A}$.

5. (QueryII-I) The step 2 is repeated in polynomial time.

6. (QueryII-II) The step 3 is repeated except for $proof_b$ as necessary in polynomial time.

7. (Guess) $\mathcal{A}$ guesses the value of b. If $\mathcal{A}$'s guess is correct, $\mathcal{A}$ wins the game. Otherwise, it loses.

Theorem 5. *(Consistent reliability of the proposed authentication with repeated requests) An adversary who knows the number of blocks within the underlying data cannot obtain any meaningful information by repeatedly running the proposed authentication protocol.*

Proof. Note that the random mask used in Step 2 presented in the Definition 3 is randomly selected in the uniform distribution for each proof generation. Although the adversary $\mathcal{A}$ can verify the validity of the received proof (by performing $proof'_i \leftarrow \Pi.\textbf{RestoreProof}(proof_i, h_i, MT_{D_i}^{root})$ and $res_i \leftarrow \Pi.\textbf{VerifyProof}(proof'_i, MT_{D_i}^{root})$ successively) in Step 3, $\mathcal{A}$ cannot distinguish whether $proof_b$ received in Step 4 is valid proof or not by Theorem 3. Furthermore, $\mathcal{A}$ cannot know the mask value used to generate the proof $proof_b$ by Theorem 2 so that the prior knowledge does not help to break the proposed authentication mechanism. In the same context, Steps 5 and 6 also only give at most negligible advantage to determine the validity of the proof $proof_b$ given by the challenge. This means the leakage resilience of the proposed authentication scheme even in the presence of adaptive adversary. $\square$

Now, we consider another adversary who has knowledge of both the number of blocks in the tree and the value of the root node in the Merkle tree. However, it make sense to assume that this kind of adversary has no knowledge of the underlying data without loss of generality (e.g., when an adversary is delegated to audit data integrity held in remote storage by a valid data owner, while the actual content is kept private). Nevertheless, even though the above two pieces of information are known to the adversary, the proposed scheme provides security that is as strong as that of typical Merkle tree-based authentication (Section 5.1).

6. Efficiency Analysis

In this section, the efficiency of the proposed scheme is evaluated based on the experimental implementation of related schemes.

6.1. Experimental Environment

According to the Commercial National Security Algorithm (CNSA) Suite [39] recommended by the National Security Agency (NSA), we used SHA-3 384-bit as a cryptographic hash function, a Diffie-Hellman key with a 3072-bit modulus, and AES-256 for key agreement and a secure encryption algorithm when implementing comparison schemes.

All experiments were performed on a single machine with a 3.5 GHz CPU (Intel i7-7800x) and 64 GB RAM (3600 MHz 4 × 16 GB) running Windows 10. Each algorithm was implemented as a singrypto version 2.6.1) [40] for AES and Diffie-Hellman key agrele-threaded 32-bit Python [41] program, using the Python cryptography toolkit (pycement and the SHA-3 wrapper for Python (Pysha3 version 1.0.2) [42] for SHA-3, respectively. In addition, the data was split into 256-byte blocks when the Merkle tree was constucted to allow for a consistent comparison.

To minimize errors caused by outliers, each experiment was repeated 1000 times in the same environment, and then the average and the standard deviation are calculated and reported. It is worth nothing that there is room for additional performance improvement because the specified libraries were used without further optimization.

6.2. Computation Overhead

The computation time for each experiment was measured based on CPU time. The performance of each algorithm for varying data sizes is analyzed and the time overhead is compared .

6.2.1. Authentication Based on Merkle Tree

Conventional online authentication applying Merkle tree guarantees neither consistent reliability nor protection from information leakage, but it was added to the experiment as a baseline indicator for efficiency. The size of the data block was fixed for the system initialization but could be varied according to the system configuration. To allow for a consistent comparison, the block size was set to 256 bytes in this and following experiments.

A comparison of the computation time required for Merkle tree-based authentication for different data sizes is presented in Figure 5. The prover constructs a Merkle tree for the possessed data and generates a proof by finding sibling nodes in the tree, while the verifier selects a random index for the leaf node (corresponding to the index for the data block) and validates the proof received from the prover by repeatedly applying a hash function for each element in the proof.

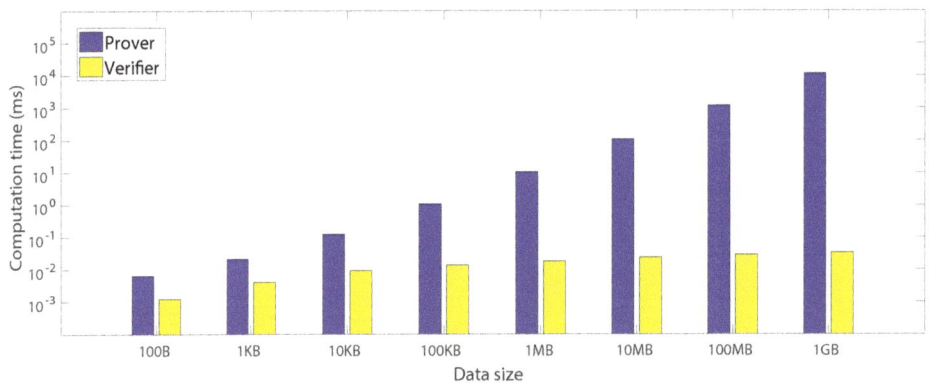

Figure 5. Computation time for Merkle tree-based authentication.

The required computation time increases as the soze of the data becomes larger. Merkle tree construction has a linear relationship with the number of blocks (i.e., leaf nodes) because the number of nodes in the tree can be at most $2n - 1$ for n blocks of data. Meanwhile, the computation time for proof generation and verification is logarithmically proportional to the number of data blocks because the number of elements is related to the tree height.

In the Merkle tree-based approach, the computation time between the prover and the verifier is not equal. This is because the prover has to construct the entire Merkle tree from the underlying data but the verifier does not. The verifier only has to apply the hash function using the received proof and compare the bitstrings of the result with locally stored information. For a data file of 1 MB, Merkle tree generation by the prover accounts for 99.9% of the computational time, which is 581.5 times longer than the verification time required for the verifier. Detailed results are summarized in Table 2.

Table 2. Average computation time (ms) for authentication based on Merkle tree by data size (standard deviation in parentheses).

	100 B	1 KB	10 KB	100 KB	1 MB	10 MB	100 MB	1 GB
Merkle tree generation	0.00461	0.01748	0.11897	1.05451	10.67579	107.07319	1134.07592	11,191.59684
	(0.00062)	(0.00080)	(0.00247)	(0.00723)	(0.09199)	(1.07465)	(19.51252)	(33.31233)
Challenge generation	0.00086	0.00086	0.00099	0.00101	0.00100	0.00100	0.00087	0.00087
	(0.00019)	(0.00017)	(0.00020)	(0.00021)	(0.00021)	(0.00023)	(0.00017)	(0.00016)
Sibling path generation	0.00201	0.00389	0.00662	0.00863	0.01101	0.01258	0.02032	0.02273
	(0.00024)	(0.00066)	(0.00048)	(0.00038)	(0.00059)	(0.00054)	(0.00159)	(0.00168)
Verification	0.00127	0.00417	0.00937	0.01398	0.01836	0.02369	0.02840	0.03294
	(0.00034)	(0.00030)	(0.00077)	(0.00075)	(0.00096)	(0.00149)	(0.00085)	(0.00096)

6.2.2. Authentication Based on the Hardcore Function

A comparison of the computation time required for hardcore function-based authentication for different data sizes is displayed in Figure 6, in which *Verifier* and *Prover* indicate the computation time required by the verifier and the prover, respectively. The verifier selects a random seed, generates a pseudorandom bitstring based on the selected seed, generates a proof using the generated bitstring, and validates the proof received from the prover by comparing it with the locally generated proof. On the other hand, the prover generates a pseudorandom bitstring based on the seed received from the verifier and generates a proof using the independently generated bitstring.

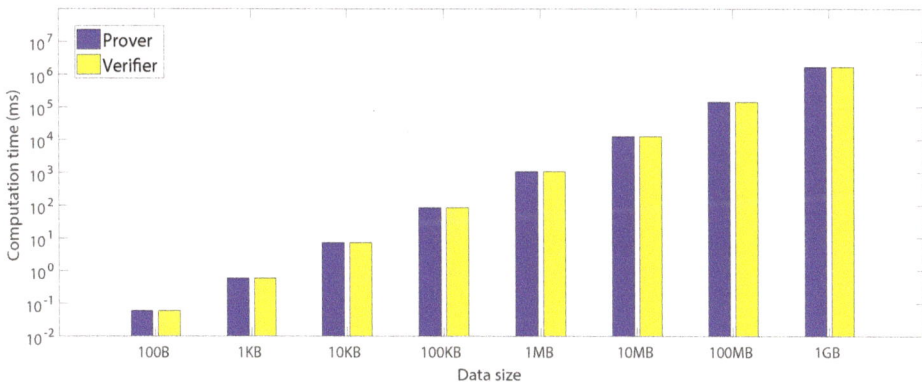

Figure 6. Computation time for hardcore function-based authentication.

The required computation time increases as the volume of data becomes larger. Seed generation time is almost constant because the size of the seed does not change. However, pseudorandom bitstring generation and verification are proportional to the logarithmic size of the data because they are closely related to the size of the proof. Proof generation is lenearly proportional to the size of the data.

There is little difference in the computation time between the prover and the verifier. This is because both the verifier and the prover generate their own proofs based on the self-generated pseudorandom bitstring, which requires the most computation time, although the verifier adiitionally conducts the initial seed generation and proof validation through a simple comparison, which requires relatively little time. For 1 MB of data, proof generation accounts for 98.2% of computation time for both the verifier and the prover. The detailed experiment results are summarized in Table 3.

Table 3. Average computation time (ms) for authentication based on a hardcore function by data size (standard deviation in parentheses).

	100 B	1 KB	10 KB	100 KB	1 MB	10 MB	100 MB	1 GB
Seed generation	0.00065	0.00075	0.00092	0.00223	0.00313	0.00420	0.00822	0.01068
	(0.00016)	(0.00018)	(0.00020)	(0.00061)	(0.00037)	(0.00054)	(0.00101)	(0.00119)
Pseudorandom bitstring	0.01106	0.01417	0.03730	0.28324	4.12432	44.29721	450.26368	4690.52351
generation	(0.00085)	(0.00133)	(0.00351)	(0.02986)	(0.79537)	(8.69847)	(89.4971)	(488.92123)
Proof generation	0.04765	0.57035	7.20359	84.95004	1058.25365	12,523.93617	139,315.96462	1,666,976.31436
	(0.00148)	(0.01614)	(0.09508)	(0.86894)	(2.32808)	(8.65474)	(236.08956)	(2663.86754)
Verification	0.00018	0.00019	0.00022	0.00032	0.00036	0.00036	0.00036	0.00100
	(0.00014)	(0.00014)	(0.00013)	(0.00013)	(0.00013)	(0.00013)	(0.00014)	(0.00023)

6.2.3. Authentication Based on Merkle Tree with Transmission in Encrypted Form

This approach requires the encryption and decryption of data transmitted between the prover and the verifier using a key agreed upon by both entities in addition to the typical Merkle tree-based authentication. Therefore, there is an additional need for a trusted authority in order to set the parameters to create an environment for key agreement. In this experiment, a Diffie-Hellman key agreement mechanism was adopted with a modulus of 3072 bits. (If communication parties require data authentication and continuous communication, the computation time for key agreement may be excluded from computation overhead. However, in this paper, we experimented with parameter setting for the same security level on the assumption that only communication for online data authentication is done.) Additionally, the data to be transmitted to the other party is encrypted with the agreed key and decrypted on the recipient's side, leaving the rest of the process the same as in the typical Merkle tree-based approach. In other words, the verifier encrypts and transmits a random challenge and decrypts the proof received from the prover. The prover decrypts the challenge received from the verifier to generate the proof, and then encrypts that proof.

If the parties communicate and perform data authentication continuously, the computation time for key agreement may be excluded from computation overhead. In this case, however, there is a possibility that an adversary can bypass authentication from eavesdropping of repeated authentication process, and there is still a leak in size information. In practice, since most of the authentication is done by a large number of independent users, individual users need to establish a new session (using a new session key) and perform authentication. Therefore, in this paper, we only measured communication overhead for online data authentication in encrypted form after a key agreement on the same security level.

A comparison of the computation time required for Merkle tree-based authentication for different data sizes is presented in Figure 7, in which *Public setting* refers to the time required for parameter generation by the trusted authority for Diffie-Hellman key agreement. As this approach is also based on a Merkle tree, the computation time increases as the volume of data increases. In addition, computational load for the prover and the verifier is also similar to that of the Merkle tree-based approach.

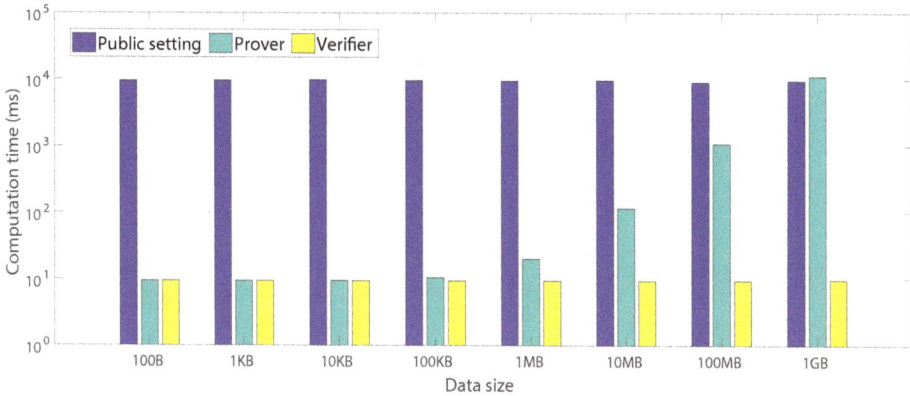

Figure 7. Computation time for Merkle tree-based authentication with encrypted communication.

However, public setting in the initial stage requires a relatively high computation time even for 1 GB of data (although it is executed only once and requires a constant amount of time). Encryption, decryption, and key agreement also increase the computation time for both the verifier and the prover for exchanges of challenges and proofs, respectively.

The detailed experiment results are summarized in Table 4. For 1MB of data, key agreement accounts for 46.9% and 99.3% of the computation time for the prover and the verifier, respectively, while encryption and decryption uses only 0.2% and 0.5%, respectively.

Table 4. Average computation time (ms) for Merkle tree-based encrypted authentication by data size (standard deviation in parentheses).

	100 B	1 KB	10 KB	100 KB	1 MB	10 MB	100 MB	1 GB
Parameter generation	9468.19790 (8493.82876)	9538.16482 (8975.20449)	9794.65946 (9175.01323)	9532.98866 (8829.14856)	9452.51795 (8759.71474)	9727.44050 (9374.65954)	9064.76798 (7981.16155)	9633.54744 (8863.74385)
Partial key generation	4.80688 (0.18776)	4.79883 (0.18395)	4.78975 (0.18719)	4.79192 (0.18367)	4.80195 (0.18475)	4.80006 (0.18142)	4.81253 (0.19221)	4.96501 (0.25943)
Key agreement	4.57732 (0.13962)	4.57399 (0.14449)	4.56398 (0.14221)	4.56755 (0.15151)	4.57577 (0.15099)	4.57745 (0.14848)	4.57501 (0.14461)	4.57433 (0.14200)
Challenge generation	0.00087 (0.00018)	0.00098 (0.00020)	0.00091 (0.00020)	0.00088 (0.00016)	0.00100 (0.00021)	0.00087 (0.00016)	0.00099 (0.00023)	0.00087 (0.00018)
Encryption of the challenge	0.02674 (0.00145)	0.02647 (0.00140)	0.02668 (0.00117)	0.02702 (0.0013)	0.03341 (0.00210)	0.03457 (0.00235)	0.03739 (0.00323)	0.05227 (0.00281)
Decryption of the challenge	0.01458 (0.00068)	0.01567 (0.00065)	0.01597 (0.00042)	0.01776 (0.00088)	0.01967 (0.00118)	0.02292 (0.00097)	0.02444 (0.00064)	0.02568 (0.00066)
Merkle tree generation	0.00324 (0.00024)	0.01570 (0.00043)	0.11664 (0.00184)	1.05112 (0.03182)	10.55169 (0.08447)	106.34833 (0.36833)	1072.881066 (6.69879)	11,196.73308 (33.06858)
Sibling path generation	0.00271 (0.00033)	0.00429 (0.00042)	0.00655 (0.00047)	0.00808 (0.00054)	0.01018 (0.00079)	0.01172 (0.00047)	0.01773 (0.00100)	0.02247 (0.00101)
Encryption of the sibling path	0.01639 (0.00061)	0.01751 (0.00043)	0.01866 (0.00089)	0.02067 (0.00117)	0.02978 (0.00323)	0.04813 (0.00177)	0.05219 (0.00118)	0.05405 (0.00357)
Decryption of the sibling path	0.01782 (0.00061)	0.01776 (0.00071)	0.01895 (0.00062)	0.01714 (0.00085)	0.01790 (0.00058)	0.01957 (0.00102)	0.02026 (0.00080)	0.02090 (0.00129)
Verification	0.00130 (0.00018)	0.00454 (0.00041)	0.00981 (0.00078)	0.01408 (0.00091)	0.01836 (0.00112)	0.02378 (0.00121)	0.02848 (0.00088)	0.03252 (0.00041)

6.2.4. Authentication Based on the Proposed Approach

Similar to Merkle tree-based authentication after encryption, the proposed mechanism obfuscates the transmitted data (i.e., challenges from the verifier and proofs from the prover). Notice that, however, unlike the other scheme, ours does not require an additional trusted authority to generate public parameters for key agreement as a preprocessing stage before the challenge-response process. In addition, the proposed scheme hides the size information by randomizing the proof size regardless of the Merkle tree structure. Specifically, the verifier requests an arbitrary proof length in terms of the hash values and the prover generates and obfuscates (and truncates if necessary) the sibling path according to the requested proof length, followed by padding with a random bitstring when the generated proof is shorter than the specified length. A comparison of the computation time required for the proposed approach by data size is illustrated in Figure 8.

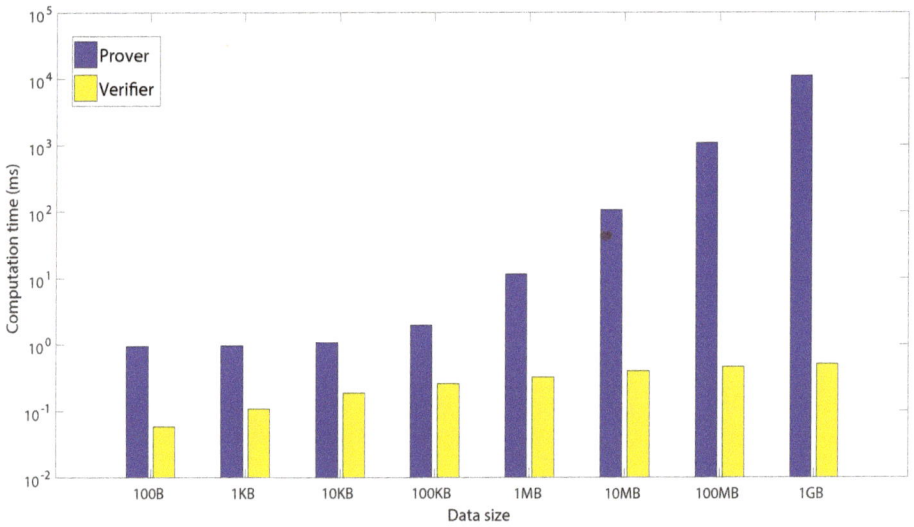

Figure 8. Computation time of the proposed authentication scheme.

The detailed experiment results are summarized in Table 5. The computation time increases as the volume of the data (consequently, the size of Merkle tree) increases, and most of the time is used to construct the tree. For 1 MB of data, Merkle tree generation requires 92.1% of the time, while the time used by the prover to obfuscate the sibling path and to add a random bitstring is just 1.6% and 6.4%, respectively. On the verifier side, mask removal is additionally performed, taking a similar amount time as the verification. However, it is logarithmically proportional to the number of data blocks and accounts for negligible amount of time.

Table 5. Average computation time (ms) of the proposed authentication scheme by data size (standard deviation in parenthesis).

	100 B	1 KB	10 KB	100 KB	1 MB	10 MB	100 MB	1 GB
Requested proof length	49.04100	50.84400	52.70700	52.58500	52.22300	51.67100	50.86000	50.81600
	(27.97269)	(28.63403)	(27.93591)	(28.83915)	(28.47626)	(28.3989)	(27.28825)	(28.78524)
Merkle tree generation	0.00299	0.01417	0.11407	1.04286	10.53377	106.27209	1081.34281	11,178.68500
	(0.00098)	(0.00324)	(0.03135)	(0.29979)	(3.02505)	(30.96492)	(320.58949)	(354.27592)
Challenge generation	0.00216	0.00226	0.00247	0.00243	0.00216	0.00240	0.00243	0.00238
	(0.00068)	(0.00067)	(0.00062)	(0.00053)	(0.00070)	(0.00056)	(0.00059)	(0.00068)
Sibling path generation	0.00000	0.00000	0.00000	0.00000	0.00002	0.00002	0.00003	0.00002
	(0.00000)	(0.00000)	(0.00000)	(0.00000)	(0.00000)	(0.00000)	(0.00000)	(0.00000)
Sibling path obfuscation	0.04547	0.06643	0.10507	0.13948	0.17899	0.23363	0.26413	0.29186
	(0.00596)	(0.00338)	(0.01607)	(0.01782)	(0.02655)	(0.03630)	(0.04581)	(0.06181)
Random bitstring padding	0.89665	0.87894	0.85868	0.78980	0.72994	0.66017	0.59333	0.55448
	(0.54933)	(0.53655)	(0.53474)	(0.53037)	(0.51176)	(0.49959)	(0.46923)	(0.47786)
Mask removal	0.02384	0.04487	0.08232	0.11417	0.14501	0.17849	0.20863	0.23202
	(0.00310)	(0.00207)	(0.01486)	(0.01715)	(0.02501)	(0.03532)	(0.04438)	(0.06090)
Verification	0.00886	0.01602	0.01936	0.02576	0.02953	0.03799	0.04319	0.04763
	(0.00358)	(0.08621)	(0.00421)	(0.05892)	(0.00504)	(0.01028)	(0.00676)	(0.00924)

6.2.5. Analysis of Computation Overhead

Based on analyses of individual algorithms, the computation overhead for the prover and the verifier is summarized in Figures 9 and 10, respectively. Although *Merkle tree-based authentication* (the first bar in the figures) does not consider information leakage, its computation overhead is used as a reference for ideal computation efficiency.

On the prover side, the operation of *authentication based on a hardcore function* (the second bar) is performed on the entire data while repeating the log of the data bit-length, leading to a computation overhead that is linearly proportional to data size. All of the other algorithms only require all of the data when constructing a Merkle tree, and the proof generation process has a relatively low overhead because it deals only with the logarithm of the data in bit-length. For'data smaller than 10 MB in size, the proposed scheme demonstrates the most efficient computation (next to the one adopting only a Merkle tree). For data over 10 MB in size, the computation overhead is very similar for the three algorithms exploiting Merkle trees. This indicates that the overhead generated by encryption/decryption and random masking in the proposed scheme is negligible.

On the verifier side, there is a relatively clear difference between the algorithms because the computation required is lower than that of the prover. Other than *authentication based on Merkle tree*, the proposed scheme exhibits the greatest efficiency, followed by authentication based on Merkle tree with encrypted transmission, with hardcore function-based authentication demonstrating the lowest efficiency. The majority of the overhead is due to key agreement stage in the algorithm requiring encrypted communication and sibling path obfuscation in the proposed scheme. However, this difference does not exceed 1ms regardless of the data size in the experimental results. Furthermore, the proposed scheme might be able to further narrow the gap by optimizing the bitwise exclusive-or (XOR) operation, which is not natively supported in Python.

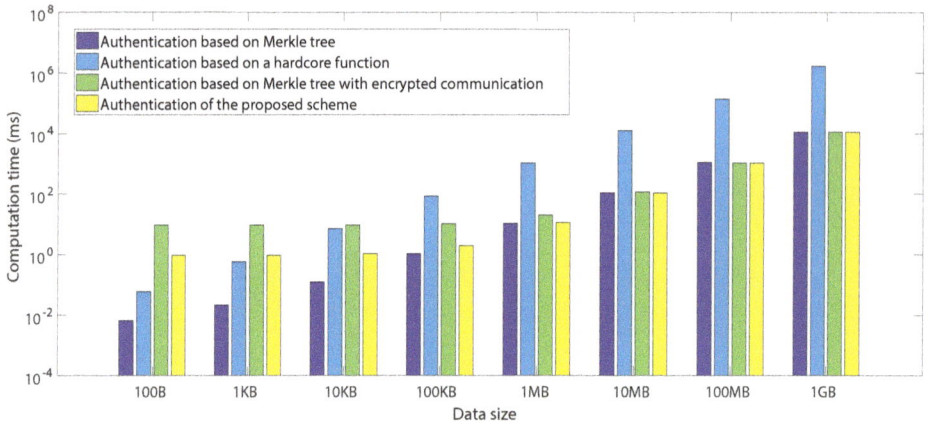

Figure 9. Comparison of computation overhead on the prover side.

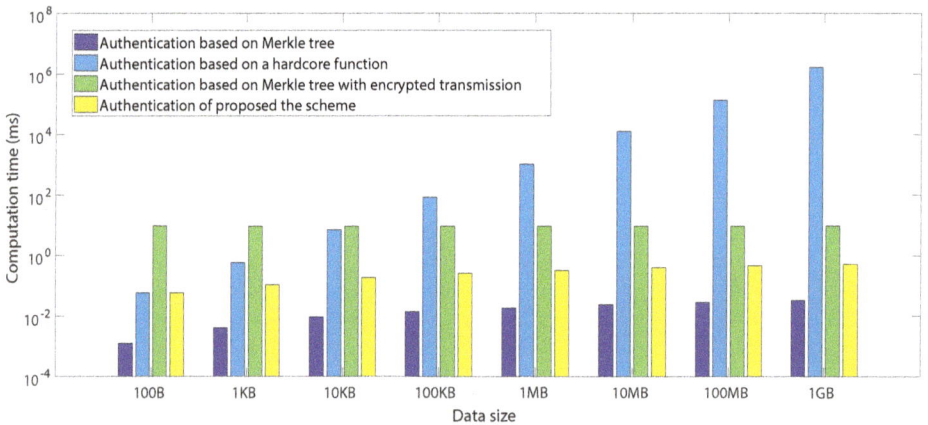

Figure 10. Comparison of computation overhead on the verifier side.

Considering the features of the related schemes summarized in Table 6, the verifier in *authentication based on a hardcore function* allows anyone to know the size of the underlying data (because the bit-length of the transmitted seed is logarithmically proportional to the data volume) even though the transmitted data is randomized. *Authentication based on Merkle tree with encrypted communication* (the third bar) is resilient to replay attacks that let the adversary reuse previously successful validation, but is still susceptible to size information leakage. In short, none of the comparison algorithmsare able to reduce information leakage to the same extent as the proposed scheme.

Nevertheless, the proposed scheme requires the least computation overhead for both the prover and the verifier (except for Merkle tree-based authentication, which does not consider information leakage).

Table 6. Average computation time (ms) for Merkle tree-based encrypted authentication by data size (standard deviation in parentheses).

Features	Authentication Based on			
	Merkle Tree	Hardcore Function	Merkle Tree with Encrypted Communication	Proposed Scheme
Resilience against size information leakage	X	X	X	O
Resilience against replay attacks	X	O	O	O
Requirement for an additional trusted authority	X	X	O	X

6.3. Communication Overhead

For all of the compared schemes, the proof is generated using all of the data, but the final proof transmitted to the verifier is proportional to the log of the data bit-length. Looking closely at the amount of data for each entity, however, there are noticible differences between approaches.

In the transmission from the prover to the verifier, *authentication based on a hardcore function* generates and sends a proof of bit-length $(|M| + \log(M) - 1)$ for data M. Therefore, the size of the generated proof becomes very small. Specifically, the proof size is only 1 Byte when the data is 100 Bytes in size, 2 Bytes for data between 1 KB and 10 KB in size, 3 Bytes for 100 KB-10 MB of data, and 4 Bytes for 1 GB of data. On the other hand, the other approaches generate and send a proof. The proof corresponds to a series of hash values and is logarithmically proportional to the number of all of the data blocks, where the size of the hash value is 384 bits (i.e., 48 Bytes). *Authentication based on Merkle tree* requires the additional transmission of a partial key generated by the prover that is 3072 bits (i.e., 384 Bytes) in size. The comparison of the data transmission from the prover to the verifier is presented in Figure 11.

Recall that the size of a proof, which is embedded in the challenge, is determined by the verifier. Therefore, the transmitted proof size is independent of the actual data size. As specified in Table 5, the average requested proof length (which is proportional to the number of hash values) of 51 is much longer than the sibling path in the Merkle tree approach. For example, 1 MB of data has a sibling path length of 13 and 1 GB of data has a sibling path length of 23. The communication overhead when the requested proof length is fixed at 25 is also illustrated as the last bar in Figure 11. In this case, the communication overhead is almost the same as that of 100 MB of data in conventional *authentication based on Merkle tree* even for 1 GB of data. This characteristic of the proposed scheme is positive in that it provides flexibility for the verifier in setting the proof length regardless of the actual data size.

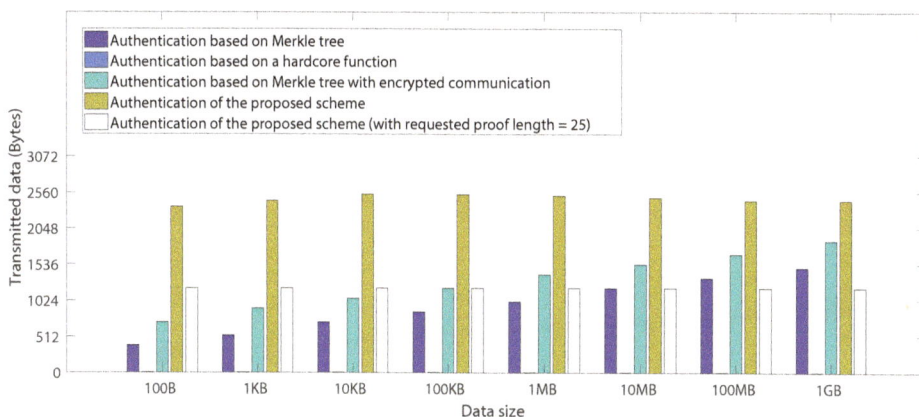

Figure 11. Comparison of trasmitted data on prover side with varying data size.

On the other hand, in the transmission from the verifier to the prover, only a constant amount of transmission is required regardless of the data size, because only the challenge is transmitted in all schemes except *authentication based on a hardcore function*. The comparison of the data transmission from the prover to the verifier is presented in Figure 12.

In terms of storage, there is no additional overhead because the random sources can be removed from the local storage immediately after the hash evaluations.

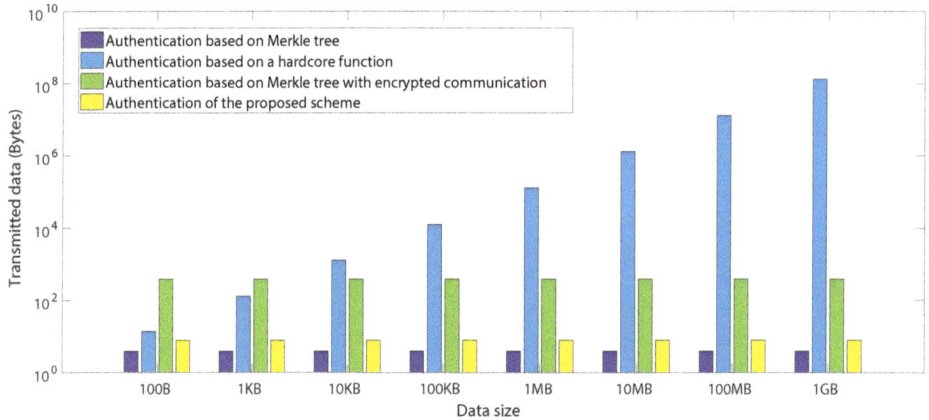

Figure 12. Comparison of trasmitted data on verifier side with varying data size.

7. Conclusions

At the present time, when data storage and maintenance costs can be reduced due to advances in information and communciation technologies, it is easy to overlook whether data is correctly and legitimated managed when outsourced to remote repositories. In this paper, we investigated the types of information leakage that can occur when data integrity is compromised between physically separate entities and reviewed representative approaches to handling this issue. A simple but efficient approach is presented to improve the security and reliability of data integrity validations, something which has been neglected in previous research. Providing rigorous security analysis, the effectiveness of the proposed scheme is examined in terms of resilience against the leakage of size information and replay attacks. Performance analysis shows that our method provides the highest efficiency in terms of computation load and improves security and reliability.

Author Contributions: D.K. contributed the ideas and wrote the paper; Y.S. and J.Y. designed and conducted the experiments; J.H. performed the security analaysis and supervised the whole paper including paper organization and proofread.

Funding: This work was supported by the National Research Foundation of Korea (NRF) grant funded by the Korea government (MSIT) (No. 2017R1C1B5077026) for Dongyoung Koo. This work was supported by Institute for Information & communications Technology Promotion(IITP) grant funded by the Korea government(MSIP) (No.2018-0-00477, Development of Malware Analysis Technique based on Deep Web and Tor) for Junbeom Hur. This work was supported by the National Research Foundation of Korea (NRF) grant funded by the Korea government (MSIP) (No.2017R1C1B5015045) for Youngjoo Shin. This research was supported by the MSIT(Ministry of Science and ICT), Korea, under the ITRC(Information Technology Research Center) support program(IITP-2018-0-01423) supervised by the IITP(Institute for Information & communications Technology Promotion) for Joobum Yun.

Conflicts of Interest: The authors declare no conflict of interest.

References

1. CPA Practice Advisor. The Top Cyber Risks to Accounting Firms Come from Inside the Firm. Available online: https://www.cpapracticeadvisor.com/news/12427308/the-top-cyber-risks-to-accounting-firms-come-from-inside-the-firm (accessed on 13 September 2018).
2. Chen, C.; Deng, I. Tencent Cloud Says Ímproper Operationsíed to Data Loss for Client as It Seeks to Implement Improvements. Available online: https://www.scmp.com/tech/article/2158785/tencent-cloud-says-improper-operations-led-data-loss-client-it-seeks-implement (accessed on 13 September 2018).
3. Zeng, K. Publicly Verifiable Remote Data Integrity. In *Information and Communications Security*; Springer: New York, NY, USA, 2008; pp. 419–434. doi:10.1007/978-3-540-88625-9_28.
4. Henry, J. These 5 Types of Insider Threats Could Lead to Costly DAta Breaches. Available online: https://securityintelligence.com/these-5-types-of-insider-threats-could-lead-to-costly-data-breaches/ (accessed on 13 September 2018).
5. Sambit.k. Global Cloud Data Loss Prevention (DLP) Market 2023 Growth Factors, Regional Analysis by Types, Applications, & Manufacturers with Forecasts. Available online: https://thetradereporter.com/global-cloud-data-loss-prevention-dlp-market-2023-growth-factors-egional-analysis-by-types-applications-manufacturers-with-forecasts/139976/ (accessed on 13 September 2018).
6. Vacca, J.R. *Cloud Computing Security: Foundations and Challenges*; CRC Press: Boca Raton, FL, USA, 2016.
7. Symantec Corporation. Symantec Data Loss Prevention. Available online: https://www.symantec.com/products/data-loss-prevention/ (accessed on 13 September 2018).
8. Trustwave Holdings, Inc. Trustwave Data Loss Prevention. Available online: https://www.trustwavecompliance.com/solutions/compliance-technologies/data-loss-prevention/ (accessed on 13 September 2018).
9. McAfee, LLC. McAfee Total Protection for Data Loss Prevention. Available online: https://www.mcafee.com/enterprise/en-ca/products/total-protection-for-data-loss-prevention.html/ (accessed on 13 September 2018).
10. Check Point Software Technologies, Ltd. Data Loss Prevention Software Blade. Available online: https://www.checkpoint.com/products/dlp-software-blade/ (accessed on 13 September 2018).
11. Digital Guardian. Digital Guardian Encpoint DLP. Available online: https://digitalguardian.com/products/endpoint-dlp/ (accessed on 13 September 2018).
12. Merkle, R.C. A Digital Signature Based on a Conventional Encryption Function. In *Advances in Cryptology—CRYPTO*; Springer: Berlin/Heidelberg, Germany, 1988; pp. 369–378. doi:10.1007/3-540-48184-2_32.
13. Swan, M. Blockchain Thinking: The Brain as a Decentralized Autonomous Corporation [Commentary]. *IEEE Technol. Soc. Mag.* **2015**, *34*, 41–52. doi:10.1109/MTS.2015.2494358. [CrossRef]
14. Liang, X.; Shetty, S.; Tosh, D.; Kamhoua, C.; Kwiat, K.; Njilla, L. ProvChain: A Blockchain-based Data Provenance Architecture in Cloud Environment with Enhanced Privacy and Availability. In Proceedings of the 2017 IEEE/ACM International Symposium on Cluster, Cloud and Grid Computing (CCGrid '17), Madrid, Spain, 14–17 May 2017; pp. 468–477. doi:10.1109/CCGRID.2017.8. [CrossRef]
15. Bitcoin.com. Bitcoin. Available online: https://www.bitcoin.com/ (accessed on 13 September 2018).
16. Halevi, S.; Harnik, D.; Pinkas, B.; Shulman-Peleg, A. Proofs of Ownership in Remote Storage Systems. In Proceedings of the 2011 ACM Conference on Computer and Communications Security (CCS), Chicago, IL, USA, 17–21 October 2011; pp. 491–500. doi:10.1145/2046707.2046765. [CrossRef]
17. Yang, C.; Ren, J.; Ma, J. Provable Ownership of Files in Deduplication Cloud Storage. *Secur. Commun. Netw.* **2015**, *8*, 2457–2468. doi:10.1002/sec.784. [CrossRef]
18. Armknecht, F.; Boyd, C.; Davies, G.T.; Gjøsteen, K.; Toorani, M. Side Channels in Deduplication: Trade-offs Between Leakage and Efficiency. In Proceedings of the 2017 ACM on Asia Conference on Computer and Communications Security (ASIA CCS '17), Abu Dhabi, UAE, 2–6 April 2017; pp. 266–274. doi:10.1145/3052973.3053019. [CrossRef]
19. Koo, D.; Shin, Y.; Yun, J.; Hur, J. An Online Data-Oriented Authentication Based on Merkle Tree with Improved Reliability. In Proceedings of the 2017 IEEE International Conference on Web Services (ICWS), Honolulu, HI, USA, 25–30 June 2017; pp. 840–843. doi:10.1109/ICWS.2017.102. [CrossRef]

20. Merkle, R.C. A Certified Digital Signature. In *Advances in Cryptology—CRYPTO*; Springer: New York, NY, USA, 1990; pp. 218–238.

21. Lamport, L. *Constructing Digital Signatures from a One-Way Function*; Technical Report, Technical Report CSL-98; SRI International Palo Alto: Menlo Park, CA, USA, 1979.

22. Kundu, A.; Atallah, M.J.; Bertino, E. Leakage-free Redactable Signatures. In Proceedings of the 2012 ACM Conference on Data and Application Security and Privacy, CODASPY '12, San Antonio, TX, USA, 7–9 February 2012; ACM: New York, NY, USA, 2012; pp. 307–316. doi:10.1145/2133601.2133639. [CrossRef]

23. Buldas, A.; Laur, S. Knowledge-Binding Commitments with Applications in Time-Stamping. In *Public Key Cryptography—PKC*; Springer: Berlin/Heidelberg, Germany, 2007; pp. 150–165. doi:10.1007/978-3-540-71677-8_11.

24. Wikipedia. Binary Tree. Available online: https://en.wikipedia.org/wiki/Binary_tree/ (accessed on 13 September 2018).

25. Ateniese, G.; Burns, R.; Curtmola, R.; Herring, J.; Kissner, L.; Peterson, Z.; Song, D. Provable Data Possession at Untrusted Stores. In Proceedings of the ACM Conference on Computer and Communications Security (CCS), Alexandria, VA, USA, 28–31 October 2007; pp. 598–609. doi:10.1145/1315245.1315318. [CrossRef]

26. Zhao, Y.; Chow, S.S.M. Towards Proofs of Ownership Beyond Bounded Leakage. In Proceedings of the 2016 International Conference on Provable Security (ProvSec), Nanjing, China, 10–11 November 2016; pp. 340–350. doi:10.1007/978-3-319-47422-9_20. [CrossRef]

27. Atallah, M.J.; Cho, Y.; Kundu, A. Efficient Data Authentication in an Environment of Untrusted Third-Party Distributors. In Proceedings of the IEEE 24th International Conference on Data Engineering, Cancun, Mexico, 7–12 April 2008; pp. 696–704. doi:10.1109/ICDE.2008.4497478. [CrossRef]

28. Atallah, M.J.; Li, J. Enhanced smart-card based license management. In Proceedings of the 2003 IEEE International Conference on E-Commerce, CEC 2003, Newport Beach, CA, USA, 24–27 June 2003; pp. 111–119. doi:10.1109/COEC.2003.1210240. [CrossRef]

29. Benjamin, D.; Atallah, M.J. Private and Cheating-Free Outsourcing of Algebraic Computations. In Proceedings of the 2008 Annual Conference on Privacy, Security and Trust, Fredericton, NB, Canada, 1–3 October 2008; pp. 240–245. doi:10.1109/PST.2008.12. [CrossRef]

30. Keelveedhi, S.; Bellare, M.; Ristenpart, T. DupLESS: Server-Aided Encryption for Deduplicated Storage. In Proceedings of the 22nd USENIX Security Symposium, Washington, DC, USA, 14–16 August 2013; pp. 179–194.

31. Xu, J.; Chang, E.C.; Zhou, J. Weak Leakage-resilient Client-side Deduplication of Encrypted Data in Cloud Storage. In Proceedings of the 2013 ACM SIGSAC Symposium on Information, Computer and Communications Security, ASIA CCS '13, Berlin, Germany, 4–8 November 2013; pp. 195–206. doi:10.1145/2484313.2484340. [CrossRef]

32. Li, H.; Lu, R.; Zhou, L.; Yang, B.; Shen, X. An Efficient Merkle-Tree-Based Authentication Scheme for Smart Grid. *IEEE Syst. J.* **2014**, *8*, 655–663. doi:10.1109/JSYST.2013.2271537. [CrossRef]

33. Rogaway, P.; Shrimpton, T. Cryptographic Hash-Function Basics: Definitions, Implications, and Separations for Preimage Resistance, Second-Preimage Resistance, and Collision Resistance. In *Fast Software Encryption*; Springer: Berlin/Heidelberg, Germany, 2004; pp. 371–388. doi:10.1007/978-3-540-25937-4_24.

34. Merkle, R.C. Protocols for Public Key Cryptosystems. In Proceedings of the 1980 IEEE Symposium on Security and Privacy, Oakland, CA, USA, 14–16 April 1980; pp. 122–122. doi:10.1109/SP.1980.10006. [CrossRef]

35. Becker, G. *Merkle Signature Schemes, Merkle Trees and Their Cryptanalysis*; Technical Report; Ruhr-University Bochum: Bochum, Germany, 2008.

36. Koblitz, N.; Menezes, A.J. Cryptocash, cryptocurrencies, and cryptocontracts. *Des. Codes Cryptogr.* **2016**, *78*, 87–102. doi:10.1007/s10623-015-0148-5. [CrossRef]

37. Bellare, M.; Pointcheval, D.; Rogaway, P. Authenticated Key Exchange Secure against Dictionary Attacks. In *Advances in Cryptology—EUROCRYPT 2000*; Springer: Berlin/Heidelberg, Germany, 2000; pp. 139–155. doi:10.1007/3-540-45539-6_11.

38. Wikipedia. Entropy (Information Theory). Available online: https://en.wikipedia.org/wiki/Entropy_(information_theory) (accessed on 13 September 2018).

39. Information Assurance by the National Security Agency. Commercial National Security Algorithm (CNSA) Suite. Available online: https://www.iad.gov/iad/customcf/openAttachment.cfm?FilePath=/iad/library/ia-guidance/ia-solutions-for-classified/algorithm-guidance/assets/public/upload/Commercial-National-Security-Algorithm-CNSA-Suite-Factsheet.pdf&WpKes=aF6woL7fQp3dJiShxsuwyRvADMxf4cwBTYEUSz (accessed on 13 September 2018).

40. Python Software Foundation. pycrypto. Available online: https://pypi.org/project/pycrypto/ (accessed on 13 September 2018).

41. Python Software Foundation. python. Available online: https://www.python.org/ (accessed on 13 September 2018).

42. Python Software Foundation. pysha3. Available online: https://pypi.org/project/pysha3/ (accessed on 13 September 2018).

![applied sciences logo] *applied sciences*

MDPI

Article

Comprehensive Evaluation on an ID-Based Side-Channel Authentication with FPGA-Based AES

Yang Li, Momoka Kasuya and Kazuo Sakiyama *

Department of Informatics, The University of Electro-Communications, 1-5-1 Chofugaoka, Chofu-shi, Tokyo 182-8585, Japan; liyang@uec.ac.jp (Y.L.); m.kasuya@uec.ac.jp (M.K.)
* Correspondence: sakiyama@uec.ac.jp

Received: 15 September 2018; Accepted: 10 October 2018; Published: 12 October 2018

Abstract: Various electronic devices are increasingly being connected to the Internet. Meanwhile, security problems, such as fake silicon chips, still exist. The significance of verifying the authenticity of these devices has led to the proposal of side-channel authentication. Side-channel authentication is a promising technique for enriching digital authentication schemes. Motivated by the fact that each cryptographic device leaks side-channel information depending on its used secret keys, cryptographic devices with different keys can be distinguished by analyzing the side-channel information leaked during their calculation. Based on the original side-channel authentication scheme, this paper adapts an ID-based authentication scheme that can significantly increase the authentication speed compared to conventional schemes. A comprehensive study is also conducted on the proposed ID-based side-channel authentication scheme. The performance of the proposed authentication scheme is evaluated in terms of speed and accuracy based on an FPGA-based AES implementation. With the proposed scheme, our experimental setup can verify the authenticity of a prover among 2^{70} different provers within 0.59 s; this could not be handled effectively using previous schemes.

Keywords: side-channel authentication; leakage model; AES; FPGA

1. Introduction

Nowadays, wearable embedded technology is being increasingly used under the rapid development of electronic devices. The users and the embedded computing systems are connected to the Internet and exposed to various security threats, such as fake silicon chips. As a fundamental method against these security threats, the authenticity of these electronic devices has to be verified carefully. As a typical authentication scenario, we focus on the case where the identity of the prover is verified based on shared secret information between the prover and the verifier.

In [1], side-channel authentication was proposed as a new authentication scheme. Side-channel leakage, e.g., power consumption and electromagnetic (EM) radiation, is the unintentional information leakage that generally exists along the device's computation. Side-channel leakage has received much attention since it can be used to perform key-recovery attacks against cryptographic implementations [2,3]. In side-channel authentication, side-channel information is constructively used as a communication channel through which certain characteristics of the performed calculation can be observed. Cryptographic hardware with a unique secret key leaks unique key-dependent side-channel information under a given challenge. The idea of side-channel authentication is to measure and analyze this side-channel information to verify whether the used secret key is the pre-shared one.

Side-channel authentication has several positive features that make it valuable to be further researched. First, the measurement of side-channel information usually requires another measurement setup which is different from the main communication. Thus, the executions of relay attacks and reply attacks are expected to become difficult. Second, side-channel information such as time, power consumption, and electromagnetic radiation generally exists during the cryptographic calculation. This side-channel information contains information about the processed data including the key-related information; this can be measured and used in the authentication. As the minimal requirement for side-channel authentication, each prover device runs a computation module that uses a pre-share key with the verifier which has measurable side-channel information during the calculation. Thus, the modification of existing prover devices could be minimal for side-channel authentication. For devices that do not have a general communication capability, side-channel authentication could still be applied by using pre-defined challenges. For example, side-channel authentication could be used for a Machine-to-Machine (M2M) authentication scenario in which the resource-restricted prover device has symmetric-key cryptographic primitives implemented. Specifically, the smart cards used in public transportation systems and the keyless entry system of vehicles could be considered to use side-channel authentication.

As the first proposed side-channel authentication scheme from [1], the 128-bit Advanced Encryption Standard (AES-128) is a cryptographic module. In order to simplify the system, a modified AES that has increased rounds is used in side-channel authentication so that a single trace of the side-channel measurement is enough for authentication. In [1], several protocols for side-channel authentication were proposed as well. According to the originally proposed side-channel authentication system in [1], the prover can be identified only with side-channel information, i.e., by deriving correlation coefficients for all of the registered devices to identify the legitimate prover. Therefore, authentication is time-consuming. The authors of [4] provided a quantitative discussion about side-channel information according to the number of distinguishable provers. However, the aspects related to the accuracy of authentication, such as the false acceptance rate and the false rejection rate, have not been discussed.

As the contribution of this paper, we propose an identification-based (ID-based) authentication scheme and perform a comprehensive evaluation with regard to the authentication speed, the authentication accuracy, and the used leakage models. The detailed contributions of this paper are summarized as follows.

1. This paper proposes the ID-based authentication scheme to mitigate the speed problem of the side-channel authentication scheme proposed in [1]. To demonstrate the advantage of the ID system for acceleration, the authentication speed and authentication accuracy are evaluated for the ID-based authentication system. We overview the technical choices for side-channel authentication schemes and compare their effectiveness based on both theoretical analysis and experiments based on field-programmable gate array (FPGA).

2. This paper evaluates the error-rate of ID-based side channel authentication in a laboratory environment. The authentication accuracy is quantitatively estimated as the false acceptance rate and the false rejection rate. First, a quantitative discussion of the side-channel information is performed according to the number of distinguishable provers. The side-channel information of the provers is experimentally obtained from AES implementations on FPGA. The histograms for rejection and acceptance trials are both approximated to a normal distribution. Based on the principle that the false rejection rate and false acceptance rate are set to be equal, the parameters in the authentication can be determined. As a result, the authentication accuracy can be determined. This part of the contribution has been partially discussed by us in [4].

3. In our evaluation, both a non-profiling leakage model and a profiling leakage model are considered for different scenarios. Similar to side-channel attacks, the leakage model describes the relations between the side-channel leakage and the processed data. Generally speaking, one can expect side-channel attacks to have a reduced data complexity with a more accurate leakage model. Specifically, we use a Hamming distance (HD) model as the non-profiling leakage model and the XOR (exclusive-or) model proposed in [5] as the representative of the profiling leakage model. It is expected that the profiling model will improve the authentication accuracy of the

system. The experiments show that the XOR model leads to a larger mean and smaller variance for the histogram of the correlation coefficients compared to that of the HD model. The authentication accuracy and the authentication time are compared between the HD model and the XOR model.

The rest of the paper is organized as follows. Section 2 reviews the previously proposed scheme for side-channel authentication. Section 3 presents the idea of an ID-based authentication system for side-channel authentication. Section 4 explains the setup for the evaluation of ID-based side-channel authentication. In Sections 5 and 6, the evaluation of the ID-based side-channel authentication system with regard to the authentication speed and the authentication accuracy is presented. Section 7 concludes this paper.

2. Preliminaries

In this section, the first side-channel authentication proposal in [1] is briefly reviewed.

2.1. n-Round AES

AES-128 has 10 rounds of operation, which usually takes 10 clock cycles to calculate in hardware implementation. Using AES-128 in side-channel authentication requires multiple traces to ensure authentication accuracy. Each execution requires a fresh plaintext. Furthermore, only the middle round of each trace is used in the authentication to prevent the security threat from conventional side-channel key recovery attacks.

An easy alternative option is to use a modified AES that has more than 10 round operations, which is called a n-round AES. Here, n should be larger than 10 and big enough, e.g., $n = 1000$, so that a single side-channel trace is enough to perform the authentication. A n-round AES could simplify the system and the modification of the AES hardware could be minimized as well. To prevent security threats from conventional side-channel key recovery attacks, several rounds, e.g., 4 rounds, near the public data are not used in the r-round AES authentication.

An illustration of 10-round AES and n-round AES is shown in Figure 1.

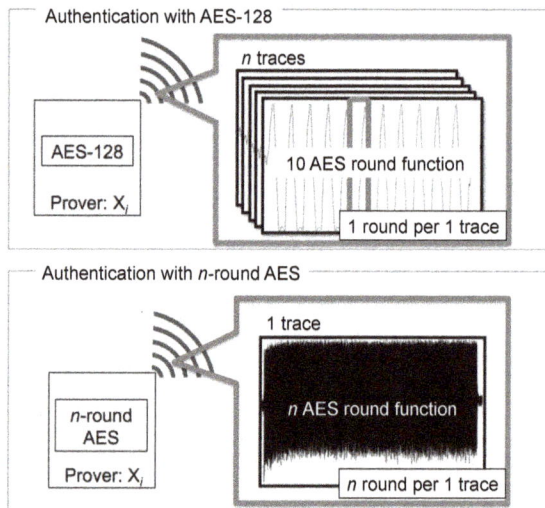

Figure 1. Two types of side-channel data with Advanced Encryption Standard (AES).

2.2. Protocol in Side-Channel Authentication

The possible protocols for side-channel authentication were discussed in [1]. In this paper, we only discuss the Challenge-S-Response authentication and the Challenge-S authentication, as shown in

Figure 2a,b. Here, challenge c and response r are the same with conventional authentication schemes, and S denotes the side-channel information.

For both schemes, prover X_i registers its secret key sk_i in the verifier before the authentication. The authentication starts when the verifier sends a challenge c to a prover X_i. Prover X_i calculates $f(c, sk_i)$ using its secret key sk_i, where $f()$ is a cryptographic calculation. The verifier measures the side-channel information S during the encryption process.

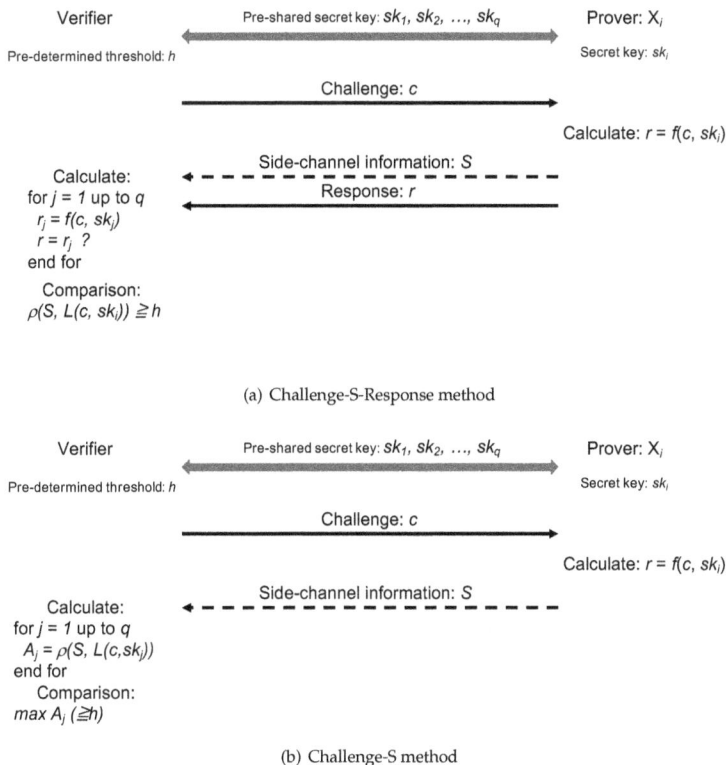

(a) Challenge-S-Response method

(b) Challenge-S method

Figure 2. Two types of authentication methods proposed in [1]. (a) Challenge-S-Response method; (b) Challenge-S method.

The major difference between these two schemes is whether the response r is sent back from the prover to the verifier to be used in the verification.

For Challenge-S-Response authentication, the response r is sent to the verifier. The identification of the prover is performed using both side-channel information S and the conventional challenge-response verification. First, the verifier searches the secret key $sk_i \in \{sk_1, sk_2, \ldots, sk_q\}$ to find the sk_i such that $f(c, sk_i) = r$. Then, the found sk_i is used with c and a leakage model to estimate the side-channel information as $L(c, sk_i)$. After that, Pearson's correlation coefficient between the measurement of real side-channel information S and the estimation $L(c, sk_i)$ is calculated and compared with a pre-determined threshold h. The authentication is passed only when both the response and the side-channel information match the expectation. This scheme is similar to conventional challenge-response authentication.

For Challenge-S authentication, only the side-channel information S is used in the authentication. The response of the encryption process is not returned to the verifier. The verifier is required to calculate the expected leakage for all registered keys as $\{sk_1, sk_2, \ldots, sk_q\}$. Then, for each registered key, a correlation calculation is conducted. The key with the largest correlation among all possible keys is compared with a pre-determined threshold h. Only when the maximal correlation is larger than the threshold, is the prover considered to be a legitimate prover. The comparison with the threshold is done to prevent a situation where the invalid keys can pass the authentication.

For Challenge-S authentication, the response r is not transmitted in the communication channel. The benefits of omitting r transmission are two-fold. First, the response r is not available for the attacker for any key recovery attack. Second, the communication for the authentication in the main channel can be reduced. Furthermore, the communication can be omitted entirely if the challenge c is predefined between the prover device and the verifier.

For the Challenge-S authentication scheme in conventional side-channel authentication [1], the correlation coefficients are calculated for all the pre-registered keys. Thus, it is expected to be time consuming when the number of registered provers is large. In this paper, we want to accelerate the Challenge-S authentication scheme of the side-channel authentication.

3. ID-Based Side-Channel Authentication System

To accelerate the authentication, we adapt a well-known ID system to the Challenge-S authentication. The idea is to reduce the amount of computation by sending an ID to the verifier before challenge-response authentication. The ID helps the verifier to quickly identify the corresponding registered key.

3.1. ID-Based Side-Channel Authentication Scheme

As shown in Figure 3, the pairs of ID and secret key of q provers, (ID_0, sk_0), (ID_1, sk_1), $\ldots$, (ID_q, sk_q) are registered in the verifier. The verifier initiates the authentication by sending an ID query to the prover, and then the verifier receives the prover's ID as ID_j. The verifier searches for the corresponding secret key and creates the corresponding leakage profile for the n-round AES. Then, Pearson's correlation coefficient ρ is calculated between the measured estimated side-channel information. Finally, the verifier confirms whether the correlation coefficient is larger than the pre-determined threshold h to decide the authentication result. If the derived ρ is larger than the pre-determined threshold h, the authentication is successful.

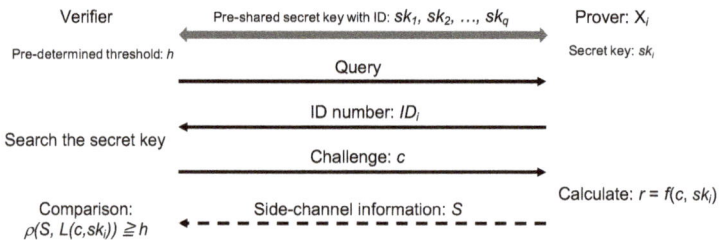

Figure 3. Proposed Authentication Method Using ID Query.

Figure 4 shows the frameworks of the conventional and the ID-based side-channel authentication systems. The major difference between the two authentication schemes is the number of correlation coefficient calculations. In previous work [1], the leakage profile and the correlation were calculated for all registered secret keys. Meanwhile, the calculation of the model and correlation coefficient is performed only once in the proposed authentication scheme as the secret key is identified using the ID sent from prover. Therefore, it is possible to authenticate much faster compared to the conventional scheme.

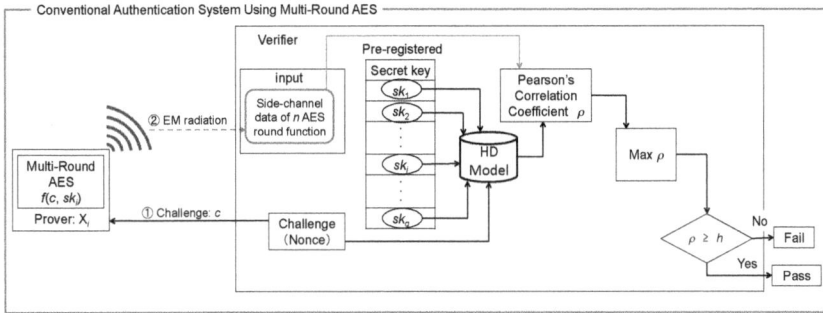

(a) Conventional side-channel authentication

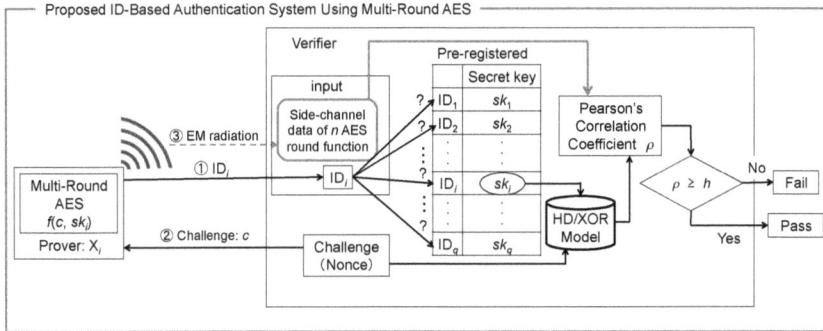

(b) Proposed ID-based side-channel authentication

Figure 4. Comparison of frameworks of the conventional (a) and proposed (b) ID-based side-channel authentication.

3.1.1. Comparison of Expected Authentication Speed

Table 1 shows the comparison of the expected authentication speed between a straightforward Challenge-S approach using normal AES and the ID-based Challenge-S using n-round AES. By denoting the clock period as T_{clk}, the acquisition time of the side-channel information for each method is $n \cdot 11T_{clk}$ and nT_{clk}, respectively. Denote the data processing time of each AES round to obtain the intermediate values as T_p; then, the total data processing time can be calculated. For the straightforward Challenge-S approach, the data processing time can be represented by $q \cdot 5n \cdot T_p$ since only 5 rounds of intermediate values are calculated. For the ID-based n-round Challenge-S approach, the data processing time is $n \cdot T_p$. The total time for authentication consists of the acquisition time and the data processing time. It can be seen that ID-based n-round approach is much more efficient.

Table 1. Comparison between two side-channel authentication schemes.

	Straightforward Challenge-S [1]	ID-Based n-Round Challenge-S
Side-channel information	n traces of 10-round AES-128	1 trace of n-round AES
Acquisition time	$n \cdot 11T_{clk}$	nT_{clk}
ID system	Not used	Used
# of trials	q (1 acceptance and $q - 1$ false trials)	1 (Only acceptance trial)
Data processing time	$q \cdot 5n \cdot T_p$	$n \cdot T_p$
Total time	$n \cdot 11T_{clk} + q \cdot 5n \cdot T_p$	$nT_{clk} + n \cdot T_p$

3.1.2. Resistance against Side-Channel Attacks

One big concern for side-channel authentication is that the shared secret key can be extracted by the attackers using the side-channel leakage. To mitigate the risk of such attacks, the system can apply the following changes. First, the side-channel information near the public data should be protected by side-channel countermeasures such as masking. Second, only the side-channel information that is far from the public data is used in the authentication. For normal AES, we only use the middle round in the authentication. Similarly, for n-round AES, several rounds near the public data are not used in the authentication.

3.1.3. Trade-Off for the ID-Based System

As for the trade-off, in conventional side-channel authentication, the verifier is pre-registered only with the secret keys. In ID-based authentication, the secret key and ID number pairs are pre-shared between the verifier and the prover. In this system, we consider a case where the ID does not contain any secret information related to the secret key. The ID works as a tag to help the verifier quickly identify the claimed secret key of the verifier. A privacy problem could also exist for the ID-based authentication system since the ID information is transmitted in air. It is possible to trace the holder of a device by tracing the ID of the device. A possible mitigation of this problem is to introduce a periodical update of the ID.

Regarding the secret of the ID and secret key, it is assumed that the registration of the ID/secret key is performed in a secure environment. As for other possible leakages of the ID and secret key, if only the ID is intercepted by a non-legitimate source, the attacker can pretend to be a certain device by using the leaked ID. However, since the secret key is unknown to the attacker, the fake device cannot pass the authentication. In the case that both the ID and secret key are intercepted and used by an attacker, the attackers can pass the authentication without any problem. As long as the verifier realizes this situation, a possible mitigation is to register the legitimate users again with new keys.

4. Evaluation Setup of ID-Based Authentication

In this work, we performed experimental evaluations of ID-based authentication using a hardware AES implemented on FPGA. This section mainly focuses on the experiment setup and leakage models used in the evaluation.

4.1. Experimental Setup on n-Round AES

In the experiment, we used ALTERA CycloneIV (FPGA) on Terasic DE0-nano (FPGA board) [6] as the prover device. A 1000-round AES modified from the 128-bit AES [7] was used as the calculation to generate side-channel information. The AES implementation uses a 128-bit data path and the composite-field S-box, which runs at 50 MHz on the DE0-nano board. The side-channel information was measured as the electro-magnetic radiation near the FPGA by the EM probe (Langer-EMV RF-U 5-2). The signal captured by the probe was recorded using an oscilloscope (Agilent Technology DSO7032A), which recorded at 1 GSa/s. Each measurement included about 21,000 samples. A photo of the experimental setup is shown in Figure 5. Note that, the measurement of side-channel information can be performed without modifying the hardware, but the probe still needs to be close to the FPGA to ensure the quality of measurements is sufficient [8,9].

On the verifier side, we used a normal PC to process the data. The correlation calculation was performed with both non-profile leakage models and leakage profiles.

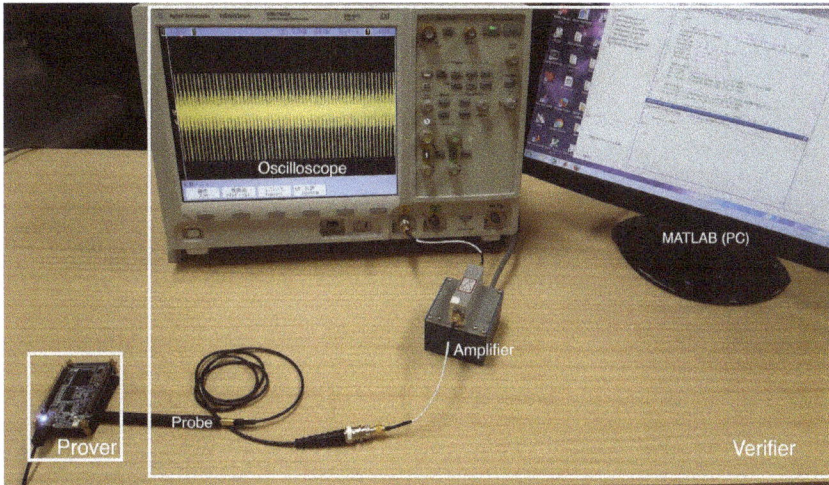

Figure 5. Experimental Environment for Side-Channel Authentication.

4.2. Leakage Model in Authentication

In this work, for both the profiling model and the general leakage model, the side-channel authentication performance was evaluated. For side-channel attacks, the leakage model describes the relations between the side-channel leakage and the processed data. An accurate leakage model could lead to better attack efficiency by side-channel attacks. Similarly, an accurate leakage model could lead to better side-channel authentication authentication efficiency. Generally speaking, a general leakage model has wide applicability but less accuracy. In contrast, a device-specific leakage model or leakage profile has better accuracy but less generality. It is well-known that side-channel attacks can be categorized into two types: non-profiling attacks and profiling attacks. In profiling attacks, the attackers have an identical device that is used to learn the leakage profile of the device so that the data complexity of the key recovery attack is reduced compared with the non-profiling attack that uses a general leakage model. Other side-channel attack techniques can be applied to side-channel authentication as well. The usage of Pearson's correlation coefficient as the distinguisher is one such example.

4.2.1. Non-Profiled Model: HD Model

As for the non-profile model, we used the well-known Hamming distance model. Since the key is known to the verifier, the Hamming distance of the 128-bit intermediate value rather than a single byte was used. For the Hamming distance model proposed in [2], side-channel information, denoted by W, is modeled as

$$W = kH(D \oplus E) + b$$

where $H(D \oplus E)$ is the Hamming distance between D and E, which are intermediate values for an AES round, and k and b are constants. For the HD model, the intermediate values are the ones stored in registers, i.e., D is stored in a register and is replaced with E after a round operation. The HD model assumes that there is a linear dependency between the side-channel leakage W and the Hamming distance value $H(D \oplus E)$.

In [1], it was shown that the 128-bit intermediate values can be used in the HD model because AES-comp implementation [7] performs each AES round in 1 cycle. In this work, we also considered the authentication using n-round AES, which is modified from the AES-comp implementation. The i-th round leakage model $\mathbf{W^i}$ and measured side-channel information $\mathbf{S^i}$ are denoted as $(W_1^i, W_2^i, \ldots, W_N^i)$, and $(S_1^i, S_2^i, \ldots, S_N^i)$, respectively. Here, N is the number of total plaintexts. The correlation coefficients are derived by $\rho(\mathbf{W}^i, \mathbf{S}^i)$ and classified into acceptance trials and rejection trials. In the acceptance trial, it is assumed that the prover who registered the pre-shared secret key in the verifier is authenticated, i.e., legitimate prover authentication. On the other hand, if it is authenticated using the unregistered secret key, it is considered to be the rejection trial.

4.2.2. Profiling Model: XOR Model

As a profiling model, we used the XOR model that was proposed in [5]. In [5], the advantage of the XOR model in correctly profiling the leakage of AES-comp implementation was shown. The XOR model leads to successful key recovery with reduced power traces compared to the HD model. In the HD model, it is assumed that the amount of bit-flipped information, i.e., the Hamming distance, is proportional to the physical information, e.g., the power consumption and the amount of EM radiation. Since the Hamming distance does not distinguish between bits, the HD model for the 8-bit intermediate value classifies the leakage into nine classes from 0 to 8. Meanwhile, in the XOR model, it is assumed that the bit reversed position affects the amount of side-channel leakage. Specifically, the XOR model classifies the side-channel leakage for 8-bit intermediate values into 256 classes ranging from 0 to 255.

The side-channel authentication is classified into a profiling phase and an authentication phase. In the profiling phase, the properties of each device are investigated in pre-processing. Specifically, the amount of EM radiation for an XORed value that changes with each product, called a model value $\mathbf{A}$, is derived using the side-channel information whose intermediate value is known. When authenticating using the XOR model for 16-byte AES, the XOR model is classified into 256×16 classes. Therefore, the model value $\mathbf{A}$ is expressed as

$$\mathbf{A} = \left(\begin{array}{ccccccc} a_{1,0} & a_{1,1} & a_{1,2} & \cdots & a_{t,r} & \cdots & a_{16,254} & a_{16,255} \end{array} \right)$$

where $a_{t,r}$ is the amount of EM radiation when the XORed value of the t-th byte is r.

In the authentication phase, the correlation coefficient is calculated using the model value $\mathbf{A}$ derived in the profiling phase. Based on the intermediate value derived from the challenge and a secret key, the amount of EM radiation is estimated using the model value $\mathbf{A}$. The process is exactly the same as that using HD model, except that the Hamming distance model is replaced with the profiling model $\mathbf{A}$. After that, the correlation coefficient is calculated between the acquired EM radiation and the estimated EM radiation. The leakage profiles for 256 classes for 16 s-boxes are obtained by solving the system of equations with the profiling measurement. Note that the authentication scheme using the XOR model was first discussed by us in [10].

5. Evaluation of the Authentication Speed

In this section, we describe the evaluation of the authentication speed using our experiment setup. Table 2 represents the difference in authentication time between the straightforward Challenge-S approach [1] and the ID-based n-round Challenge-S approach. As for the acquisition time, our setup takes 43 s to measure 1000 EM traces and 0.5 s to measure a 1000-round EM trace. This shows that the n-round approach could largely reduce the data acquisition time.

As for the data processing time, both the n-round AES and the ID system have advantages. In the data processing of the straightforward Challenge-S approach, 5 AES rounds have to be calculated for each EM trace to estimate the side-channel information. Since there are, in total, 1000 traces, 5000 AES rounds must be calculated using 1000 different plaintexts. In contrast, the 1000-round AES calculates 1000 intermediate values in total. Moreover, the 5000 AES rounds of calculation need to be performed for each register key without the ID system. Using the ID system, only the claimed register key is compared with the observed side-channel information. As shown in Table 2, the data processing time is $0.34 \cdot q$ for the straightforward Challenge-S approach, while the ID-based n-round Challenge-S approach requires less than 0.1 s. For both the HD model and the XOR model, the leakage profile is prepared before the processing the measurement. Therefore, both models will be able to authenticate in a short time period.

Table 2. The difference in authentication time in seconds.

Used Model	1000 Traces of AES-128 [1] HD model [1]	1000-Round AES (1000 Round Function Calls) HD Model	XOR Model [2]
Acquisition	43	0.50	
Data Processing	$0.34 \cdot q$	0.083	0.086
Total	$43 + 0.34 \cdot q$	0.583	0.586

[1] Hamming distance model, [2] XOR (Exclusive-or) model.

It is reasonable to expect the acceleration of authentication when the ID system is applied to side-channel authentication. With the performed experiments and the time measurements, the acceleration can be understood more clearly since both the decomposition of the consumed time and the contributions of each techniques are clear.

Note that, the time required for the pre-authentication processes is similar for both authentication schemes. The pre-authentication processes consist of the key registration part and the leakage profiling part. As for the key registration part, the ID-based side-channel authentication scheme is the same as the conventional scheme except that a device ID is additionally registered together with the secret key. For the leakage profiling part, only the scheme using the XOR model requires the leakage profiling, which has negligible time consumption, since the profile only needs to be performed once for each type of prover device.

6. Evaluation of Authentication Accuracy

In this section, the parameters and the performance of the side-channel authentication are discussed. First, we define several parameters that are related to the perforation evaluation. Then, we discuss how to obtain reasonable choices for these parameters. Then, based on our laboratory setup, we calculate the optimal parameters for both the non-profiling model and profiling model. Finally, we apply these parameters and evaluate the error rate for several variations.

6.1. Accuracy-Related Parameters

As for the authentication accuracy, we refer to the error rate as the false acceptation rate and false rejection rate. After the setup is fixed, we consider that there are two related system parameters: the number of the total provers M and the threshold h. We consider parameter M to be the maximum number of authentication trials that enables an authentication system to operate without producing false errors. M is the number of devices that can be used in the system. The error rate is likely to be increased along the increase in M.

6.2. Relationship Among M, n, and False Errors

Following the approach in [4], the relationship between M, n, and the false acceptance and false rejection errors $S_1(h)$ and $S_2(h)$ can be visualized when changing h, as shown in Figure 6. The threshold h was set in the range from -1 to 1.

There are two major differences between the work of [4] and the proposed scheme. One is that the parameter M is regarded as the number of authentication trials in this paper, although it was previously taken to mean the number of provers in [4] by assuming that each prover was only accessed once. That is, M corresponded to the number of distinguishable provers. In contrast, in this study, it is assumed that M fake provers access the authentication system as well as M legitimate provers. In total, $2M$ trials are assumed when estimating the false errors, whereas M^2 trials were used in [4]. This assumption affects the variance parameter and the mean values of approximated normal distributions because the number of samples is different. The other difference relates to the parameter n, which is defined as the number of rounds of 128-bit AES in this paper, whereas it was defined as the number of traces of 128-bit AES encryption in [4].

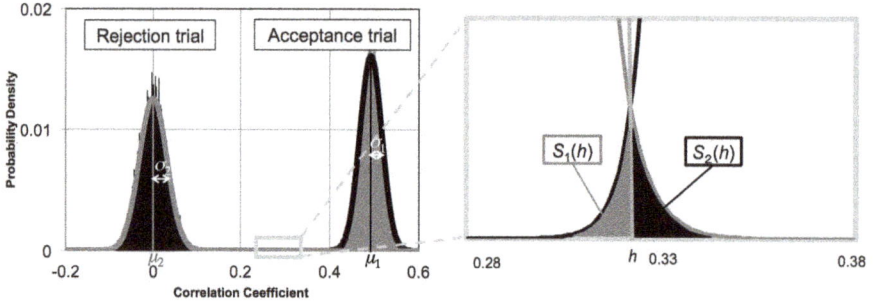

Figure 6. Conceptual diagram of the normal distribution derived from the correlation coefficients.

To derive the histogram of acceptance and rejection trials, Fisher z-transformation was applied to achieve approximation. After that, we verified the validity with the Jarque–Bera test. Since the histogram can approximate a normal distribution, it was found that the correlation coefficients dependent on secret keys and plaintext were not derived. Then, the histograms of the acceptance and rejection trials were approximated to follow normal distributions (see Figure 6), respectively, as $\mathcal{N}(\mu_1, \sigma_1^2)$ and $\mathcal{N}(\mu_2, \sigma_2^2)$ where the variances σ_1^2 and σ_2^2 are described with n as

$$\sigma_1^2 = \frac{\beta_1}{n}, \quad \sigma_2^2 = \frac{\beta_2}{n} \tag{1}$$

where β_1 and β_2 are constants that are experimentally determined with the correlation coefficients between the observed n-round side-channel information and the leakage model. Therefore, with the threshold, defined as h, the probability of a false rejection ratio $S_1(h)$ and false acceptance ratio $S_2(h)$ are represented by

$$S_1(h) = \frac{1}{2}\mathrm{erfc}\left(\frac{\mu_1 - h}{\sqrt{2\sigma_1^2}}\right), \tag{2}$$

$$S_2(h) = \frac{1}{2}\mathrm{erfc}\left(\frac{h - \mu_2}{\sqrt{2\sigma_2^2}}\right). \tag{3}$$

As the total number of false errors should be equal to or less than one for $2M$ trials, we have

$$MS_1(h) + MS_2(h) \le 1. \tag{4}$$

Therefore, the total number of trials M can be derived from

$$M \le \frac{1}{S_1(h) + S_2(h)}. \tag{5}$$

By increasing the number of rounds, the system is capable of distinguishing many provers.

6.3. Formulation of n under Equal Error Rate

In our method, h is determined such that the probabilities of false acceptance and false rejection rates occurring are equal, i.e., the error rate was required to be equal. Thus, the equal error rate adopted in our authentication system assumes that false acceptance and false rejection occur with the same probability. Therefore, in the case of $S_1(h) = S_2(h)$, the threshold h is expressed as

$$h = \frac{\sqrt{\beta_1}\mu_2 + \sqrt{\beta_2}\mu_1}{\sqrt{\beta_1} + \sqrt{\beta_2}}. \tag{6}$$

Hence, the maximum number of total trials is expressed as

$$\begin{aligned} M &= \left(\mathrm{erfc} \; \frac{\mu_1 - \frac{\sqrt{\beta_1}\mu_2 + \sqrt{\beta_2}\mu_1}{\sqrt{\beta_1} + \sqrt{\beta_2}}}{\sqrt{2\frac{\beta_1}{n}}} \right)^{-1} \\ &= \frac{1}{\mathrm{erfc} \; \alpha\sqrt{n}} \end{aligned} \tag{7}$$

where the constant α is

$$\alpha = \frac{\mu_1 - \mu_2}{\sqrt{2\beta_1} + \sqrt{2\beta_2}}. \tag{8}$$

Accordingly, the number of AES round function calls is represented using M as

$$n = \left(\frac{\mathrm{erfc}^{-1} \frac{1}{M}}{\alpha} \right)^2. \tag{9}$$

6.4. Parameters with Different Settings

We derived the relationship between n and M using two datasets corresponding to two authentication schemes discussed throughout this work.

- Dataset A: n EM traces of AES-128
- Dataset B: one trace for n-rounds of AES

Table 3 summarizes the parameters that were experimentally obtained, which are necessary for approximating a normal distribution.

Table 3. Experimentally obtained parameters: mean values μ_1 and μ_2 and constants of proportionality β_1 and β_2; α, and h.

Dataset	Leakage Model	μ_1	μ_2	β_1	β_2	α	h
A	HD Model	0.57	0.00	0.98	1.00	0.20	0.29
B	HD Model	0.541	0.00	1.05	1.00	0.19	0.27
B	XOR Model	0.718	0.00	1.06	1.06	0.25	0.36

6.5. Experimental Results

Figure 7a shows the relationship between the number of EM traces and the number of distinguishable provers when the previous authentication scheme was used with Dataset A. In the case of $n = 1171$, i.e., using 1171 EM traces, $M = 2^{50}$, which indicates that 2^{50} provers were distinguishable from the previous authentication scheme. Contrary to the above result, the results obtained with the proposed scheme using Dataset B show that when $n = 908$, i.e., 908-round AES, 2^{50} was obtained (see Figure 7b) which means that false errors do not occur even if 10 million provers are authenticated twice a day for 100 years. Furthermore, it should be noted that the authentication time was 0.58 s. In addition, Figure 7c shows that $M = 2^{70}$ was obtained with 759-rounds of AES. The summaries of these figures are listed in Table 4.

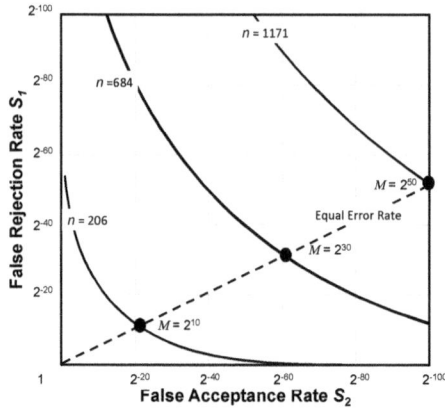

(a) Standard AES-128

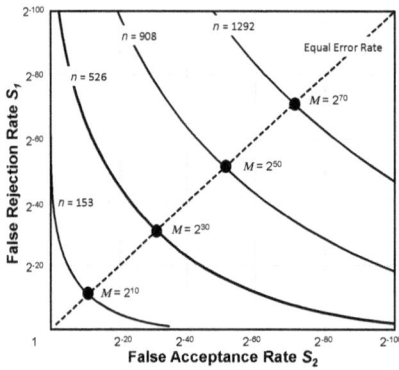

(b) *n*-rounds of AES with the HD model

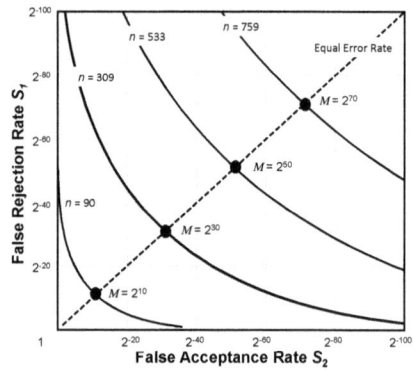

(c) *n*-rounds of AES with the XOR model

Figure 7. Relationship between the number of rounds and the number of distinguishable provers for Datasets A and B.

Table 4. Summary of Figure 7: the relationship between M and n.

M	2^{10}	2^{30}	2^{50}	2^{70}
AES-128	206	684	1171	-
HD Model	153	526	908	1292
XOR Model	90	309	533	759

7. Conclusions

In this work, an ID-based authentication scheme was adopted for side-channel authentication. In addition, the performance of the side-channel authentication was evaluated in terms of the authentication speed and authentication accuracy. In the performance evaluation, this work overviewed several technical choices for side-channel authentication to compare them. Based on both theoretical analysis and FPGA-based experiments, it is clear the ID-based scheme can accelerate the authentication speed, and the profiling model leads to better data complexity. The results showed that our experimental setup is a possible way to check the authenticity of a prover among 2^{70} different provers within 0.59 s using 759 AES round function calls, which demonstrates the feasibility for side-channel authentication to be used as a future practice. In order to apply the side-channel authentication in a specific scenario, further optimization and field tests are considered as future works.

Author Contributions: Conceptualization, M.K. and K.S.; Investigation, M.K.; Supervision, K.S.; Writing–original draft, M.K.; Writing–review & editing, Y.L.

Funding: This research was funded by Japan Society for the Promotion of Science (JSPS) Grants-in-Aid for Scientific Research (KAKENHI) grant number 15K12035.

Acknowledgments: We are grateful to the editor and the anonymous reviewers for various constructive comments.

Conflicts of Interest: The authors declare no conflict of interest.

References

1. Sakiyama, K.; Kasuya, M.; Machida, T.; Matsubara, A.; Kuai, Y.; Hayashi, Y.I.; Mizuki, T.; Miura, N.; Nagata, M. Physical Authentication Using Side-Channel Information. In Proceedings of the 2016 4th International Conference on Information and Communication Technology (ICoICT), Bandung, Indonesia, 25–27 May 2016.

2. Brier, E.; Clavier, C.; Olivier, F. Correlation Power Analysis with a Leakage Model. In Proceedings of the Cryptographic Hardware and Embedded Systems—CHES 2004, Cambridge, MA, USA, 11–13 August 2004; pp. 16–29.

3. Kocher, P.; Jaffe, J.; Jun, B. Differential Power Analysis. In Proceedings of the Advances in Cryptology—CRYPTO' 99, 19th Annual International Cryptology Conference, Santa Barbara, CA, USA, 15–19 August 1999; pp. 388–397.

4. Kasuya, M.; Machida, T.; Sakiyama, K. New Metric for Side-Channel Information Leakage: Case Study on EM Radiation from AES Hardware. In Proceedings of the 2016 URSI Asia-Pacific Radio Science Conference (URSI AP-RASC), Seoul, South Korea, 21–25 August 2016.

5. Clavier, C.; Danger, J.L.; Duc, G.; Elaabid, M.A.; Gérard, B.; Guilley, S.; Heuser, A.; Kasper, M.; Li, Y.; Lomné, V.; et al. Practical improvements of side-channel attacks on AES: Feedback from the 2nd DPA contest. *J. Cryptogr. Eng.* **2014**, *4*, 259–274. [CrossRef]

6. Terasic Inc. DE0-Nano Development and Education Board. Available online: http://www.terasic.com.tw/en (accessed on 29 March 2018).

7. Tohoku University. Cryptographic Hardware Project. Available online: http://www.aoki.ecei.tohoku.ac.jp/crypto/ (accessed on 29 March 2018).

8. Gandolfi, K.; Mourtel, C.; Olivier, F. Electromagnetic Analysis: Concrete Results. In Proceedings of the Cryptographic Hardware and Embedded Systems—CHES 2001, Paris, France, 14–16 May 2001; pp. 251–261.

9. De Mulder, E.; Örs, S.B.; Preneel, B.; Verbauwhede, I. Differential Power and Electromagnetic Attacks on a FPGA Implementation of Elliptic Curve Cryptosystems. *Comput. Electr. Eng.* **2007**, *33*, 367–382. [CrossRef]

10. Kasuya, M.; Sakiyama, K. Improved EM Side-Channel Authentication Using Profile-Based XOR Model. In Proceedings of the International Workshop on Information Security Applications (WISA 2017), Jeju Island, Korea, 24–26 August 2017.

![applied sciences logo] *applied sciences*

MDPI

Article

Using Ad-Related Network Behavior to Distinguish Ad Libraries

Ming-Yang Su *, Hong-Siou Wei, Xin-Yu Chen, Po-Wei Lin and Ding-You Qiu

Department of Computer Science and Information Engineering, Ming Chuan University, Taoyuan 333, Taiwan; kanzaki0421aria@gmail.com (H.-S.W.); best01010193@gmail.com (X.-Y.C.); vongolax343@gmail.com (P.-W.L.); eggegg234567@gmail.com (D.-Y.Q.)
* Correspondence: minysu@mail.mcu.edu.tw

Received: 24 August 2018; Accepted: 1 October 2018; Published: 9 October 2018

Abstract: Mobile app ads pose a far greater security threat to users than adverts on computer browsers. This is because app developers must embed a Software Development Kit (SDK), called an ad library or ad lib for short, provided by ad networks (i.e., ad companies) into their app program, and then merge and compile it into an Android PacKage (APK) execution file. The ad lib thus becomes a part of the entire app, and shares the whole permissions granted to the app. Unfortunately, this also resulted in many security issues, such as ad libs abusing the permissions to collect and leak private data, ad servers redirecting ad requests to download malicious JavaScript from unknown servers to execute it in the background of the mobile operating system without the user's consent. The more well-known an embedded ad lib, the safer the app may be, and vice versa. Importantly, while decompiling an APK to inspect its source code may not identify the ad lib(s), executing the app on a simulator can reveal the network behavior of the embedded ad lib(s). Ad libs exhibit different behavior patterns when communicating with ad servers. This study uses a dynamic analysis method to inspect an executing app, and plots the ad lib behavior patterns related to the advertisement into a graph. It is then determined whether or not the ad lib is from a trusted ad network using comparisons of graph similarities.

Keywords: mobile ads; software development kit (SDK), android package (APK), ad lib; ad libraries; ad networks; graph; graph similarity

1. Introduction

According to a report released by the Interactive Advertising Bureau (IAB) of the United States on 26 April 2017 [1], the trend of digital advertising has transferred from personal computers to mobile devices. In 2016, the annual revenue of digital advertising in the United States was USD 72.5 billion, of which revenue from mobile ads exceeded 50% for the first time, reaching USD 36.6 billion. The Mobile Application Industry Report for 2015 [2] revealed more about the popularity and importance of mobile advertising: 82% of app developers made a profit by advertising, and 91% were still using banner ads. Obviously, consumers were not willing to pay to download apps, so app developers turned to free apps and used ads to make profit.

The report on malicious mobile software evolution released by Kaspersky in February 2017 [3] listed 8,526,221 detected malicious apps in 2016, which was three times as many as that in 2015. Increasingly more information security reports related to mobile ads have since been conducted. A report by Trend Micro in June 2017 [4] showed that a Trojan Android ad program called Xavier could steal users' personal information and transmit it to somewhere without user permission whenever users downloaded the embedded app. According to the Trend Micro data, over 800 Google Play Android apps contained the Trojan ad lib, which had been downloaded millions of times. These apps included utility apps such as photo editing apps, desktop and ringtone change apps. Xavier had a self-protection mechanism to avoid detection, and also downloaded and executed other malicious codes.

Doctor Web, an anti-malware company, indicated that in June 2016 a Trojan called Android.Spy.305, had been embedded in 155 Google Play apps, and estimated that more than 2.8 million people had downloaded and installed them [5]. The new Trojan, Android.Spy.305.origin, originally put into an ad lib, was embedded in apps when some developers used this ad lib to generate advertising revenue. It was known that 155 kinds of apps made by 8 app development companies had been infected. Once mobile device users had installed the embedded Android.Spy.305.origin module with the ad lib, it then connected to a Command and Control server to download the additional Android.Spy.306.origin module. The additional module would then begin to steal personal data, including, Google account E-mail logins and passwords, installed app lists, system languages, mobile brands, device names, IMEI numbers, OS versions, screen resolution, telecom operators, etc. In addition, third party apps would be installed during the app installation, which would then display various malicious advertisements from time to time. Many researchers have noticed the security issues caused by ad libs and a lot of efforts have been made in recent years to address this. Some of them are introduced below.

Athanasopoulos et al. [6] estimated that more than half of the apps available on Google Play contained ad libs linked to third party advertisers, posing a significant security risk to mobile app users. They therefore proposed the Native Code Isolation for Android Applications (NaClDroid) architecture to separate the program code of an ad lib from that of an app, thus preventing permission sharing. Kumar et al. [7] noted that many ad libs required too many privileges or used privileges for which they did not have authority. Some observed apps could also sniff network traffic to obtain package content across the ad requests of multiple ad networks, making a user's personal information more easily accessible. They also discussed how a few notorious ad libs used online third parties to stealthily transfer personal information to an unknown server. Gao et al. [8] noted that because ad libs and apps were compiled after their merging, it was impossible to prevent the ad lib from using unauthorized permissions that exceeded the ad lib instructions. They therefore designed the Permission Supervision for Android Advertising (PmDroid) system to block ad libs' unauthorized use of permissions to transfer information. PmDroid employed a graphical interface to present the seriousness of any unauthorized usage. To understand the actual actions of these SDKs, Gao et al. wrote 53 different apps, each with a different ad lib embedded. The apps did not do anything, but announced all the privileges of the Android system to which they had access. The packet traffic of the apps was then recorded in order to understand how the ad libs abused permissions. Because the apps themselves did nothing, all network traffic was the result of ad lib activity. The authors concluded that unauthorized use of permissions by ad libs was very serious.

Narayanan et al. [9] observed that it was difficult to judge ad lib behavior using only the ad lib program code due to the widespread use of modern obfuscation tools. They used 26 different ads in their experimental dataset in order to test such obfuscation tools. They then proposed the AdDetect framework to assist in detecting ad libs and their behaviors in apps. AdDetect used semantic analysis to check ad libs, and used a support vector machine (SVM) to make classification judgments. Liu et al. [10] proposed their system, called PEDAL, to de-escalate privileges for ad libs in mobile apps. The study reported that, even if ad libs used obfuscation tools, PEDAL had a 98% accuracy in detecting them. Yan et al. [11] designed a new Android model, RTDroid, which basically modified the internal components of the Android operating system, and made use of a real-time Virtual Machine (VM) instead of the original Android Dalvik VM. This ensured that the execution of any app and its ad lib had greater predictability.

Book and Wallach [12] noted that, while ad libs could use the privileges of the host app to secretly transmit data, the host app could also use the privileges of the ad lib to engage in extra, unauthorized actions. That is, app developers and ad networks were colluding to carry out aggressive activities. The authors collected 114,000 apps, and collected statistics for the 20 most frequently used advertisers, identifying a total of 64,000 apps using those 20 ad libs. By observing the behavior of the 64,000 apps, they concluded that app developers often actively collected too much personal user information to supply to ad networks in pursuit of high advertising profits. In addition, they found that the greater the popularity of an app, the easier it was to engage in such behavior, since as the number of users of an app increased, so did the motivation for advertisers to engage in such profit-seeking actions. Ruiz et al. [13] discussed the problems caused by ad lib updates. According to their experiment data, over 90% of apps were free, and advertising was the only income for these app developers, so it was very important to ensure that ad libs embedded in apps could bring the expected profit. If ad libs didn't achieve the expected profit, they were replaced or updated. The authors collected 13,983 versions of 5937 apps, and found that nearly 50% of these apps had changed their ad libs within 12 months by increase, removal or update. Ad lib maintenance was thus a burden on app developers. Su et al. [14] developed a data exploration method for HTTP dataflow. The features adopted were quantitative, timing and semantic. The authors claimed that their traffic identification of malicious ad libs could achieve an accuracy of 95% in their experiments. Kuzuno and Magata [15] used the difference of HTTP online traffic to identify ad libs. They adopted 1000 known advertising pictures to identify others. The experiment results exhibited a 76% detection rate for known advertisement maps and 96% for manual sorting advertisement maps.

Kajiwara et al. [16] observed that ad libs periodically used ad request packets to transfer personal information to ad servers, and received ad reply packets from ad servers. These reply packets were mainly advertisement pictures which appeared on the apps, changing the window screen. It was thus possible to estimate whether an app had an embedded ad lib by mathematically processing the HTTP frequencies online and screen changes. Crussell et al. [17] focused on the issue of MAdFraud, wherein app developers used background processing to connect to ad servers and ask for advertisements for profit, without users' knowledge, or have the program automatically click ads, thereby deceiving the ad networks. The PrivacyGuard system proposed by Song and Hengartner [18] had a number of functions which could not only track the flow of sensitive information, but also handled sensitive information to protect against illegal access. Backes et al. [19] noted the trend of ads being embedded in free apps, but those released apps often using an old version of an ad lib, thus hiding security weaknesses. The authors designed a system to help users check whether the ad libs contained in downloaded apps had security concerns, including whether there were malicious behavior instructions for obfuscation. Lee et al. [20] proposed the use of Contextual and Semantic perspectives to distinguish between app behavior and ad behavior. Tang et al. [21] carried out a static analysis of 10,710 apps, and found that 76.08% of them had obvious unauthorized use problems, and of those, 424 apps' sensitive permissions were only used by ad libs, instead of the host apps. This study also deals with the abuse of permissions by ad libs in a semantic way. Liu et al. [22] discussed the possibility that analytics libraries were more likely to leak users' personal information than ad libs. Analytics libraries are the mechanisms for tracking ad presentation and ad clicking on mobile phones.

Stevens et al. [23] evaluated 13 well-known ad networks and found that some ad libs had significant problems. For example, Mobclix used some permissions unrelated to displayed ads, such as Send SMS and Read Calendar. Through testing, this ad lib adopted 7 undeclared privileges, including four very aggressive permissions: Read Calendar, Write Calendar, Read Contacts and Write Contacts. In addition, 7 of 13 ad libs analyzed included JavaScript interface, which indicated that these modules could perform external JavaScript. The external malicious JavaScript could be embedded into four of seven ad libs (Mobclix, Greystripe, Mocean, InMobi) and executed; the behavior of which was as follows. Mobclix modified the user's calendar, contacts, message and image files, and opened or closed the camera. Greystripe obtained or set cookies, which could include account passwords or credit card numbers. Mocean sent newsletters and e-mails, made calls, added calendar items, and obtained users' locations and any network requests. InMobi sent short messages (SMS) and e-mails, made phone calls and modified users' calendars.

Today, what users most want to know in this context is what ad libs are embedded in their downloaded apps. If an ad lib is well known, it may be relatively safe, and vice versa. This study runs apps on an emulator, and analyzes their network behaviors related to advertising. Most ad libs exhibit different behavior patterns, which are plotted into graphs to determine whether an ad lib comes from a trusted advertising company, using similarities between the graphs. The remainder of this paper is arranged as follows. Section 2 describes the operation of ad libs and related knowledge. Section 3 describes the proposed method of graph drawing according to an ad lib's network behavior patterns. Section 4 presents experimental results, and Section 5 offers conclusions.

2. The Operation of Ad Libs and Their Security Issues

Since 2010, mobile networks have undergone rapid growth and development. The boom included the adoption of mobile devices to quickly and reliably send messages. This sudden ubiquity of mobile devices resulted in a new mobile advertising market worth thousands of millions of U.S. dollars each year. The revenue of Internet advertising took 23 years to catch up with that of TV advertising, but the income of mobile advertising took only 6 years to surpass that of computer ads in 2016 [24]. However, as shown in a Purdue University and Microsoft report [25], the cost of using these free apps, which depended on ads for income, was the power consumption of the mobile phone and leakage of users' personal information. The surprise finding was that up to 75% of mobile device electrical power was used for advertising services, or tracking and uploading the relevant information of the user. There were already a variety of proper solutions for the advertising problems caused by websites, which could be addressed by computers via browsers. However, the information security problems caused by the mobile app ads had not yet been completely solved.

The reason these mobile advertising security-related problems were so difficult to work out was that ad libs were embedded in host apps and compiled together into APK execution files. That is, the ad lib had become a part of the entire app and could use all the permissions granted to the app. For example, an ad lib could claim to only use permissions P1 and P2, while the host app claims permissions P3, P4 and P5. Once merged and complied, the ad lib would be able to use all permissions, i.e., P1, P2, P3, P4 and P5. When installing this app, the Android system only informs the user that it will use P1, P2, P3, P4 and P5, and once the app is installed, the system does not distinguish between the host app and ad lib permission use, and does not prevent the ad lib from using all privileges belonging to permissions P3, P4 and P5.

Ad libs are SDKs (Software Development Kits). In order for ad networks to target users with appropriate ads, the embedded ad libs collect user information such as age, gender, income, etc. However, a number of ad libs exceed their permissions, and collect the sensitive user data, web use habits or track a user's position, etc. The ad libs connect to an ad server autonomously and employed ad request messages to send out the collected personal information without the awareness of the user. Meanwhile, the ad reply messages from the ad server are received, which may include JavaScript. Thus, ad libs carry out unsanctioned, malicious actions such as collecting contact information, audio and image files, sending messages and emails, or stealing mobile users' cookies to obtain user accounts and passwords. The ad replies may also require the app to connect to a third-party server to download malware and automatically install and execute it in the background. This kind of app itself does not contain malicious codes, and all security risks occur in the advertising behavior during execution, so that general anti-virus software is unable to effectively detect them. Figure 1 shows the flowchart of app advertising processing.

The AppBrain website [26] listed the top five hundred ad networks around the world, of which the Google's AdMob was the most popular. According to the latest information released by this website in December 2017, 61.52% of all apps installed had the AdMob ad lib embedded. The second most popular network was Unity, with 18.73% of all apps installed having this ad lib embedded. Third was Chartboost, with 14.00% of installed apps using this ad lib. The less well-known ad networks may offer higher advertising profits to attract app developers, but their security risk is higher. Ad networks provided documentation on their official websites, but some collect more personal information than their permissions allow, and app developers are not aware of it, or do not mind, because of their desire for profit. This is because, if an ad lib cannot collect enough personal information to include in ad request messages, ad servers determine that it is unable to provide effective advertising images to potential customers, and will thus not reply to the ad request messages. This means that app developers lose financially on the app, as free app developers rely on ads being sent to the users' phones for income.

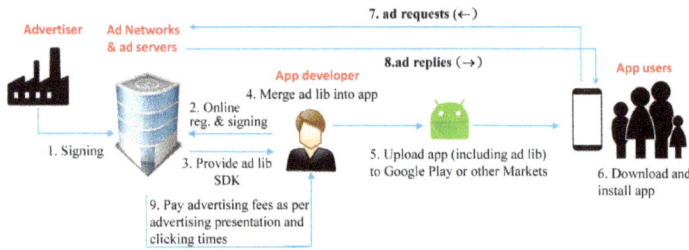

Figure 1. Ad lib operation flowchart.

According to Ruiz et al. [27], as a result of a large increase in free apps, the reply rate to ad requests of the Top 40 ad networks is lower than 18%. Therefore, free app developers must turn to the less known ad networks and (or) embed multiple ad libs from different companies in an app at the same time so as to increase the possibility of getting advertising pictures. The authors also collected more than 625,000 apps on these ad networks. After analysis, it was found that 34.88% of those apps had two or more ad libs embedded. A small number of apps had as many as 28 ad libs embedded.

Wei et al. [28] found that even apps with a good reputation and flagged as normal by anti-virus software were likely to be connected to malicious websites during their operation. They combined static (decompiling and checking program code) with dynamic (running apps for two hours and clicking as many links as possible through the tools) methods to observe who an app would communicate with. They collected 13,500 normal and popular apps, and found that in the course of their execution, these apps were connected to 254,022 URLs. In addition, 1,260 known malicious apps were collected, and it was found that they were connected to 19,510 URLs in their execution processes. According to the check returns based on Web-Of-Trust (WOT) [29] and VirusTotal [30], the authors divided all the above URLs into four categories: good websites, low-reputation websites, bad websites and malicious websites. Of the normal and popular apps, 8.8% of them were connected to malicious websites, 15% were connected to bad websites, and 73% to low-reputation websites. A total of 74% were connected to websites unsuitable for children. Of the known malicious apps, the situation can be expected to be worse. But otherwise, the authors found that the online URL distribution was similar to that of normal apps. This paper revealed an important point: even thorough and effective anti-virus software cannot guarantee that a certain app is safe, because the problem may not lie in the app itself, but with the website associated in the execution process. If connected to malicious or bad websites, a normal app could cause unimaginable damage.

The above authors [28] also found that only static decompilation of apps was not sufficient to achieve an effective full check by examining all possible online URLs, because the website could reconnect to other URLs through HTTP redirect mechanisms. Such problems are more difficult to predict because of the embedded ad libs of the apps. In fact, online advertising companies could resell ad slots to other ad networks (usually less known) through Ad exchange [31] so as to maximize advertising profits. This increases the advertising security risk, as the website could connect to multiple URLs when, for example, a free online game app is executed. Aside from the game server(s), the site could connect to ad server(s), redirect or unnamed server(s) by an ad resale mechanism.

3. The Proposed Method

Unlike static analysis, dynamic analysis focuses on the behavior of program execution, by analyzing the behavior of an app in an emulator. In some cases, better results may be obtained by dynamic analysis because it is resistant to obfuscation tools. Some researchers have emphasized the importance of dynamic analysis [32–36] for this reason. Since ad messages are carried out through HTTP packets, an understanding of HTTP is necessary to study the behavior pattern analysis of an ad lib, including the meaning of each field, and the information contained in it, so that the required data used in this research can be obtained.

In this study, an app was executed in an emulator and the packets of all network behavior were recorded, from which the packets related to the advertisement were picked using the proposed method. The tools used in this study were BlueStacks, TCP DUMP, ADB and self-created software. The emulator, called BlueStacks, used TCP DUMP to record network traffic from the virtual network adapter. The Android deb bridge (ADB) tool could directly access the Android emulator. The "logcat" instructions therein produced the required record files and the "pull" instructions exported the packet files (PCAP format) in the virtual machine. Because the captured packets were extremely large and messy, a program was designed to filter the packets related to the advertisement. Figure 2a shows a part of the proposed program, Figure 2b shows an ad request message, and Figure 2c shows an ad reply message.

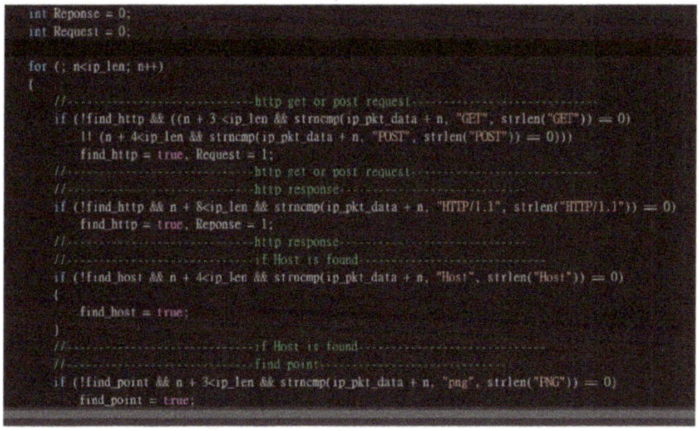

(a)

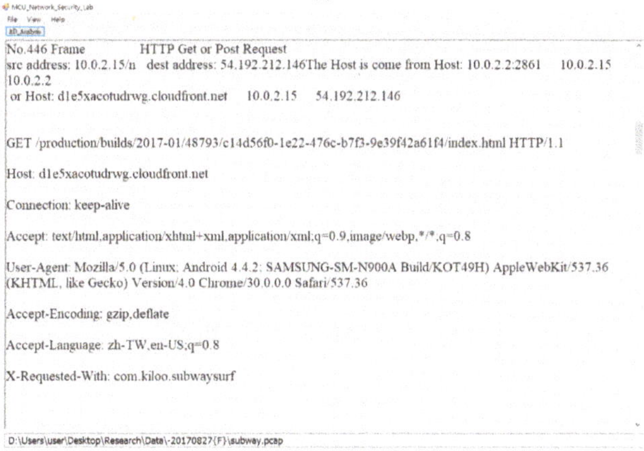

(b)

Figure 2. *Cont.*

(c)

Figure 2. The proposed tool for obtaining advertising behavior: (**a**) Partial program; (**b**) Ad request; (**c**) Ad reply.

In this study, the interaction between ad lib and ad server was presented by a graph according to their HTTP connections. A series of contents of ad replies from the server were observed, which were driven by the ad request from the ad lib. There were basically 3 kinds of content types sent back to the app by the ad server: HTML, JavaScript, and IMG, of which IMG could have different picture formats, such as PNG, JPG and gif.

The proposed graph-based method first identified the main behaviors of the ad lib, each of which was expressed by one vertex in the graph. All vertices were connected according to the proposed algorithm, and then an undirected graph was constructed to represent the network behavior of the ad lib. The PChome [37] ad lib was taken as an example to illustrate as follows. Figure 3a shows the main behavior related to ads extracted from the packet traffic of the emulator by the proposed program. In order, HTML -> JavaScript (JS) -> IMG -> IMG -> html, where the upper-case HTML represented ad requests, and the lowercase html indicated ad replies. This array of ad behavior pattern [HTML, JavaScript (JS), IMG, IMG, html] was taken as input, and an undirected graph, as in Figure 3b, was constructed using the algorithms, as shown in Figure 4.

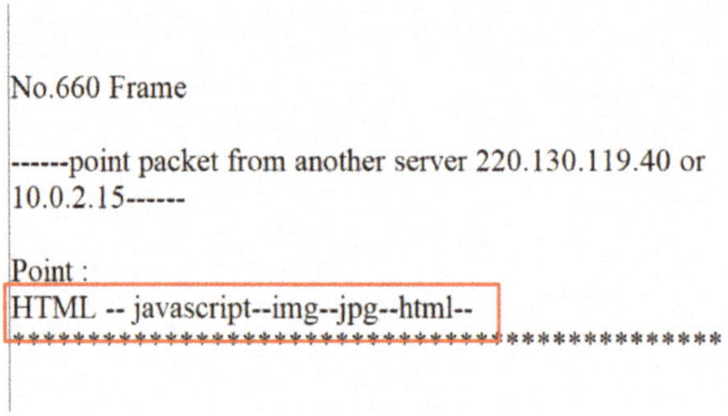

(**a**)

Figure 3. *Cont.*

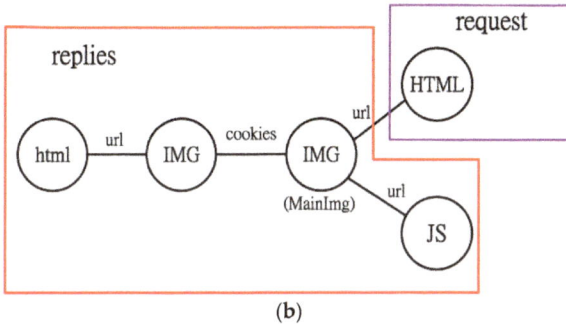

(b)

Figure 3. Graph construction of PChome ad lib: (**a**) online advertising behavior and (**b**) graph representation.

INPUT: An array of Actions //for exmaple: [**HTML, JavaScript (JS), IMG, IMG, html**]
//types: HTML, IMG, Javascrip(JS)
//HTML means coming from ad request, html means coming from ad replies

OUTPUT: a graph

Algorithm Graph_Construction()
{
  Scan the input array from left to right to locate the first IMG, also called MainImg.
  If there is no IMG, then exit;
  //supposedly at least an ad pictutre has been delivered in the ad activities
  else Creat a vertex for the MainImg in the graph; //this is the first vertex created in the graph//

  for every element on the leftside of the MainImg in the input array
  {
    Create a vertex for the element and put the vertex on the right-hand of the MainImg in the graph;
    //the current element is either html(HTML) or JS//
    Link the current vertex to the MainImg and label the edge as "url";
  }

  for every element on the rightside of the MainImg in the input array
  {
    if it is IMG
    {
      Create a vertex for the IMG; //current vertex
      Put the current vertex on the left-hand of the MainImg in the graph;
      Link the current vertex to the MainImg and label the edge as "cookies";
      //cookies used to link two image vertices
    }
    else //either JS ot html
    {
      Create a vertex for the element; //current vertex
      Find the closest IMG on the leftside in the input array; //related IMG. It could be MainImg or regular IMG
      Put the current vertex on the left-hand of the related IMG vertex in the graph;
      Link the current vertex to the related IMG vertex and label the edge as "url";
    }
  }
}

Figure 4. Algorithm of graph construction by ad lib network behavior.

According to the algorithm in Figure 4, the input array of actions is first checked from left and right so as to find the first IMG, which was taken as the first IMG point (vertex) in the graph, also called the MainImg in the algorithm. Each element of the input array on the left of the first IMG formed its own vertex, which was drawn on the right side of the MainImg and linked to the MainImg by the edge marked "url". Then the element on the right of the first IMG in the input array was processed. The element on the right of the first IMG was checked in the input array, and each IMG on the right formed its own vertex, which was drawn on the left side of the MainImg and linked to the MainImg by the edge marked "cookies". If there was no IMG (either HTML or JS), one vertex was formed, which was connected to the vertex formed by the nearest IMG on the left side of the element in the input array with the edge marked "url".

4. Experiment Results

This section gives more examples to demonstrate the effectiveness of the proposed approach. In Figure 5, on the left of each subgraph is the main ad behavior of an ad lib obtained from the network traffic of the emulator, and on the right side of each subgraph is the undirected graph based on the algorithm, as shown in Figure 4.

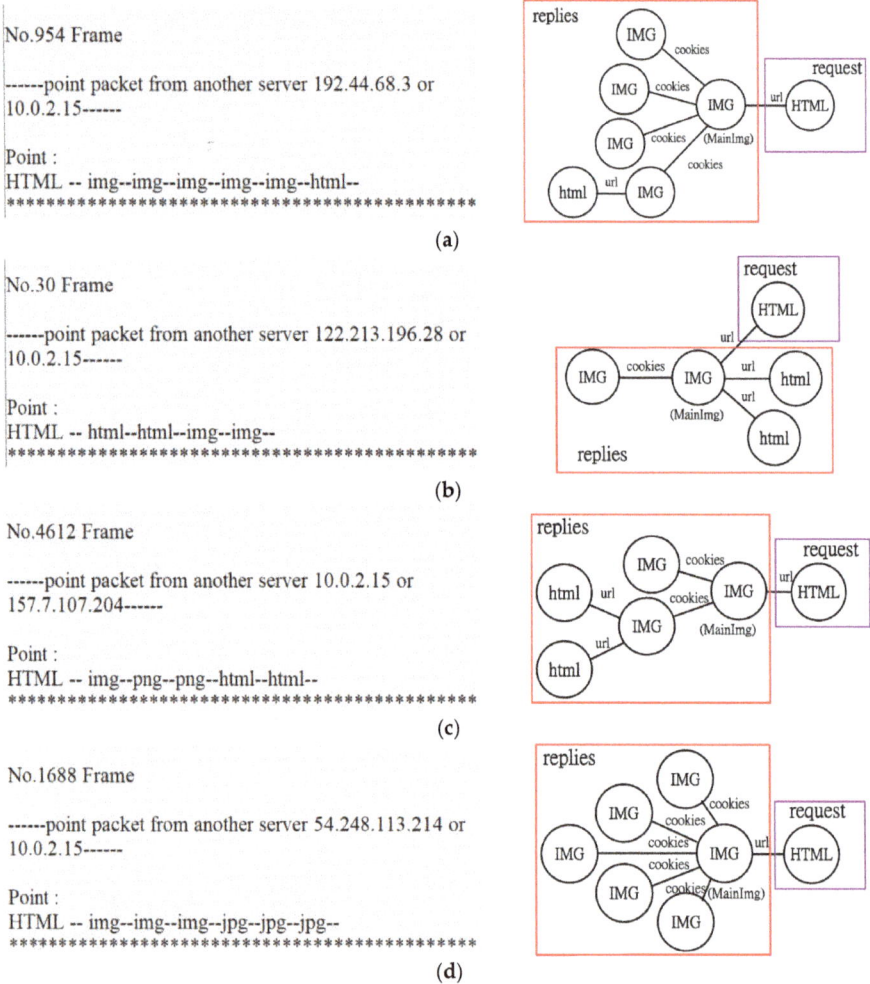

No.954 Frame

------point packet from another server 192.44.68.3 or 10.0.2.15------

Point :
HTML -- img--img--img--img--img--html--
****

(a)

No.30 Frame

------point packet from another server 122.213.196.28 or 10.0.2.15------

Point :
HTML -- html--html--img--img--
****

(b)

No.4612 Frame

------point packet from another server 10.0.2.15 or 157.7.107.204------

Point :
HTML -- img--png--png--html--html--
****

(c)

No.1688 Frame

------point packet from another server 54.248.113.214 or 10.0.2.15------

Point :
HTML -- img--img--img--jpg--jpg--jpg--
****

(d)

Figure 5. *Cont.*

No.46 Frame

------point packet from another server 122.213.196.28 or
10.0.2.15------

Point :
HTML -- img--img--
****

(e)

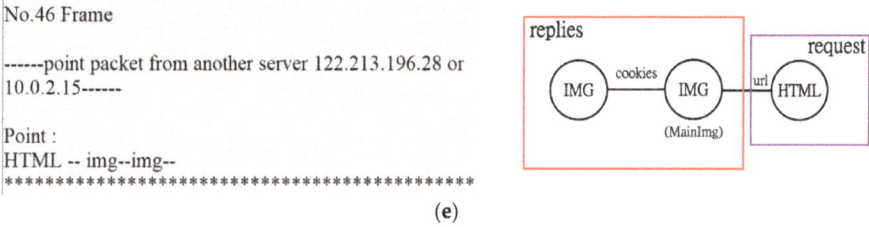

Figure 5. Ad libs' network behaviors and their graph representations (**a**) MoPub, (**b**) Mmate, (**c**) Hap-game, (**d**) Millennial Media, (**e**) AppLovin.

Some of the advertising companies listed on the AppBrain website [26] and their ad libs were chosen and processed according to the above process, with the results presented in Table 1. For some ad libs, only one type of advertising graph was observed in the experiment, but most ad libs exhibited different advertising behavior patterns because of different advertising types, such as banner or full page, or version difference. Most of the javascripts found in this study were used to deal with the ad pictures, for instance adjusting the size of picture presented on the screen.

Table 1. Advertising behavior graphs of ad libs (with ad lib as index).

Table 1. *Cont.*

appservestar

campaign.ad-brix

cdn.unityads.unity3d.com

cloudfront.herocraft.com

hap-game.sub.jp

JScount	hap-app.net	magic.cmcm.com

mydass

rayjump

Table 1. *Cont.*

yahoo		
IMG —Url— HTML	html—Url, JS—Url → IMG —Url— HTML	

Startapp		
IMG—Cookies, IMG—Cookies → IMG —Url— HTML	IMG —Url— HTML	IMG —Url— HTML
supersonicads-a.akamaihd.net	surpax	wpc.32DF9.rhocdn.net
IMG —Url— HTML	IMG —Url— HTML	IMG —Url— HTML
ying	mysearch-online	omax.admarvel.com

Millennial Media	moreadexchange	AdMob
IMG, IMG, IMG (Cookies), IMG (Cookies), IMG — IMG —Url— HTML	IMG, IMG, IMG, IMG, IMG, IMG (Cookies) — IMG —Url, Url → HTML, JS	IMG —Cookies— IMG —Url— HTML
IMG —Url— HTML	IMG —Url— HTML	IMG —Url— HTML
alog.umeng	app.hap.ne.jp	b.scorecardresearch.com
IMG —Url— HTML	IMG —Url— HTML	html —Url— IMG —Url— HTML
cdnicons.pluginmanagerconfig1.info	entity3.com	gamepromote

Table 1 uses ad lib as the index, while Table 2 uses the graph as the index. In Table 2, all graphs are categorized into different types, from A to P.

Table 2. Graph types and their corresponding ad libs (with graph as index).

A		Adecosystems, ads.mopub.com, AppLovin, appservestar, campaign.ad-brix, cdn.unityads.unity3d.com, cloudfront.herocraft.com, hap-game.sub.jp, JScount, hap-app.net, magic.cmcm.com, mydass, rayjump, Startapp, surpax, wpc.32DF9.rhocdn.net, ying, mysearch-online, omax.admarvel.com, alog.umeng, app.hap.ne.jp, b.scorecardresearch.com, entity3.com, cdnicons.pluginmanagerconfig1.info,
B		Adecosystems, app.mmate.jp, AppLovin, AdMob, cdn.unityads.unity3d.com
C		ads.mopub.com
D		ads.mopub.com, yahoo, PChome
E		app.mmate.jp, appservestar, cdn.unityads.unity3d.com
F		Hap-game
G		campaign.ad-brix, rayjump, yahoo
H		cdn.unityads.unity3d.com, cloudfront.herocraft.com, gamepromote
I		mydass
J		mydass

Table 2. *Cont.*

K		mydass
L		Startapp
M		supersonicads-a.akamaihd.net
N		Millennial Media
O		moreadexchange
P		app.mmate.jp

Because of the variety of graphs, a formula was designed to quantize the figures in order to make the graph classification easier later. This formula is mainly based on the vertices. The larger the IMG number, the lower its value; on the other hand, the larger the number of JavaScript and HTML, the lower the value. However, the differentia was not big. That is, the impact caused by JavaScript and HTML was relatively small compared to the number of IMG.

$$\frac{100}{2^{\#IMG-1}} - 2\#JavaScript - 3\#HTML + 5$$

Using this formula, the corresponding values of different graphs could be obtained, and the corresponding values of some graphs are shown in Table 3. If an unknown pattern of ad lib produced the advertising behavior shown in Figure 6a, for example, i.e., the vertices were HTML, HTML, IMG, IMG, IMG and JavaScript, according to the algorithm in Figure 4, the graph was generated as Figure 6b. The value obtained by the suggested formula is 22, according to the numbers of different types of vertices. However, there was no matched graph value in Table 3, which indicated that it was a newly found advertising behavior model. Therefore, the content of the ad packet needed to be further analyzed, and it was found that this was the behavior of Mydas ad lib, shown in Figure 6c. Finally, the newly acquired information was added to Tables 1–3, in order to expand the content of known advertising patterns. The method proposed in this paper made it possible to more quickly classify the ad lib in an app. Some advertisers or app developers may deliberately hide the Host name. In this situation, the ad lib could still be classified by checking the ad behavior graph. If two or more ad libs shared the same graphs, the range of candidates was significantly reduced because of the classification.

No.468 Frame

------point packet from another server 216.157.12.18 or
192.168.137.70------

Point :
HTML -- html--img--img--img--javascript--
****

(a)

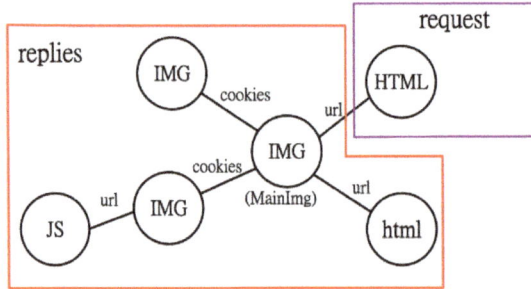

(b)

No.468 Frame   HTTP Response
src address: 216.157.12.18/n   dest address: 192.168.137.70The Host is come
from Host: androidsdk.ads.mp.mydas.mobi | 192.168.137.70   216.157.12.18
or Host: androidsdk.ads.mp.mydas.mobi   192.168.137.70   216.157.12.18

(c)

Figure 6. An ad lib discovery, for example (**a**) network patterns (**b**) corresponding graph (**c**) packet details.

Table 3. Values of the different types of graphs.

Graph Type	#IMG	#html	#JS	Value
A	1	1	0	102.00
B	2	1	0	52.00
C	5	2	0	5.25
D	2	2	1	47.00
E	2	2	0	49.00
F	3	3	0	21.00
G	1	1	1	100.00
H	1	2	0	99.00
I	3	4	1	16.00
J	2	3	1	44.00
K	3	1	1	25.00
L	1	2	1	97.00
M	3	1	0	27.00
N	6	1	0	5.13
O	7	1	1	1.56
P	2	3	0	46

5. Conclusions

Mobile security issues today are not limited to malicious apps. This is because some security risks do not lie in an app itself, but in the advertising network behavior carried out in the course of the app execution. Well known advertising networks pose fewer security risks in this regard, but as their popularity increases and more instances of an app are installed, the number of ad requests sent to ad servers increases, resulting in fewer replies to ad requests by individual apps. App developers therefore often embed several ad libs into an app in order to increase advertising profits. Less well-known advertisers sometimes may offer better profits in order to attract app developers, and developers may increase profits at a higher risk to user security without users' knowledge. When advertisers hide their identity by removing their brand names or specific symbols, decompiling an app will still not identify which ad libs are embedded in the app. Therefore, the purpose of this study was to transform the network behaviors of ad libs into graphs, and then identify ad libs through the comparison of the graph similarities. Letting users know about the embedded ad(s) in an app is a further step to protect users' smartphones. If an ad lib cannot be identified, possibly from unknown ad networks, the user would be informed.

Author Contributions: M.-Y.S. conceived and organized the research work; H.-S.W., X.-Y.C., P.-W.L. and D.-Y.Q. conducted the experiments and analyzed data; M.-Y.S. wrote the paper; H.-S.W. and X.-Y.C. checked and verified the paper. All authors reviewed the paper.

Funding: This research was funded by the Ministry of Science and Technology, Taiwan, grant numbers MOST 106-2221-E-130-002 and MOST 107-2221-E-130-003. The APC was funded by the latter.

Conflicts of Interest: The authors declare no conflicts of interest.

References

1. IAB Internet Advertising Revenue Report—2016 Full Year Results. Available online: https://www.iab.com/wp-content/uploads/2016/04/IAB_Internet_Advertising_Revenue_Report_FY_2016.pdf (accessed on 7 October 2018).
2. Millennial Media: State of the Apps Industry Snapshot 2015. Available online: http://visionmediainteractive.com/millennial-medias-state-of-the-apps-industry-snapshot-2015/ (accessed on 7 October 2018).
3. Mobile Malware Evolution 2016. Available online: https://securelist.com/analysis/kaspersky-security-bulletin/77681/mobile-malware-evolution-2016/ (accessed on 7 October 2018).

4.   Trend Micro Detects Xavier Android Malware in 800 Mobile Apps Downloaded from Google Play Store. Available online: https://cio.economictimes.indiatimes.com/news/digital-security/trend-micro-detects-xavier-android-malware-in-800-mobile-apps-downloaded-from-google-play-store/59284441 (accessed on 7 October 2018).

5.   Dr. WEB Anti-Virus, New Trojan Found in 155 Apps on Google Play: 2.8 Million Mobile Devices Already Infected. Available online: https://news.drweb.com/show/?i=10115&lng=en (accessed on 7 October 2018).

6.   Athanasopoulos, E.; Kemerlis, V.P.; Portokalidis, G.; Keromytis, A.D. NaClDroid: Native Code Isolation for Android Applications. *LNCS* **2016**, *9878*, 422–439.

7.   Kumar, P.; Singh, M. Mobile Applications: Analyzing Private Data Leakage Using Third Party Connections. In Proceedings of the IEEE Proceedings of International Conference on Advances in Computing, Communications and Informatics (ICACCI), Kochi, India, 10–13 August 2015; pp. 57–62.

8.   Gao, X.; Liu, D.; Wang, H.; Sun, K. PmDroid: Permission Supervision for Android Advertising. In Proceedings of the 34th Symposium on Reliable Distributed Systems, Montreal, QC, Canada, 28 September–1 October 2015; pp. 120–129.

9.   Narayanan, A.; Chen, L.; Chan, C.-K. AdDetect: Automated Detection of Android Ad Libraries using Semantic Analysis. In Proceedings of the IEEE Ninth International Conference on Intelligent Sensors, Sensor Networks and Information Processing (ISSNIP), Singapore, 21–24 April 2014; pp. 1–6.

10. Liu, B.; Liuy, B.; Jin, H.; Govindan, R. Efficient Privilege De-Escalation for Ad Libraries in Mobile App. In Proceedings of the 13th Annual International Conference on Mobile Systems, Applications, and Services, Florence, Italy, 18–22 May 2015; pp. 89–103.

11. Yan, Y.; Cosgrove, S.; Anand, V.; Kulkarni, A.; Konduri, S.H.; Ko, S.Y.; Ziarek, L. RTDroid: A Design for Real-Time Android. *IEEE Trans. Mob. Comput.* **2016**, *15*, 2564–2584. [CrossRef]

12. Book, T.; Wallach, D.S. A Case of Collusion: A Study of the Interface between Ad Libraries and their Apps. In Proceedings of the International Workshop on Security and Privacy in Smartphones and Mobile Devices (SPSM), Berlin, Germany, 8 November 2013; pp. 79–85.

13. Ruiz, I.J.M.; Nagappan, M.; Adams, B.; Berger, T.; Dienst, S.; Hassan, A.E. On Ad Library Updates in Android Apps. *IEEE Softw.* **2017**. [CrossRef]

14. Su, X.; Liu, X.; Lin, J.; He, S.; Fu, Z.; Li, W. De-cloaking Malicious Activities in Smartphones Using HTTP Flow Mining. *KSII Trans. Internet Inf. Syst.* **2017**, *11*, 3230–3253.

15. Kuzuno, H.; Magata, K. Detecting Advertisement Module Network Behavior with Graph Modeling. In Proceedings of the Ninth Asia Joint Conference on Information Security, Wuhan, China, 3–5 September 2014; pp. 1–10.

16. Kajiwara, N.; Kawamoto, J.; Matsumoto, S.; Hori, Y.; Sakurai, K. Detection of Android Ad Library Focusing on HTTP Connections and View Object Redraw Behaviors. In Proceedings of the IEEE International Conference on Information Networks (ICOIN), Cambodia, 12–14 January 2015; pp. 104–109.

17. Crussell, J.; Stevens, R.; Chen, H. MAdFraud: Investigating Ad Fraud in Android Applications. In Proceedings of the 12th International Conference on Mobile Systems, Applications, and Services, Bretton Woods, NH, USA, 16–19 June 2014; pp. 123–134.

18. Song, Y.; Hengartner, U. PrivacyGuard: A VPN-based Platform to Detect Information Leakage on Android Devices. In Proceedings of the ACM Workshop on Security and Privacy in Smartphones and Mobile Devices, Denver, CO, USA, 12 October 2015; pp. 15–26.

19. Backes, M.; Bugiel, S.; Derr, E. Reliable Third-Party Library Detection in Android and its Security Applications. In Proceedings of the 23rd ACM Conference on Computer and Communications Security, Vienna, Austria, 24–28 October 2016; pp. 356–367.

20. Lee, J.-H.; Jun, S.-Y.; Park, S.-J.; Kim, K.-M.; Lee, S.-K. Demo: Mobile Contextual Advertising Platform based on Tiny Text Intelligence. In Proceedings of the 15th ACM International Conference on Mobile Systems, Applications, and Services (MobiSys), Niagara Falls, NY, USA, 19–23 June 2017; p. 181.

21. Tang, J.; Li, R.; Han, H.; Zhang, H.; Gu, X. Detecting Permission Over-claim of Android Applications with Static and Semantic Analysis Approach. In Proceedings of the IEEE Trustcom/BigDataSE/ICESS, Sydney, NSW, Australia, 1–4 August 2017.

22. Liu, X.; Zhu, S.; Wang, W.; Liu, J. *Alde: Privacy Risk Analysis of Analytics Libraries in the Android Ecosystem*; Springer: Cham, Switzerland, 2017; pp. 655–672.

23. Stevens, R.; Gibler, C.; Crussell, J.; Ericksonand, J.; Chen, H. Investigating User Privacy in Android Ad Libraries. In Proceedings of the IEEE Mobile Security Technologies (MoST), San Francisco, CA, USA, 24 May 2012.

24. Mobile Web Usage Overtakes Desktop for First Time. Available online: https://www.telegraph.co.uk/technology/2016/11/01/mobile-web-usage-overtakes-desktop-for-first-time/ (accessed on 8 October 2018).

25. Without Ads, Android Apps Could Be More Than Twice as Power-Efficient. Available online: http://www.theverge.com/2012/3/19/2884902/android-apps-battery-efficiency-study (accessed on 7 October 2018).

26. AppBrain, Android Ad Networks. Available online: http://www.appbrain.com/stats/libraries/ad (accessed on 7 October 2018).

27. Ruiz, I.J.; Nagappan, M.; Adams, B.; Berger, T.; Dienst, S.; Hassan, A.E. Impact of Ad Libraries on Ratings of Android Mobile Apps. *IEEE Softw.* **2014**, *31*, 86–92. [CrossRef]

28. Wei, X.; Neamtiu, I.; Faloutsos, M. Whom Does Your Android App Talk To? In Proceedings of the IEEE Global Communications Conference (GLOBECOM), San Diego, CA, USA, 6–10 December 2015.

29. Web of Trust. December 2017. Available online: http://www.mywot.com/ (accessed on 7 October 2018).

30. VirusTotal. December 2017. Available online: https://www.virustotal.com/en/#url (accessed on 7 October 2018).

31. Ad Exchange. Available online: https://en.wikipedia.org/wiki/Ad_exchange (accessed on 7 October 2018).

32. Enck, W.; Gilber, P.; Chun, B.; Cox, L.P.; Jung, J.; McDaniel, P.; Sheth, A.N. TaintDroid: An information-flow tracking system for realtime privacy monitoring on smartphones. In Proceedings of the USENIX Symposium on Operating Systems Design and Implementation (OSDI), Vancouver, BC, Canada, 4–6 October 2010.

33. Blasing, T.; Batyuk, L.; Schmidt, A.D.; Camtepe, S.A.; Albayrak, S. An Android application sandbox system for suspicious software detection. In Proceedings of the 5th International Conference on Malicious and Unwanted Software (Malware 2010), Nancy, France, 19–20 October 2010.

34. Liu, J.; Liu, J.; Li, H.; Zhu, H.; Ruan, N. Who Moved My Cheese: Towards Automatic and Fine-Grained Classification and Modeling Ad Network. In Proceedings of the Global Communications Conference (GLOBECOM), Washington, DC, USA, 4–8 December 2016.

35. Chan, J.; Keng, J.; Jiang, L.; Wee, T.K.; Balan, R.K. Graph-aided directed testing of Android applications for checking runtime privacy behaviours. In Proceedings of the 11th International Workshop on Automation of Software Test, Austin, TX, USA, 14–15 May 2016; pp. 57–63.

36. Biswas, S.; Haipeng, W.; Rashid, J. Android Permissions Management at App Installing. *Int. J. Secur. Appl.* **2016**, *10*, 223–2322. [CrossRef]

37. PChome. Available online: http://show.pchome.com.tw/pfb/ (accessed on 7 October 2018).

MDPI

St. Alban-Anlage 66

4052 Basel

Switzerland

Tel. +41 61 683 77 34

Fax +41 61 302 89 18

www.mdpi.com

Applied Sciences Editorial Office

E-mail: applsci@mdpi.com

www.mdpi.com/journal/applsci